Underground Transmission of Electric Power

Underground Transmission of Electric Power

B. M. WEEDY

Department of Electrical Engineering,
The University of Southampton,

A Wiley–Interscience Publication

JOHN WILEY & SONS

CHICHESTER NEW YORK BRISBANE TORONTO

British Library Cataloguing in Publication Data:

Weedy, Birron Mathew
Underground transmission of electric power.
1. Electric lines–Underground
I. Title
621.319'23 TK3251 79–42896

ISBN 0 471 27700 2

Typeset by Preface Ltd, Salisbury, Wiltshire and printed
by Page Bros. (Norwich) Ltd, Norwich

To
Betty, Kate, and
Laurence

Preface

Increasing public pressure against overhead circuits has resulted in intense interest in the further improvement of conventional cables and research and development into new forms of underground circuits. I have attempted in this book to give the scientific basis underlying the design and installation of cable systems and to apply this knowledge to practical schemes. To a large extent it is aimed at users, e.g. utilities and consultants, rather than at manufacturers. Hence no attempt is made to describe in detail the complexities of cable manufacture. It is hoped that the book will serve as a useful introduction for engineers involved in electric transmission and distribution. In particular, the text has been developed to provide reasonable coverage of postgraduate courses in electric power, and worked examples and problems for student solution are included.

At the present time the long-term future for underground transmission is unclear. Reduced rates of increase in electricity consumption in recent years have probably put back the time when very high power cables such as the superconducting type will be required, especially in Europe. However, as oil and natural. gas become scarcer, in the long term an increasing use of electricity must result. This in turn will entail the use of underground transmission in ever larger amounts.

My thanks are due to friends in the electricity supply industry and with manufacturers who have encouraged my interest in underground transmission over many years.

Southampton 1979

B. M. WEEDY

Contents

Symbols Used

A = Area
a–b–c = Phase rotation (alternatively R–Y–B)
a and b = Temperature and stress coefficients of d.c. resistivity
a = Operator $1\angle 120^\circ$
C = Capacitance
C = Thermal capacity (J/deg C)
C = Cost
c_p = Specific heat (J/kg per deg C)
d, D = Diameter
$\mathbf{E}$ = Electric stress (MV/m or kV/mm)
f = Frequency (Hz)
g = Thermal resistivity (deg. C-m/W)
a = Acceleration due to gravity (m/s^2)
h = Heat transfer coefficient (Watts/m^2 per deg C)
$\mathbf{I}$ = Current (A)
j = $1\angle 90^\circ$ operator
$\mathbf{J}$ = Current density (A/m^2)
k = Thermal conductivity (W/m-deg C)
L = Inductance (H)
L.F. = Load factor
L = Insulation life (years)
L.L.F. = Load loss factor
l = length (m)
M = Mutual inductance (H)
P = Power
P = Pressure (Pa)
p.f. = Power factor
Q = Reactive power (VAr)
q = Loss dissipated as heat (Watts)
R = Resistance (ohms) also thermal resistance (deg C/W)
$\mathbf{S}$ = Complex power = $P \pm jQ$
s = spacing between cables
T = Absolute temperature (K)

t = Time
Δt = Interval of time
U = Velocity
V = Voltage
W = Volumetric flow of coolant (m^3/s)
$\mathbf{Y}$ = Admittance (p.u. or Ω^{-1})
$\mathbf{Z}$ = Impedance (p.u. or ohms)
$\mathbf{Z}_0$ = Characteristics or surge impedance (ohms)
β = Coefficient of volumetric expansion (deg C^{-1})
γ = Kinematic viscosity (m^2/s)
ϵ = Permittivity
μ = Viscosity (kg/m-s)
θ = Temperature rise (deg C) above reference or ambient
ρ = Electrical resistivity (Ω-m)
ρ = Density (kg/m^3)
τ = Time constant
ϕ = Angle between voltage and current phasors (power factor angle)
ω = Angular frequency (pulsatance) rad/s
σ = Electrical conductivity
σ = Stefan Boltzmann Constant (5.7×10^{-8} W/m^2 K^4)
N_{Re} (or Re) = Reynolds Number $\left(\frac{dU\rho}{\mu}\right)$ or $\left(\frac{dU}{\gamma}\right)$
N_{Gr} (or Gr) = Grashof Number ($g\beta d^3 \Delta\theta/\gamma^2$); values at NTP are: air 7.5×10^3, water 2.4×10^5
N_{Pr} (or Pr) = Prandtl Number ($\mu c_p/k$); values at NTP are: air 0.7, water 6.5

I

Introduction

Experience indicates that, dependent on the voltage, the capital costs of underground electric power circuits are several times those of the overhead line of equivalent capacity. However, comparisons are based on particular schemes and circumstances, and should be treated with some degree of caution. One reason for this cost disadvantage is not difficult to see: with the overhead line the insulation is mainly air, which could still at the time of writing be considered an inexpensive commodity. In the underground cable, on the other hand, continuous man-made insulation must be applied.

In an underground cable the high-voltage conductor is contained in as small an insulating envelope as possible, for reasons which will become apparent. This represents a manufacturing task of the highest quality and hence of considerable expense. Also, the installation costs of undergrounding are high. Nature conspires against the conventional underground circuit in almost every way. The heat evolved in the conductor has to traverse paths in series comprising insulation and soil, both materials possessing high thermal resistance. Also, the effective life span of conventional insulations becomes uneconomically short if the maximum temperature caused by the losses exceeds a certain value. This restricts the permissible loss and hence the current. With the overhead line, on the other hand, the heat is transferred directly to the air much more effectively.

In the technology of underground cables it is necessary to provide the following:

(*a*) Adequate insulation as economically as possible and such that the cable may be transported on a reel or drum, i.e. possess a viable radius of curvature.
(*b*) Adequate current-carrying capacity which is dependent on installation techniques and heat-transfer properties.
(*c*) Schemes, whilst probably still more expensive than equivalent overhead lines, which are attractive enough for utilities to install them on environmental grounds and also to reduce costs in urban areas where underground circuits must be used.

Whereas a single overhead-line circuit striding over the countryside may have a certain appeal, especially to electrical engineers and their dependants, this appeal soon diminishes with multi-circuit routes and, to the more sensitive, converts to horror at the sight of lines converging on a large substation which are visible over an angle of 360°. This may be considered a personal matter, and it has been suggested, somewhat frivolously, that in the 'all underground era ahead', redundant towers will be cherished and maintained and sought after by conservation groups as old windmills are at the present time in Britain and elsewhere. However, the public rejection of unsightly overhead circuits, especially in suburban areas, is now so strong and well organized as to be a major factor in the provision of transmission circuits. At the present time, underground circuits are used mainly in short lengths in urban and suburban areas and for special requirements such as water crossings, airports, and under highways. In Britain, where a relatively large amount of undergrounding has taken place, underground rings round major cities exist at major transmission voltages from which the lower-voltage cables into the city (e.g. 66 kV in London) are fed.

In the future development of major power systems, power generation may be concentrated in very large units or parks (over 10 GW) remote from large urban areas, resulting in an environmental need for very high capacity underground connections from the power plant to the overhead-line system some distance away. Conversely, urban and suburban areas will be supplied underground from the major overhead system. Eventually underground circuits of 5 to 10 GW capacity for lengths of up to 80 km (50 miles) and over may be used. It is probable that eventually major power pools will be connected by d.c. links (to avoid stability and other problems) several hundred kilometres long and of several Gigawatts, capacity.

The environmental problems of overhead lines at the ultrahigh-voltage levels required for future loads are well known, and it may well be that with limited right of way or wayleave corridors available sophisticated underground circuits will have to be used. There will be increasing problems in transferring power across great conurbations with progressively less space available under pavements for cable circuits as other utilities, e.g. gas and water, must increase their capacity also. The extent of this problem may be seen by the fact that at the present time the installed capacity is about 20 GW in New York City. Although in the 1970s the growth rate for electricity has lessened, in the long term considerable increases in growth are likely to result from shortages of oil and natural gas. The widespread use of air-conditioning in North America has resulted in peak transmission loads at high ambient temperatures, the worst possible combination of events.

It is the object of this book to discuss the technical problems associated with the bulk transfer of electric energy by underground circuits. Transmission by direct current will be confined to Chapter 5, the other chapters referring (unless otherwise stated) to alternating current. Due attention will be given to conventional systems, i.e. those well tried and commercially available. Also, new forms of cable such as compressed-gas-insulated and cryogenic will be discussed.

1.1 HISTORICAL[1]

The first use of insulated wires was for telegraph purposes and dates back to the first half of the nineteenth century. For the conveyance of power underground, insulated wires were first used in the 1880s at about the same time by Edison (in the USA) and Ferranti (in London), both for electric lighting systems. The cables were rigid, consisting of copper rods insulated with a jute wrapping. Reliability was resonable and most problems arose from the large number of joints or splices necessary in a rigid system in which the conductor could not be coiled on a reel.

In a subsequent scheme Ferranti decided on a new form of insulation – paper impregnated with Ozokerite wax (a by-product from the manufacture of candles). In this cable the modern paper-taped version was born. The Ferranti 10 kV cable consisted of two concentric tubes (conductors) insulated from each other by the impregnated paper. The conductors were brazed copper tubes of 20.7 mm and 49.5 mm in diameter. To satisfy the communication authorities Ferranti was forced to cover the outer tube (sheath) with further layers of paper and the cable was then installed in an iron pipe. The route length was 7.5 miles, and four cables were laid involving over 7000 joints. Over the service life of 42 years relatively few joint failures occurred and the cable was only superseded because of a need for higher current in the circuit. This rigid conductor was followed rapidly by the development of a flexible stranded conductor cable, and by 1898 the insulation maximum stress was 2.1 kV/mm (single-phase cable). Many insulation materials were investigated and used, including paper, cotton, gutta percha, vulcanized bitumen, and rubber. 11 kV rubber vulcanized cables were installed in Buffalo (N.Y.) in 1897 and 25 kV cables in St. Paul and Minneapolis in 1900. Paper impregnated with oil using vacuum drying and hot-oil impregnation was available at 10 kV (single-phase circuits) in 1895.

In the early days in the USA cables were often laid in ducts because of legislation which allowed only one authority to open up trenches in a city. Electrical utilities hired a passage through the duct ways provided. In this environment single-core rubber insulated cables were used because of their greater flexibility compared with the paper–oil type, and this influence remains today in the great usage of single-core elastomeric insulation in the USA compared with paper–oil cables in Britain which are direct-buried in trenches.

At these lower voltages oil-impregnated (mass impregnated) paper-insulated cables (solid type) were used, often with the three conductors contained in a single sheath. The three conductors were stranded and insulated separately and then laid up spirally together. The space between and around the insulated conductors was packed with paper or jute to form a circular surface which was then further wrapped with insulation. This is called the 'belted' type of cable (see Figure 1.1) and it may have steel-wire armouring over the sheath because only small induced eddy currents are produced in the armouring wires, whereas with single-conductor cables severe loss and increased impedance would result. With the three-core cable, high electric stresses are set up tangentially to the paper insulation surfaces, in which direction the insulation strength is weakest. To overcome this each core is wrapped with a conducting layer of metallized paper which converts the cable electrically into three single-core cables with the electric stress completely in the radial direction. This form of construction was originated by Hochstadter and is known as the 'H' type. As system voltages increased above 33 kV the solid-type

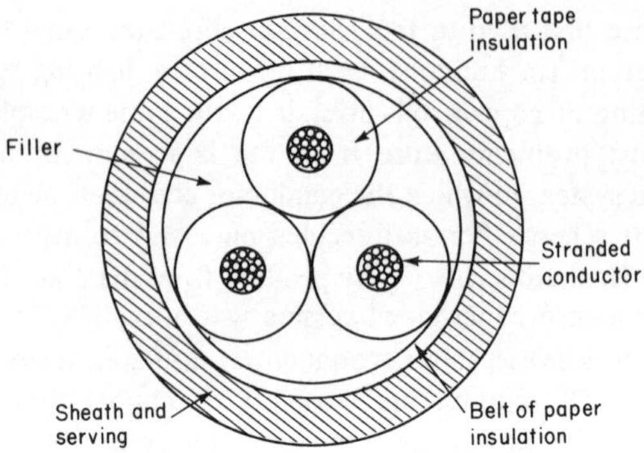

Figure 1.1 Belted type – 3 core cable.

paper–oil cable became prone to breakdown because of the voids (small pockets of air or gas) formed in the insulation when constituent parts of the cable expanded and contracted to different extents with the heat evolved on load cycles. The stress across these voids is high and local discharges occur, creating heat which chars the paper; eventually complete breakdown results.

1.2 CONVENTIONAL CABLE SYSTEMS

The cable must fit naturally into the transmission system and is subject to the network power and voltage levels. A network at a given voltage may be extended to cope with greater load demand, but eventually a limit arises because of the resulting higher short-circuit levels and the inability of appropriate plant to cope. Hence a new voltage is introduced with a corresponding network overlapping the original one. A further factor causing the introduction of higher voltages is the increasing difficulty in obtaining 'rights of way' for circuits. As much more power can be transmitted by a line (or cable) at the higher voltage then fewer circuits are needed (see Section 1.7).

In Britain the main transmission voltage is 400 kV with some circuits at 275 kV. The extensive 132 kV network has recently been designated as distribution rather than transmission and would probably in some countries be called subtransmission. For cables, 66 kV is sometimes used, particularly for the London network. In most of Europe the main transmission voltage is 400/420 kV with circuits also at 225/245 kV, 145 kV, and 72.5 kV. In North America the highest transmission voltage is 765 kV followed by 500 kV, 345 kV, 230 kV, and 138 kV.

The difficulties outlined in the previous section with the mass-impregnated cable were overcome by the filling of the cable with oil under pressure. In 1926, Emanueli[2] in Italy introduced the self-contained low-pressure oil-filled (LPOF) cable which is still extensively used today. A cross-section of a 275 kV high-capacity version of this cable is shown in Figure 1.2 and a section of a through-joint or splice in Figure 1.3.

In the oil-filled cable the hollow centre of the conductor is filled with insulating oil maintained under pressure by reservoirs feeding the cable along the route. As the

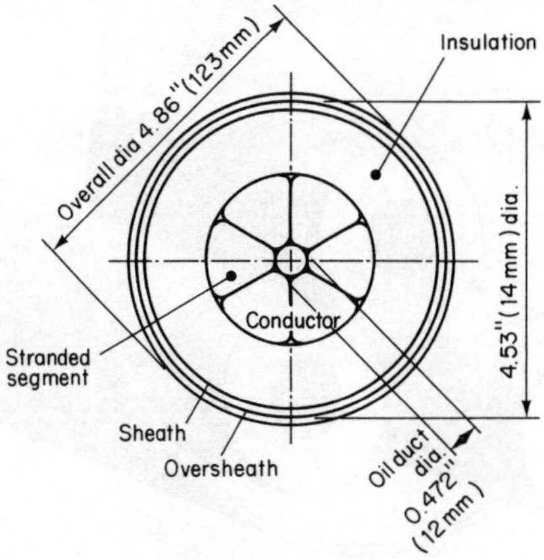

Figure 1.2 Cross section of a 2000 mm² (3 in²) copper conductor, self-contained low-pressure oil-filled cable.

cable heats on load, oil is driven from the cable in the reservoirs, and vice versa, and hence the creation of voids is avoided. In gas-pressure installations, nitrogen at several atmospheres maintains a constant pressure on an inner sheath compressing the dielectric and so preventing the formation of voids.

In the USA the paper/oil insulated cores are installed in a rigid pipe containing the insulating oil, and the conductor oil-duct is not required. The difficulties in the USA of keeping a trench open for a time sufficient for the direct burial of the self-contained (LPOF) cable have led to the use of the pipe-type oil-filled cable, in which the three paper-tape insulated cores are pulled into a previously buried pipe via manholes at intervals along the route. (For cross-sections see Figures 1.4 and 1.5). After core installation the pipe is filled with insulating oil which is maintained under a pressure of 1.38 MN/m^2 (200 psi). This is known as the high-pressure oil-filled pipe-type cable (HPOF) and is widely used in the USA and elsewhere. The pipe is generally provided in 20 m (60 feet) lengths which are welded together on site using backing rings. During installation, extreme care is taken to prevent moisture entering the pipe, and at the end of a day's operations the pipe is sealed and filled with a dry gas. The reels containing the insulated conductors are

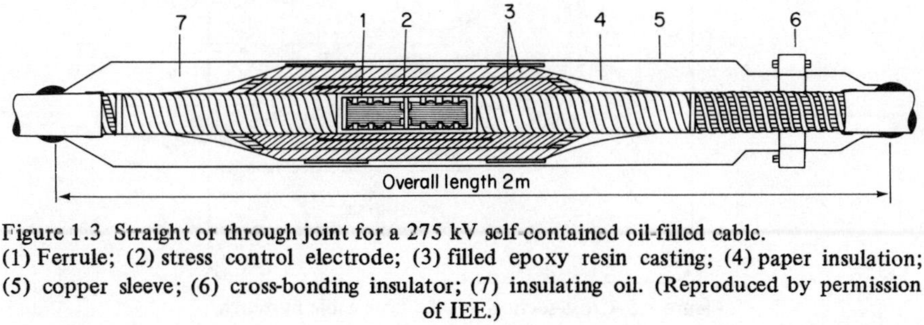

Figure 1.3 Straight or through joint for a 275 kV self contained oil-filled cable.
(1) Ferrule; (2) stress control electrode; (3) filled epoxy resin casting; (4) paper insulation; (5) copper sleeve; (6) cross-bonding insulator; (7) insulating oil. (Reproduced by permission of IEE.)

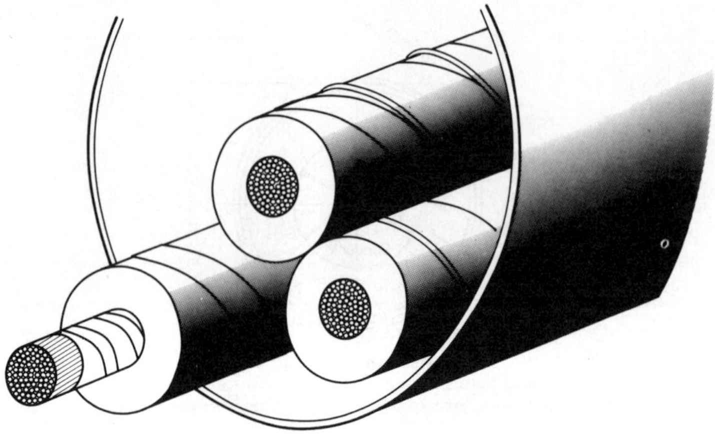

Figure 1.4 View of high-pressure oil-filled pipe-type cable.

hermetically sealed when shipped from the factory. The installation of the pipe-type cable must take place on a dry day and the splicing operations call for precision and cleanliness. Strict temperature and humidity limits are observed.

The self-contained LPOF cable[4] is completely finished and sealed at the factory and is transported containing oil under a pressure of about one atmosphere.

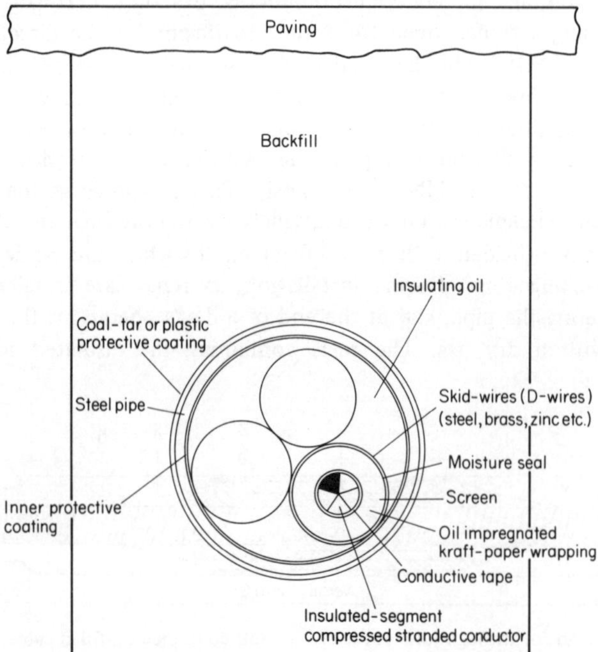

Figure 1.5 Cross-section of pipe-type cable in trench.

A further method of preventing cavity of void formation is the use of an inert gas, e.g. nitrogen, which exerts a pressure on the paper dielectric. There are two basic types:

(*a*) External compression cable in which the gas is separated from the oil-impregnated dielectric by a diaphragm, normally a lead sheath though one design has used polyethylene. The cable is contained in a pressure vessel which may consist either of a steel pipe or a lead sheath with metallic reinforcement.
(*b*) Internal gas-pressure cable, in which the gas is contained within a lead sheath and is in contact with the dielectric thus suppressing ionization.

For both types the nominal gas (nitrogen) pressure is 1.38 MPa (200 psi) with a maximum pressure of 1.725 MPa (250 psi). The maximum conductor temperature for both types is 85 °C and the maximum stress 11 kV/mm (280 V/mil). These cables have been used up to 275 kV system voltage.

In Europe the low-pressure oil-filled (LPOF) system is widely used mainly in the form of three, separate, self-contained cables. Occasionally, at 132 kV and below, three insulated cores in one sheath are used. These cables are buried in trenches which are refilled around the cables with special backfill and the trench refilled with local soil. The maximum voltage at which LPOF cables are in use is 500 kV.[6] In high-capacity installations water cooling is incorporated to increase the current rating. The percentage of British transmission circuit lengths installed underground (mainly LPOF) is shown in Figure 1.6.

In the USA the pipe-type cable has virtually superseded the self-contained oil-filled type, as will be seen in Figure 1.7.[7] The first 345 kV pipe type cable was installed in New York City in 1964, being 24 km in length. An advantage of the HPOF cable is the longer lengths of insulated cores that can be placed on a reel, e.g. at 138 kV typically 800 to 1000 m, and at 345 kV, 600 m.

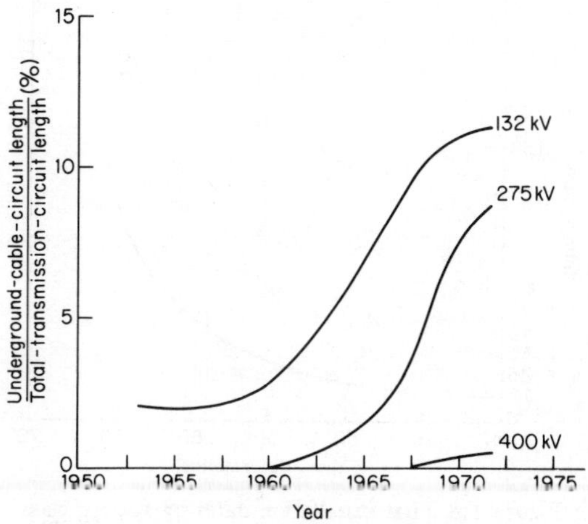

Figure 1.6 Percentage of British Transmission circuit length. (Reproduced by permission of IEE.)

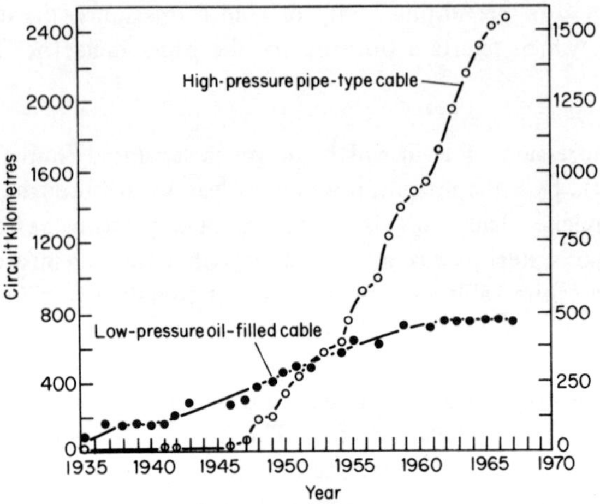

Figure 1.7 Underground transmission cable installed in the USA (69 kV to 345 kV).[7] (Reproduced by permission of IEEE.)

A further type of cable, in limited use at the lower transmission voltages (138 kV) uses solid extruded polyethylene insulation. This has the advantages of simpler splicing, ease of handling, and no requirement for expensive auxiliary equipment. The introduction of progressively higher voltages with this insulation is shown in Figure 1.8 (for the USA).[8] Unfortunately, dielectric problems have slowed progress to higher voltages. The difficulty has been that the extrusion of the insulation creates voids and inclusions of foreign material which precipitate breakdown. Intensive work is in progress to extend the voltage to 230 kV and above.

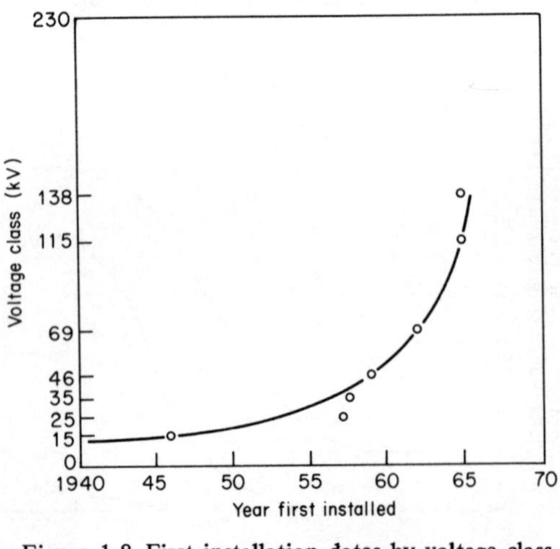

Figure 1.8 First installation dates by voltage class of polyethylene-insulated power-cable.[7] (Reproduced by permission of IEEE.)

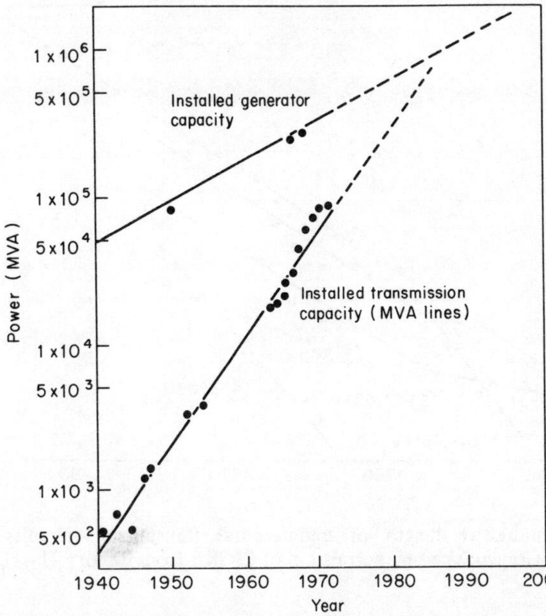

Figure 1.9 Underground transmission (MVA x number of lines) and generation capacity compared by year (Western hemisphere).[8] (Reproduced by permission of CIGRE from Report 21–11, 1976).

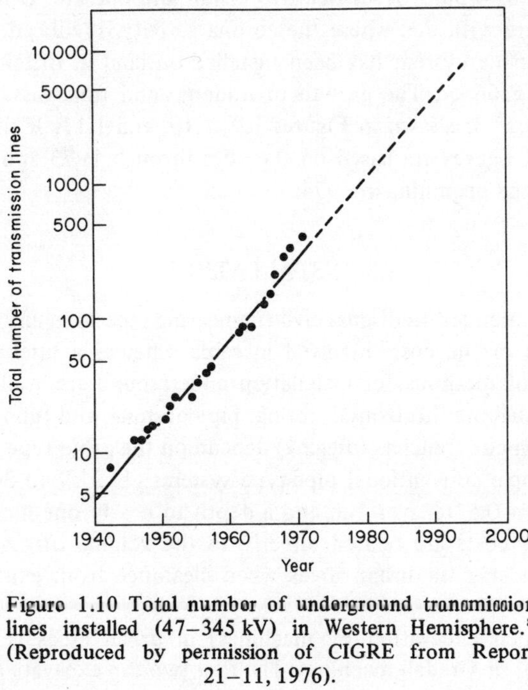

Figure 1.10 Total number of underground transmission lines installed (47–345 kV) in Western Hemisphere.[8] (Reproduced by permission of CIGRE from Report 21–11, 1976).

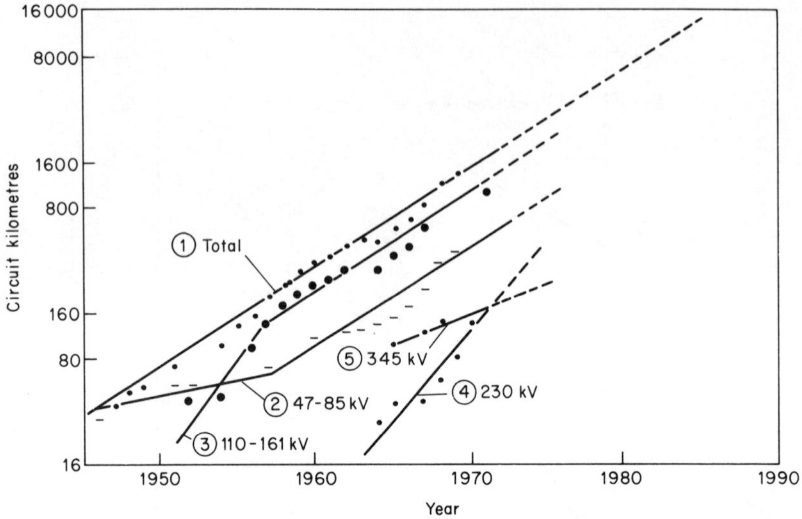

Figure 1.11 Cumulative length of underground transmission circuits – Western hemisphere[8] (Reproduced by permission of CIGRE from Report 21–11, 1976).

Although the mass-impregnated (nondraining) solid-type paper/oil cable has great limitations because of partial discharge (corona) activity in cavities, it is still used extensively for distribution networks and for underwater connections where the oil-filled systems would be difficult to install and operate. Often today these underwater links are with d.c. where the corona activity is reduced. At the time of writing, one d.c. cable system has been installed on land in Britain; this uses the LPOF system at ±266 kv. The growth of underground transmission lines in the Western Hemisphere[8] is shown in Figures 1.9, 1.10, and 1.11. It should be noted that the projected figures are based on statistics through 1973 and do not reflect reductions in demand beginning in 1974.

1.3 INSTALLATION

Although it is not intended to discuss civil engineering aspects in any detail, because of the importance of the costs involved in cable schemes a brief outline will be given. A number of methods for installation underground are available, e.g. open cut, trenching, ploughing, horizontal boring, pipe pushing, and tunnelling.[9,10] The dimensions of open-cut trenches (digging) depend on the cable type and number of circuits. For example conventional pipe-type systems (115 kV to 345 kV) require trench widths from 0.5 to over 1 m and a depth to nearly one metre. In general, open-cut trenching costs are related directly to the volume of material handled. Complications can arise in urban areas when clearance from existing utilities is difficult.

Trenching (trench excavation by machines) in urban areas is performed by backhoe, Dynahoe, or Gradall machines. The first two can excavate a width of over

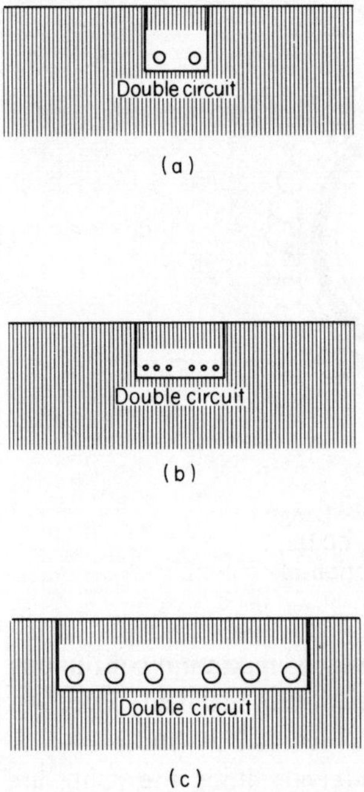

Figure 1.12 Typical cable installations below ground double circuits: (a) HPOF pipe-type; (b) extruded and LPOF self-contained direct buried; (c) C.G.I.

1 m in one pass, the latter machine a width of 2 m (6ft). Maximum rates of advance are about 25 m per day in cities and 60–100 m in suburban areas. Trenching machines are at best in soil with few obstacles and usually require level-ground clearance of several feet on both sides of the trench. Typical cross-sections of buried installation are shown schematically in Figure 1.12.

In urban areas especially in North America, the use of buried ducts is widespread. Into these are pulled the lengths of self-contained cables, either LPOF or extruded dielectric types. The use of ducts permits the minimum length of trench to be open at a given time because the road, etc., above the duct can be reinstated before cable installation commences. Ducts are made of tile, fibre, asbestos, cement, concrete, or plastic. At the present time plastic ducts, e.g. of polyethylene, are the most popular. The minimum space from the bottom of the trench to the cables is normally about 150 mm and the cover over a buried cable system about one metre.

Tunnels of small diameter, say up to 1 m, are used for cable installations under roads and railways and in special circumstances. Large-diameter tunnels (say 3 m diameter) have been purpose built for crossings under rivers and estuaries. This method of installation is costly unless combined with other utilities or for transportation. Small sections of cable circuits are sometimes installed above ground on such structures as bridges and on utility-owned land.

Underwater crossings of bays, estuaries, etc., are often essential and are frequently of considerable length, e.g. 30 km. The cables are laid on the bottom or in cut-and-cover trenches. HPOF pipe-type cables require weighting to reduce the

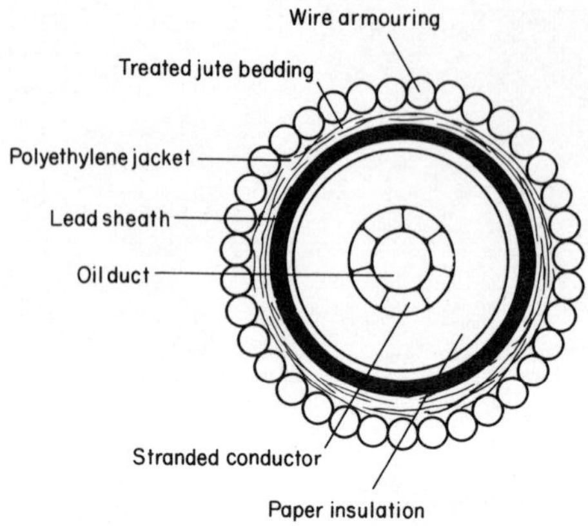

Figure 1.13 Underwater cable link across Long Island Sound using 138 kV, 300 MVA, 60 Hz, LPOF, self-contained cables – cross-section of cable.

buoyancy of the pipes before the pulling in of the cables; a maximum continuous core length in the order of 2000 m can be pulled in the installed pipe. An underwater connection using LPOF cables is shown[11] in Figures 1.13 and 1.14.

Underground vaults, manholes, or joint bays at intervals along the route are required for installation and jointing. Manholes are constructed of concrete and reinforcing steel, and must be able to support the pressures resulting from the soil and traffic above. Typical dimensions are: height 2 m, internal width up to 2 m, and length up to 10 m. A schematic diagram of a manhole is shown in Figure 1.15. The essential components in the installation of the pipe-type system are shown in Figure 1.16. An extensive treatment of the environmental impact of installation techniques is given in Reference 9.

1.4 CONDUCTORS AND SHEATHS

Conductors

Various conductor materials and their properties are given in Table 1.1. It will be seen at first glance that sodium has much to recommend it; this will be referred to later. However, copper and aluminium are universally used at the present time. The large powers of interest involve large conductors and the continued use of copper. Comparative shapes and sizes of 507 mm^2 (10^6 circular mil) conductors are shown in Figure 1.17. For the LPOF cable the conductor has a central duct (see Figure 1.2) through which the oil flows.

With large conductors carrying a.c. the skin effect has a significant influence on the effective a.c. resistance (R_{ac}) of the conductor at power frequencies. With 2000 mm^2 conductors for LPOF cables a ratio of R_{ac} to R_{dc} of up to 1.2 has been measured. This increases the conductor loss and reduces current capability. Considerable development work has been undertaken to reduce the skin effect and for

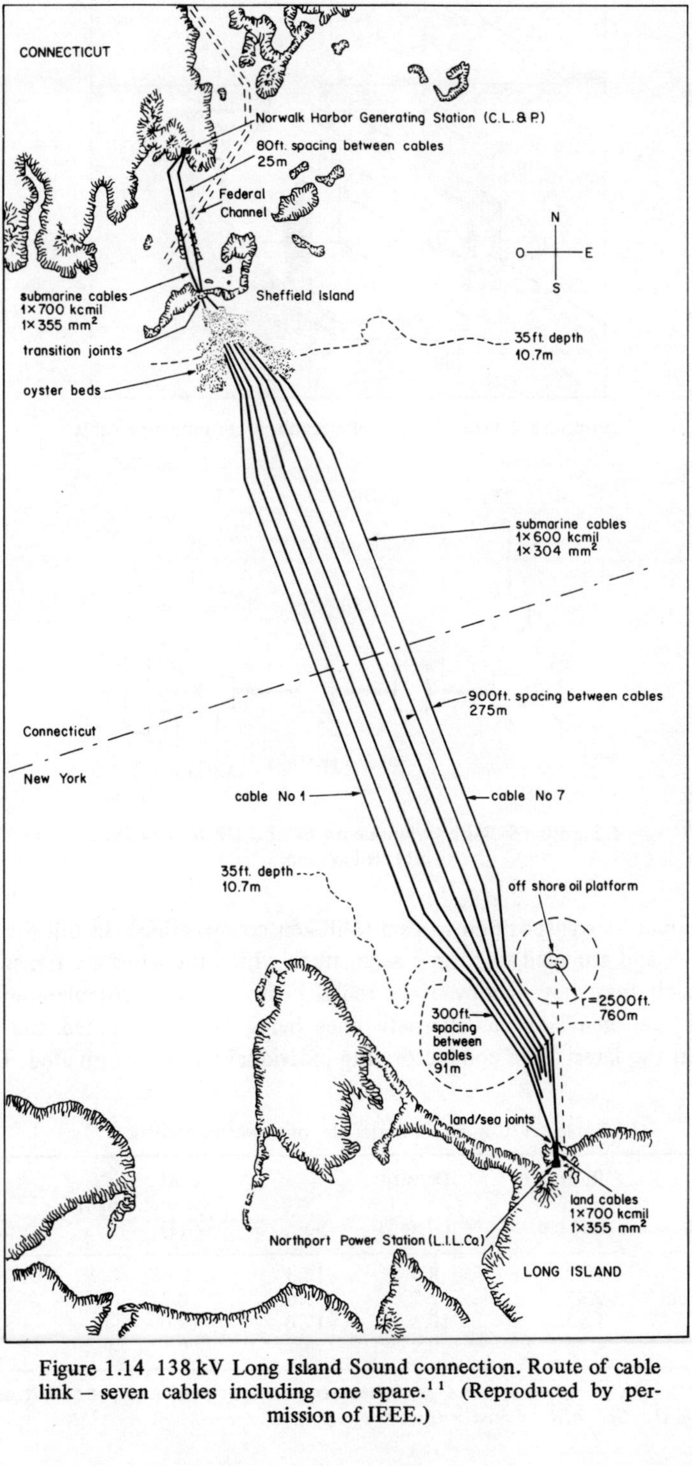

Figure 1.14 138 kV Long Island Sound connection. Route of cable link – seven cables including one spare.[11] (Reproduced by permission of IEEE.)

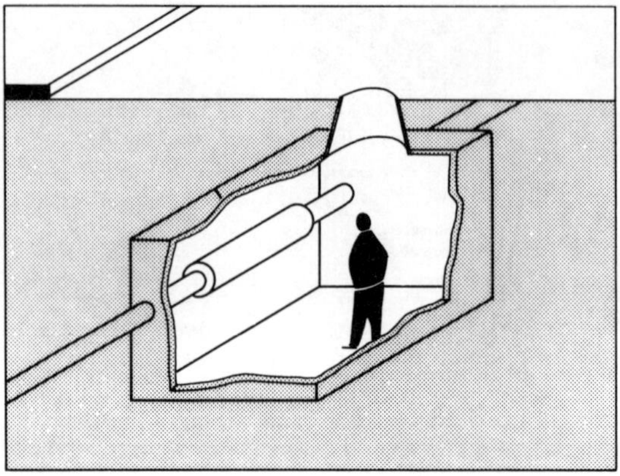

Figure 1.15 Schematic view of a manhole for pipe-type cable.

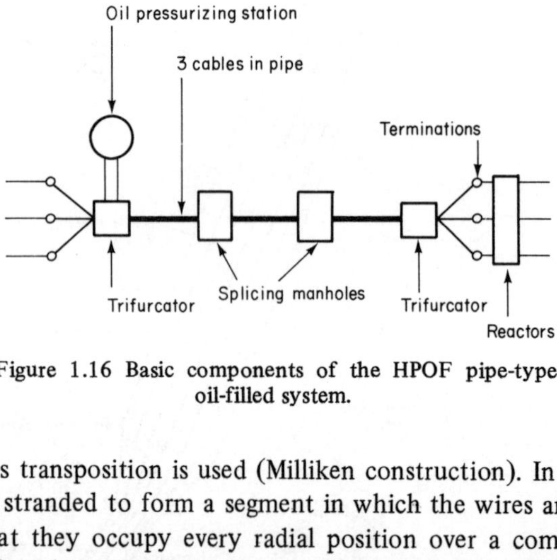

Figure 1.16 Basic components of the HPOF pipe-type oil-filled system.

large conductors transposition is used (Milliken construction). In this the wires are compacted and stranded to form a segment in which the wires are transposed, i.e. wound such that they occupy every radial position over a complete lay. Several segments are then laid up, alternate ones being lightly insulated and screened overall. In the latest large conductors the individual wires are insulated. The effect

Table 1.1 Conductor materials for power transmission.[17]

Material	Resistivity σ ($\mu\Omega$ cm)	Density ρ (g/cm^3)	$\rho\sigma$	Cost per pound weight C($)	$C\sigma\rho$
Copper	1.72	8.9	15.3	0.43	658
Aluminium	2.82	2.7	7.62	0.31	236
Silver	1.62	10.5	17.0	19.00	32300
Sodium	4.3	0.97	4.17	0.17	71.0

If l = length, A = area and R = permitted resistance of conductor, $A = \sigma l/R$ and total cost = $C\rho Al = C\rho l^2\ \sigma/R$ = Const. × $C\rho\sigma$.

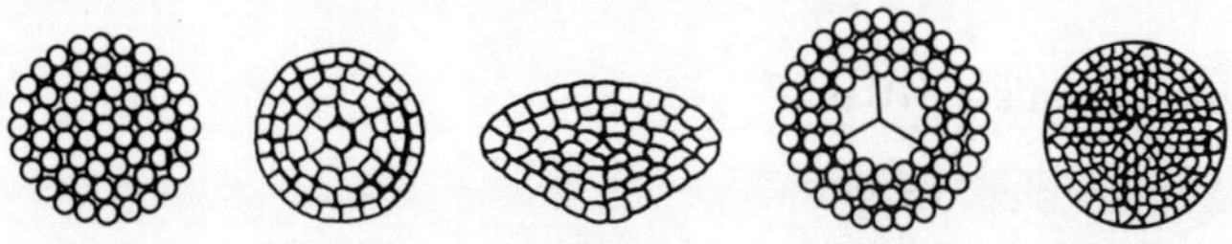

Figure 1.17 Shapes and sizes of conductors used in cables. Effective area 507 mm.[2] Types (a), (b), and (e) used for single-conductor cables. Types (a) and (c) used for three conductors in one sheath. Types (d) and (e) used for high currents (lower skin effect) – in (e) segments are insulated from each other.

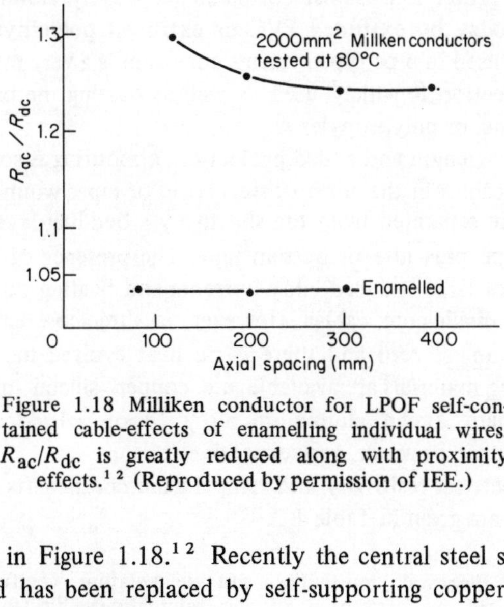

Figure 1.18 Milliken conductor for LPOF self-contained cable-effects of enamelling individual wires. R_{ac}/R_{dc} is greatly reduced along with proximity effects.[12] (Reproduced by permission of IEE.)

of this is shown in Figure 1.18.[12] Recently the central steel spiral on which the wires are formed has been replaced by self-supporting copper strips forming an annulus of the required rigidity.

A considerable development in recent years has been the widespread use of aluminium as both a conductor and sheath material. In the corrugated form for large LPOF cables it has largely superseded lead as a sheathing material and is quickly gaining ground as a conductor. A major problem with aluminium conductors has been making of effective joints which are in practice subject to thermal cycling and creep. Many techniques have been used, and today complete confidence is placed in the available jointing techniques. One reason for the use of alumium is its price stability compared with copper, which undergoes severe price fluctuations.

Conductor cross-sectional areas are described in different units in various countries. In the USA the circular mil is used (area enclosed by a circle of 0.001 in diameter) and in Europe the square millimetre. Relevant conversion factors are:

To convert mm^2 to circular mils (CM) multiply by 1970
To convert in^2 to circular mils multiply by 1.273239×10^6

e.g.

$$10^6 \text{ circular mils} = 510 \text{ mm}^2$$

$$1 \text{ in}^2 = 645 \text{ mm}^2.$$

Sheaths

For self-contained systems using impregnated paper as insulation it is essential to form a seal over the insulation in the form of a sheath. Up to recent years, lead and lead alloys have been very widely used because of ease of application, flexibility, and corrosion resistance. Aluminium is superior to lead alloy in resistance to creep, fatigue, and vibration and the absence of intercrystalline cracking. Cable sheaths must be afforded protection against corrosion (especially aluminium) and this is largely provided today by extruded PVC or extruded polyethylene over-sheaths. For the steel pipes used in pipe-type systems corrosion is a very major consideration and cathodic protection is widely used as well as coating the pipe with asphaltic mastic, polyethylene, or polypropylene.

For mechanical strength and added protection, armouring is sometimes provided for self-contained cables in the form of steel wires or tapes wound helically around the cable. These are separated from the sheath by a bedding layer comprising two layers of paper tape plus jute or hessian tape. The presence of steel wire or tape armouring produces large induced eddy currents and heating due to the magnetic field produced in single-core cables. However, in three-core cables the resultant magnetic field is almost zero and there is no heat evolved in the armour. Non-magnetic armouring materials are available, e.g. copper–silicon–manganese alloy (ρ of 24.6 $\mu\Omega$cm), hard drawn aluminium alloy (ρ of 3.1 $\mu\Omega$cm). This type of nonmagnetic armour is, however, very expensive.

Values[13] of electrical resistivity and temperature coefficients of metals used in cable construction are given in Table 1.2.

Table 1.2 Electrical resistivities and temperature coefficients.[13] (Reproduced by permission of the International Electrotechnical Commission, to whom the copyright belongs).

Material	Resistivity (ρ) ohm m at 20 °C	Temperature coefficient (α_{20}) per deg C at 20 °C
Copper	1.7241 10^{-8}	3.93 10^{-3}
Aluminium	2.8264 10^{-8}	4.03 10^{-3}
Lead or lead alloy	21.4 10^{-8}	4.0 10^{-3}
Steel	13.8 10^{-8}	4.5 10^{-3}
Stainless steel	70 10^{-8}	Negligible
Aluminium	2.84 10^{-8}	4.03 10^{-3}

Effective Resistance of Conductors

The d.c. resistance (R_{dc}) of a conductor, is given by

$$R_{dc} = R_0 [1 + \alpha_{20} (\theta' - 20)]$$

where

R_0 = d.c. resistance at 20 °C

α_{20} = temperature coefficient at 20 °C

θ' = operating temperature (°C).

The conductor cross-section normally quoted is the effective area to be used for resistance calculations. The actual physical area is larger because of the stranding or space factor. The skin effect at power frequencies becomes more pronounced as the conductor cross-section increases. Proximity effect in a conductor results from the flux from a neighbouring conductor which causes greater eddy-current induction over one half of the cross-sectional area than over the other. Although normally small this can become significant when large cables are in touching formation.

The effective a.c. resistance (R_{ac}) of a conductor is given by:

$$R_{ac} = R_{dc}(1 + y_s + y_p)$$

where y_s and y_p are the skin and proximity effect factors respectively.

The mathematical treatment of skin and proximity effects is complex and involves the use of Bessel Functions. These treatments are widely reported in a number of tests on both circuits and field theory.

The skin effect factor,[13]

$$y_s = \frac{x_s^4}{192 + 0.8x_s^4} \tag{1.1}$$

where

$$x_s = \frac{8\pi f}{R_{dc}} \times 10^{-9} k_s.$$

k_s is tabulated in Table 1.3.[13] Also,

$$y_p = \frac{x_p^4}{192 + 0.8x_p^4} \left(\frac{d_c}{s}\right)^2 \left[0.312\left(\frac{d_c}{s}\right)^2 + \frac{1.18}{x_p^4/(192 + 0.8x_p^4) + 0.27}\right] \tag{1.2}$$

where

$$x_p^2 = \frac{8\pi f}{R_{dc}} \times 10^{-9} k_p$$

s = spacing between cable areas (cm)

d_c = conductor diameter (cm).

k_p is tabulated in Table 1.3.

For pipe-type cables the following formula is often used for the conductor a.c. resistance

$$R_{ac} = R_{dc}[1 + 1.7(y_s + y_p)].$$

This is an empirical relation for existing sizes of cable, and caution should be exercised when considering higher-voltage systems.

The analysis of skin and proximity effects for solid conductors is well documented, but for stranded conductors it becomes more complex and uncertain. Recourse is made in practice to the measurement of a.c. resistance. For large conductor sizes, considerable care is needed to obtain realistic measurements and appropriate techniques have been reported.[14,15,16]

Typical values of a.c. resistance for typical sizes of conductor in pipe-type and self-contained cables are as follows (US sizes):

(*a*) Pipe-type (cables in steel pipe and in trefoil formation and occupying 60% of pipe space)

Area (mm²)	507	761	1014	1269
Area (CM)	10^6	1.5×10^6	2×10^6	2.5×10^6
R_{ac}(Ω/phase/km)	0.058	0.0475	0.0425	0.04

(*b*) Self-contained (round concentric stranded copper) – not in a pipe

Area (mm²)	507	761	1014	1269	2000 (LPOF)
Area (CM)	10^6	1.5×10^6	2×10^6	2.5×10^6	–
R_{ac}/R_{dc}	1.068	1.145	1.239	1.336	–
R_{ac}(65 °C) (Ω/phase/km)	0.04375	0.03125	0.0256	–	0.0124 (85 °C)
Conductor diameter (mm)	29.2	35.8	41.4	46.3	57
Conductor duct diameter (mm)	–	–	–	–	16.5

Table 1.3 Experimental values of k_s and k_p for copper conductors.[13] (Reproduced by permission of the International Electrotechnical Commission to whom the copyright belongs).

Type of conductor	k_s	k_p
Round, stranded	1	0.8
Round, compact	1	0.8
Round, segmental[a]	0.435	0.37
Hollow	b	0.8
Sector-shaped	1	0.8

[a]Conductors having 4 segments (with or without central duct) and sectional areas up to 1 500 mm².

[b]The following formula should be used for k_s:

$$k_s = \frac{d'_c - d_i}{d'_c + d_i}\left(\frac{d'_c + 2\,d_i}{d'_c + d_i}\right)^2$$

where:

d_i is the inside diameter of the conductor (central duct), cm
d'_c is the outside diameter of the equivalent solid conductor having the same central duct, cm

1.5 JOINTS AND TERMINATIONS

The maximum length between joints is determined by the size of the reel or drum; for most cable circuits several lengths must be joined. At high voltages the design of the joint is complex and critical, especially as the splice must be applied by hand *in situ* and hence does not possess the electric strength of the cable insulation because of the presence of moisture. A straight-through joint (connecting two lengths) is shown in Figure 1.3. Self-contained oil-filled systems need special joints to hydraulically isolate sections of cable and also to allow oil from the reservoirs to enter the cable (feed joints). Both from an electrical and thermal standpoint, the joint represents a crucial part of a cable system.

At the input and output ends of a high-voltage cable the conductor emerges from the cable insulation and enters the atmosphere and thence is connected to an

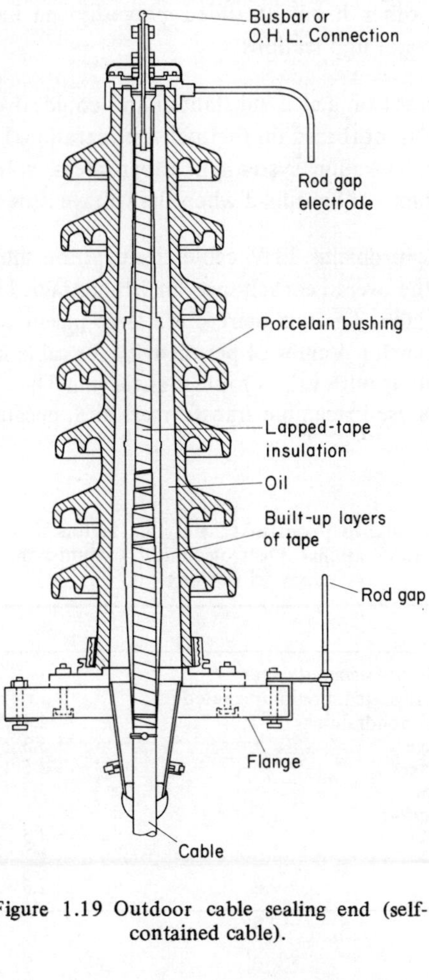

Figure 1.19 Outdoor cable sealing end (self-contained cable).

overhead line, switch, or transformer. The effective containment of the nonradial electric field in these terminations calls for devices with special insulation systems known as sealing ends or pot-heads. The outer stress containment is provided by a porcelain housing which is shaped to minimize external surface-leakage currents. Normally termination of the cable circuit or transition to an overhead line is accomplished in a substation. Typical areas required at 345 kV are 745 m^2 (8000 ft^2) for a single-circuit termination and 372 m^2 (4000 ft^2) for a single-circuit transition to an overhead line. The heights of typical terminations are as follows: 138 kV, 4.9 m; 345 kV, 6.7 m; and 500 kV, 8.5 m. An outdoor sealing end for a LPOF cable is shown in Figure 1.19.

1.6 ELECTRICAL PERFORMANCE OF CABLES IN A POWER SYSTEM

The conventional cable constitutes a very good capacitor. The relative permittivity (ϵ_r) of paper/oil insulation is of the order of 3.5, resulting in a high capacitance per unit length. On the other hand, extruded polyethylene has an ϵ_r of 2.2. This capacitance has two major implications:

(*a*) The charging current on a.c. is substantial and could, over reasonable lengths, equal the rated current (based on thermal considerations).
(*b*) In the largely underground systems in urban areas, voltage-control problems result mainly at times of light load when the voltage rises are considerable.

To overcome these problems EHV cable circuits have shunt inductive reactors connected to reduce the overall capacitance. In self-contained systems this increases the cost by roughly 20%. There is therefore a large incentive to use insulation of low ϵ_r, e.g. gases or plastics. Values of permittivity for cable insulation systems are given in Table 1.4[13] along with values for loss angle (tan δ).

The shunt reactors used resemble transformers and, because of size, are located

Table 1.4 Values of relative permittivity and loss factors.[13] (Reproduced by permission of the International Electrotechnical Commission, to whom the copyright belongs).

Type of cable	ϵ_r	tan δ
Cables insulated with impregnated paper		
Solid type, fully-impregnated, pre-impregnated or mass-impregnated nondraining	4	0.01
Oil-filled, low-pressure	3.3	0.004
Oil-filled, high-pressure	3.5	0.0045
Oil-pressure pipe-type	3.7	0.0045
Other kinds of insulation		
Butyl rubber	4.5	0.05
PVC	8	0.1
Polyethylene	2.3	0.0004

above ground normally adjacent to the cable terminations. Ideally, at intermediate points along the route separate reactance stations are required although seldom provided. Space requirement for a 345 kV single-circuit reactor in an existing substation is of the order of 930 m^2 (10 000 ft^2) and for a single-circuit reactor station over one acre.

In Table 1.5 are shown the reactive power requirements for LPOF cables at European voltage levels. Reactive power generation for pipe-type cables is shown in Table 1.6.

The capacitance of a concentric conductor–screen arrangement is (outer and inner radii R and r respectively) is given by:

$$C = 2\pi\epsilon_0\epsilon_r/\ln\left(\frac{R}{r}\right) \quad \text{(F/m)} \tag{1.3}$$

The total three-phase reactive power

$$= \sqrt{3}VI_c = 3V\,2\pi f\left(\frac{V}{\sqrt{3}}\right)\left(\frac{2\pi\epsilon_0\epsilon_r}{\ln(R/r)}\right)$$

$$= \frac{4\pi^2V^2\epsilon_0\epsilon_r \times 10^{-6}}{\ln(R/r)} \quad \text{(MVAr/m).} \tag{1.4}$$

This can only be reduced by reduction in ϵ_r, e.g. from the paper/oil value of about 3.5 to the value for polyethylene of 2.2 or even better, for gas insulation with ϵ_r of around 1.0.

A further result of the cable capacitance is the magnitude of the charging current. Consider a cable fed from one end. The charging current per metre length, $I_c = (V/\sqrt{3})2\pi fC$, where V is the line–line voltage. When the length of the cable is such that the current on no-load at the sending end (the maximum value) equals the rated current (based on temperature rise) the length is termed critical (l_{cr}), there being no possibility of passing any load current through the cable without excessive temperature rise. For example, a 345 kV HPOF cable (ϵ_r = 3.5) with an insulation wall of 25 mm thickness has a critical length of 42 km (26.4 miles). It should, however, be remembered that well before this length is achieved the current rating

Table 1.5 Reactive power for LPOF cable systems

	MVA Natural cooling	Thickness of dielectric (mm)	MVAr absorption	
			No load	Full load
*400 kV, 2 x 2000 mm^2, 16 km	2220	24.5	−540	−487
275 kV, 1 x 2000 mm^2, 16 km	760	20.3	−170	−155
132 kV, 1 x 350 mm^2, 8 km	160	10.7	−14.5	−13.6

*Double circuit (E_m 15 kV/mm).

Table 1.6 Losses and reactive power (MVAr) – HPOF pipe-type cables (1020 mm² (2 x 10^6 circular mils) conductors, naturally cooled, conductor temperature 75 °C, g_s 0.90 K-m/W, ambient 25 °C) diameter over conductor 41.4 mm.

Voltage (kV)	138	230	345	500	765 (1270 mm² conductor)
Pipe size (mm)	203 (8 in)	254 (10 in)	254	305 (12 in)	356 (14 in)
Insulation thickness (mm)	12.8	21.2	26.0	31.8 (E_m = 16 kV/mm)	39.7 (E_m = 19 kV/mm)
Capacity (MVA) – naturally cooled	224	344	453	571	264
Dielectric loss (W per conductor m)	2.32	4.4	8.6	12.6	27.6
Resistive loss (W per conductor m)	24.4	22.3	17.4	13.9	11.7
3-phase reactive loss (MVAr per circuit km)	2.94	5.58	10.8	19.9	43.5
Typical reactive loss for equivalent overhead line MVAr per circuit km	0.067	0.19	0.53	1.0	–

of the cable is becoming very restricted. Cost advantages of decreasing the thickness of insulation are offset by the increased capacitance and dielectric loss.

A further consequence of the capacitance and hence ϵ_r concerns the dielectric loss and thermal rating. At high voltages the dielectric loss becomes a substantial part of the total loss in conventional oil-filled cables and reduction of the product $\epsilon_r \times \tan\delta$ is of great importance. (*Note*: dielectric loss = $V^2\omega C\tan\delta$; see Chapter 2.)

The interaction of voltage levels and flow of reactive power is well known.[17] The voltage drop across a transmission link of resistance R and reactance X with capacitive var absorption Q_c (see Figure 1.20) is

$$V_S - V_R = \Delta V \doteqdot \frac{PR + XQ_T}{V_R}.$$

Assuming that the sending end is an infinite busbar (V_S), then the transferred vars $Q_T = Q_L - \frac{1}{2}Q_c$, assuming Q_c to be lumped at each end of the cable.

If R = O, then $\Delta V = X(Q_L - \frac{1}{2}Q_c)/V_R$ and if $\frac{1}{2}Q_c > Q_L$ then ΔV is negative and there will be a voltage rise at the receiving end. Under no-load conditions, Q_L = O and $\Delta V = -\frac{1}{2}Q_cX/V_R$. Between no-load and full loads there exists therefore a considerable variation in voltage for system operators to handle.

The above treatment may be extended to a network consisting wholly or partially of underground circuits as would be found in urban–suburban areas. The obvious solution is the connection wherever convenient of shunt inductive reactors to neutralize the generated reactive power of the cables. In large urban systems the problem is aggravated by the relatively poor ability of large turbo-generators to deliver leading reactive power to the network. A 500 MW machine of low short-circuit ratio can absorb relatively much less reactive power than smaller generators.

The Consolidated Edison system in New York[18] illustrates the situation. It has a backbone of 345 kV pipe-type cables and within the city limits the distribution circuit (at 27 kV and 13.8 kV) is completely underground. Although at peak-load periods the associated capacitance is useful, at light loads it poses a serious problem. The minimum daily load during the off-peak season is about 20% of the weekday peak, and on an average day the ratio of minimum to peak is about 30%. Generally the excess vars may be absorbed as follows:

(1) Underexcited generator operation.
(2) Use of shunt reactors.
(3) Circulation of vars in selected circuits.
(4) Switching out of selected cable circuits.

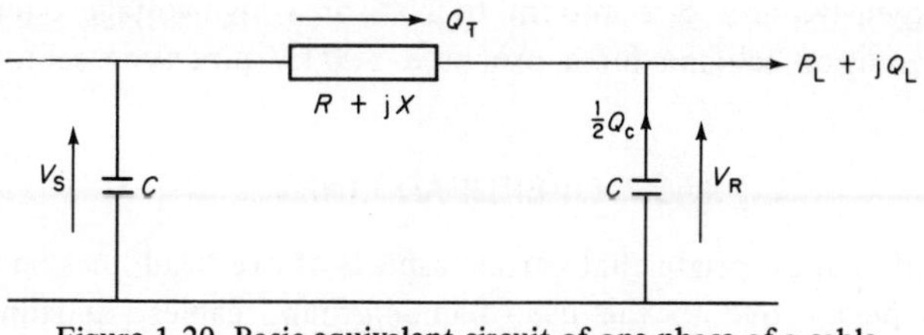

Figure 1.20 Basic equivalent circuit of one phase of a cable.

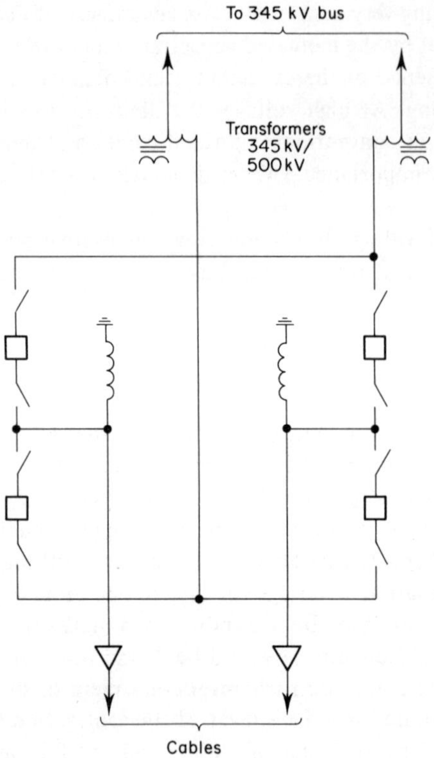

Figure 1.21 Reactive compensation – arrangements for 500 kV double circuit cables.

The generator problem has been previously mentioned. In addition to the limited ability to absorb vars there is the problem of sets being removed from service at periods of light loads for economic and maintenance reasons. These machines, if in operation, could help with the surplus var situation. With the two largest machines out of service and perhaps other machines for maintenance, it would be necessary to switch out of service selected circuits if voltage is to be adequately controlled. By the use of tap changing on transformers it is possible to force more current to flow than would do so otherwise in certain circuits and hence increase the var absorbtion via the inductive losses (known as tap-stagger). The switching out of cable circuits has certain problems due to switching surges. These may be modified by resistance switching in the circuit breakers. The installation of reactors is expensive and can amount to 20% of a high-voltage circuit costs. A typical compensation scheme for a proposed 500 kV pipe-type cable is shown in Figure 1.21.

1.7 OVERHEAD LINES

At this point it is appropriate that certain aspects of overhead lines be summarized to give some perspective to the use of underground cables. At subtransmission voltages and above, suspension-type insulators are used to support the conductors and are made of toughened glass in Europe and porcelain in North America. At the higher voltages the movement of conductors is restricted by the use of 'V' insulator strings. Conductor corona which is determined by the electric stress is reduced for a given voltage by the use of several subconductors per phase. These are separated mechanically by spacers along the span but are electrically at the same potential and are known as bundle conductors.

Regulations exist governing the minimum heights of line conductors above the ground and it is necessary to calculate the sag in the line and its variation with temperature. Although the power transmitted by a line may be limited by electrical considerations, e.g. stability, often for shorter lines around cities it is the current which is limited, this being decided by allowable temperature. Rating temperatures in the range 60° to 75° are in use and the current will depend on the ambient temperature. Hence summer, normal, and winter ratings are used. In addition the thermal time-constants of the conductors are utilized to give short-term emergency ratings. Although load currents of a cyclic nature with time attain the rated value for only a few hours per day, because of the relatively short temperature time-constant (up to 20 minutes) the line must be rated for the full-load current.

Line Route Selection

Because of the visual impact of towers and lines the selection of routes is of great importance. Attempts are made to modify the designs of towers to give reduced weights and dimensions and by the use of guyed structures. Route design must consider natural landscapes and the geology, flora, and fauna of the land. Also industrial and residential constraints must be met. Computer programs now exist which develop 'suitability' maps with suggested right of way (wayleave) corridors based on minimal environmental impact.

The effective use of a right of way (ROW) by a given line or lines may be expressed by an index defined as follows:

(Line Loading)/(Width of right of way) (Tower Height)

Values of this index for lines in the USA are given in Table 1.7. On a right of way

Table 1.7 Line parameters.

Voltage (kV)	Loading MVA	Right of way (m)	Tower height (m)	Index value	D.C. line occupying same ROW Voltage (kV)	MW
345	500	45.8	27.4	0.037	± 500	2500
500	1200	61.0	36.6	0.050	± 700	4500
765	2500	76.2	41.2	0.074	±1000	8000
1200*	7500	91.5	50.4	0.152	–	–

*Proposed design.

for one single 1200 kV line the following number of lines (circuits) are required to transmit 7500 MW: 345 kV, 15 (index 0.058), 765, 3 (index 0.085). The improvement in corridor use by increase in voltage is very evident.

D.C. Lines

The reduced insulation requirements for direct voltages and the reduced number of conductors results in a smaller, more compact, and more economic line for a given power transfer. Other factors are different from a.c., e.g. corona appears to be reduced in wet weather, audible noise levels are relatively low, and radio interference levels are less (but RI emanating from converter stations must be considered). A comparision of a.c. and d.c. line capabilities is given in Table 1.7.

Mechanical Design

For low-voltage lines all-aluminium conductors are used and the span restricted to less than 100 m because of unfavourable creep performance. For subtransmission and above the conductor, Aluminium Conductor Steel Reinforced (ACSR), is formed of strands of aluminium wire wound around a core of steel strands, the latter providing the mechanical strength. At lower voltages copper, cadmium copper, aluminium, and aluminium alloy, all hard drawn, are used. A typical cross-section of an ACSR conductor is shown in Figure 1.22.

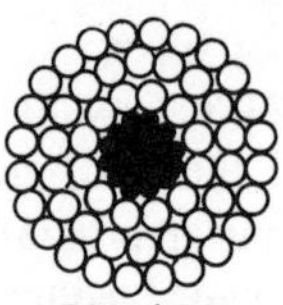

54Al/19St.

Figure 1.22 Overhead line conductors (ACSR).

An alternative to the ACSR is one using aluminium alloy instead of steel as the core with an aluminium conductor (ACAR). Advantages claimed are better corrosion resistance, lighter weight, and simpler splices. The maximum operating temperature of a conductor is governed by a number of factors; ground clearance, energy losses, run-out of grease (grease used to prevent corrosion), loss of strength due to annealing, and creep. Typical overhead line conductors in use over the range of voltage are given in Table 1.8. When determining the rating of a line solar radiation and wind velocity are included in the calculation.

Environmental Criteria for Overhead Lines

Although the visual impact of lines comes immediately to mind other aspects have become of importance for EHV lines and will become more so for UHV. These are:

(*a*) Audible noise emission (AN).
(*b*) Emission of ozone and oxides of nitrogen.
(*c*) Radio and television interference (RI).
(*d*) Safety and comfort problems caused by electrostatic fields.

Table 1.8 Typical conductors for a range of line voltages.[20] (Reproduced by permission of IEE.)

Line voltage (kV)	Conductors: Material	Number per bundle and nominal area (mm^2)	diameter (mm)
		a.c. lines	
low voltage	Al	25–100	7.2 –13.2
11–33	ACSR	25–200	7.08–19.3
220	ACSR	2 x 300	25.97
		2 x 400	28.6
275	ACSR	2 x 175	19.6
		2 x 400	28.6
330	ACSR	2 x 350	26.73
400	ACSR	4 x 400	28.6
	alloy	2 x 800	37.26
500	ACSR	2 x 800	40.70
		4 x 445	29.52
735	ACSR	4 x 689	35
765	ACSR	4 x 765	38
1100	ACSR	6 x 765	38
1500	ACSR	12 x 765	38
		d.c. lines	
±250	ACSR	2 x 930	38.43
±375	ACSR	2 x 1165	45.7
±450	ACSR	2 x 933	43.2

Criteria used in the design of lines must meet various central and local government agencies with jurisdiction over the land over which the line is constructed.

Audible noise, radio noise, and ozone emission are caused by corona and are reduced by any measures which will reduce the field strength at conductor surface. Electric fields from EHV lines induce charge on well-insulated objects, e.g. cars, trucks, etc., close to the line. The national safety code (US) recommends a limit of 5 mA for the maximum short-circuit current from the largest anticipated vehicle. Recommended criteria for EHV, a.c. lines are given in Table 1.9. The line design requires correct selection of subconductor size, phase to ground and phase to phase clearances, and mechanical loading. The size of bundle and subconductor are most crucial for the effects indicated above. The basic configuration and environmental data for 500 kV and 765 kV a.c. lines (USA) are given in Table 1.10 and for d.c. lines in Table 1.11. In EHV/UHV lines apart from visual impact the most severe design constraint is audible noise on wet conductors.

1.8 FUTURE DEVELOPMENT

The problems associated with conventional cables have been outlined and will be treated in detail in later sections. From these discussions it is evident that reduction in cost is essential, and to achieve this the development of new types of cable must

Table 1.9 Recommended electrical environmental criteria for EHV a.c. lines

Criteria	Magnitude	Justification
ES Ground gradient	Gradients not exceeding max. 13 kV/m	5 mA let-go-current criteria in forest areas and areas non-accessible by large vehicles
	Maximum 7.5 kV/m	Farmlands and areas accessible to heavy machinery 5 mA let-go-current criteria
	Around 2 kV/m at edge of right-of-way	Most lines meet this criterion
Radio interference fair weather	37 dB above 1 μV/m at the edge of right-of-way in residential areas or as legally required	IEEE recommendations of 27 dB S/N ratio, for Grade B quality reception of Class 1 stations
Audible noise wet-conductor	52 dB(A) at edge of right-of-way or as legally required for residential areas	Bonneville Power Administration data based on public response to a survey
Ozone	0.080 ppm maximum 1 hour concentration	US Environmental Protection Agency recommendation
Nitrogen oxides	0.050 ppm annual arithmetic mean or as legally required	

Table 1.10 Data for a.c. lines. (Reproduced by permission of IEEE.)

Voltage level (kV)	Bundle conductor size	Configuration	RI (WC) dBμ 2500 (MVA)	RI (WC) dBμ 5000 (MVA)	ES (ROW) (kV/m) 2500 (MVA)	ES (ROW) (kV/m) 5000 (MVA)	AN (WC) dB (A) 2500 (MVA)	AN (WC) dB (A) 5000 (MVA)	ROW widths in m 2500 (MVA)	ROW widths in m 5000 (MVA)
500	Four 45/7 Std 'Bunting' ACSR	457mm 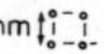	32	38	1.91	1.99	24	27	61	95
500	Three 45/7 Std 'Bunting' ACSR	457mm 	59	60	1.77	1.79	43	49	95	162
765	Four 84/19 Std 'Chukar' ACSR	457mm 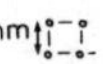	60	63	1.53	1.59	49	55	88.5	137
1100	Eight 45/7 Std ACSR	914mm  Sp	53	53	1.92	1.92	52	52	122	122

Table 1.11 Data for d.c. lines. (Reproduced by permission of IEEE).

Voltage (kV)	Line length (Miles)	Power transfer (MW)	Pole bundle design	Configuration	Max. cond. surface gradient (kV/cm)	Electric field (kV/m)		ROW width (m)
						At edge of ROW	Maximum	
± 400	200	2500	3-Bluebird		16.9	2.1	12.1	53
± 400	200	5000	6-Bluebird	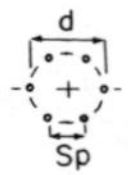	11.0	2.6	12.3	53
± 800	200	5000	6-Pheasant		23.7	2.7	12.6	111
±1200	200	5000	8-Chukar	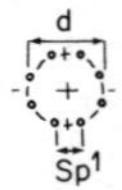	23.8	2.7	12.4	122

Sp = 457 mm Sp[1] = 350 mm d = 914 mm.

have the following objectives:

(*a*) Use of dielectrics of low relative permittivity and power factor, and of low thermal resistivity (to aid heat dissipation).
(*b*) Reduction of conductor and other losses.
(*c*) Effective removal of heat evolved by losses by special installation techniques and artificial cooling.

The pursuit of these objectives will be discussed in detail later and will only be summarized here as follows:

(*a*) Development of plastic–oil taped insulation of low ϵ_r and tan δ.
(*b*) Use of water cooling for self-contained systems and oil circulation for pipe-type system.
(*c*) Use of gas, e.g. SF_6, as insulation either in tube spacer designs or as the impregnant in plastic tapes.
(*d*) Development of extruded polyethylene solid-wall insulation for higher voltages, i.e. above 138 kV.
(*e*) Use of superconductors or low resistance conductors, involving the use of cryogenic liquids.
(*f*) Further development of high-voltage d.c. cables.
(*g*) Simplification of splices or joints.

A possibility which at the present time is considered to involve technical difficulties of such a scale as to not warrant development resources is the use of *waveguides.* The waveguides would use oversized circular pipes and work in TE_{01} mode at frequencies from 3 to 10 GHz. Attenuation of the TE_{01} mode in a circular waveguide decreases with increasing radii of pipe and increasing frequency. The variation of the specific attenuation (α) per kilometre and the maximum power

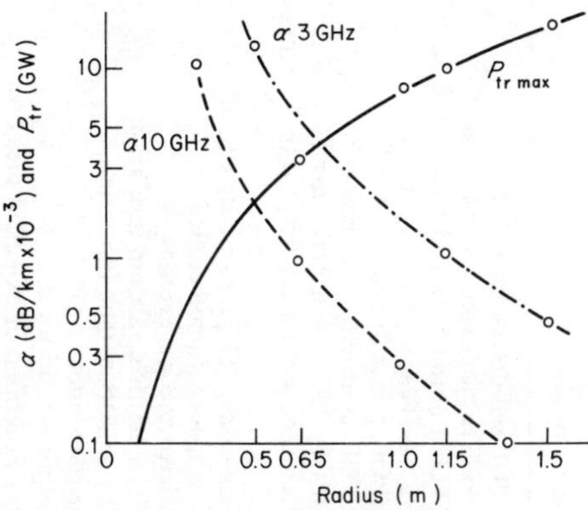

Figure 1.23 Attenuation (α) and power handling (p_{tr}) of a circular waveguide.[19] (Reproduced by permission of IEE.)

Table 1.12 Summary of existing and future cables.

Cable type	Best power level (MW)	Further development economic incentive	Comments
LPOF self-contained cable[b]	1200	–	No advantage > 400 kV
LPOF cable forced cooled (water)[a,b]	2600	–	Test cable at 765 kV
LPOF internally forced cooled with oil[a,b]	5000	Moderate	Advanced development in U.K.
HP oil-filled, pipe type, paper tape[b]	200–600	–	No advantage in voltages > 345 kV
HPOF, paper forced cooled[a,b]	500–1500	Moderate	Pay off small at lower voltages
HPOF, synthetic, laminate (PPC) tape[a,b]	300–700	Moderate	No advantage in voltages > 500 kV
HPOF, PPC or synthetic, forced cooled[a,b]	800–2000	Moderate	Pays off at higher voltages
Extruded solid dielectric, Cu or Al conductor[c]	200–600	Large	Available now at lower voltages, dielectric breakdown poorly understood
Gas-spacer (SF_6), naturally cooled	500–1500	Moderate	Compatible with overhead
Gas-spacer, forced cooled	1000–5000	Large	Very large capacity
Microwave waveguides	Enormous	None	State of the art not yet ripe
Cryoresistive, vacuum insulation	800–2000	Dubious	
Cryoresistive, tape insulation	1500–5000	Dubious	
Superconducting a.c.	1500–5000	Moderate	Prototype cables installed in 1980
Superconducting d.c.	>5000	Moderate	
D.C. oil-filled or extruded dielectric	1000–5000	Large	Use limited by converter costs

[a] Can be laid in short lengths.
[b] Length limited by charging current and compensation.
[c] Most advantageous when laid in long lengths; compatible with integrated laying techniques.

transmittable with radius are shown in Figure 1.23. For 4 GW (4000 MW) of power to be transmitted over 1000 km with the same loss as with a 50 Hz EHV, a.c., line (about 1 dB) the waveguide diameter will be at least 1.15 m at 3 GHz and 0.65 at 10 GHz, giving power transfers of 10.6 and 3.4 GW, respectively. The relative merits of such a system have been discussed by Paul.[19] A summary of present and future underground systems and their capabilities is given in Table 1.12.[10]

1.9 SUMMARY

In this chapter the conventional cable systems have been briefly summarized and the major problems identified. Crucial areas have been seen to involve the dielectric and heat transfer. New systems which have large power carrying capability have been outlined ranging from some already being used on a limited scale to others requiring several years of research and development.

The paramount problem with both novel and established methods for undergrounding is, however, cost. All schemes are more expensive than the overhead equivalent. Technical advantages in undergrounding are freedom from direct lightning strikes and elimination of hazards caused by vehicles colliding with poles, etc. Against this cables are injured by digging operation and the location of faults and subsequent repairs take a considerable time. The overwhelming argument for undergrounding is on grounds of visual amenity. In urban areas this is accepted, but in suburban and rural areas is more debatable. However, it would seem that over part of the routes carrying the large transfers of power envisaged in the last decades of this century undergrounding will be inevitable and great efforts will be required to achieve this at reasonable cost.

REFERENCES

1. Barnes, C. C., 'Power Cables Old and New', *I.E.E. Electronics and Power*, 1976.
2. Emanueli, L., *High Voltage Cables*, Chapman and Hall, London, 1926.
3. *Underground Systems Reference Book*, Edison Electric Institute, 1957.
4. Avila, C. F., Status Report, 'Underground Transmission Technology', *Power Engineering (USA)*, March, 1969, pp 35–38.
5. Barnes, C. C., *Electric Cables*, Pitman, London, 1964.
6. Ray, J. J. *et al.*, '525 kV Self Contained Oil Filled Cable Systems for Grand Coulee Third Power Plant Design and Development', *I.E.E.E. Trans. Paper T73*, 492–6, 1973.
7. Avila, C. G. and Corry A. F., Underground Transmission in the United States, *I.E.E.E. Spectrum*, March 1970.
8. *Cable Development in the USA*, CIGRE, Report 21.11, 1976.
9. *Study of Environmental Impact of Underground Electric Transmission Systems*, EPRI Report 7826, ERDA E(49-18)-1608, May 1975 (USA).
10. *Underground Power Transmission*. Study by Arthur D. Little Inc. for the Electric Research Council, ERC Publication No. 1–72 (USA).
11. Gazzana-Priaroggia, P. *et al*, 'The Long Island Sound Submarine Cable Interconnection', *Trans I.E.E.E. PAS-90*, 1971, pp 1863–1873.
12. Miranda, F. J. and Gazzana Priaroggia, P. 'Recent Advances in Self-contained Oil-filled Cable Systems', *I.E.E. Electronics and Power*, Feb. 1977, pp 134–140.

13. *Calculation of the Current Rating of Cables (100% load factor)*, I.E.C. Report 287, 1969.
14. Ball, E. H. and Maschio, G., 'The A.C. Resistance of Segmented Conductors as used in Power Cables', *I.E.E.E. Trans. PAS-87*, 1968.
15. 'Problems in the Measurement of A.C. Resistance and Reactance of Large Conductors', *I.E.E.E. Trans.*, Vol. 67, 1948, pp 1390–1397.
16. Wiseman, R. J., 'Resistance of Large Size Conductors in Steel Pipe or Conduit', *I.E.E.E. Trans.*, Vol. 7, 1948, pp 1745–1758.
17. Weedy, B. M., *Electric Power Systems*, Third Edition, Wiley, Chichester, 1979.
18. Balet, W. J. and Webb, R. L., 'Reactive Power and its Control in a large Metropolitan Electric Supply System.', *I.E.E.E. Trans. PAS-87*, 1968, pp 49–52.
19. Paul, H., 'Power Transmission of the future', *I.E.E. Electronics and Power,* 1970, p. 171.
20. Ash, D. O. *et al.*, Conductor systems for overhead lines, *Proc. I.E.E.*, **126**, 333–339, (1979).

2

High-Voltage Insulation

2.1 INTRODUCTION

It is intended here to outline the various systems of materials in common usage for cables for bulk electic power transmission. The forms of insulation may be broadly divided into paper or plastic tapes in combination with insulating oil or gas and solid wall extruded insulation. Gas and low-temperature insulation will be dealt with separately in Chapters 6 and 7 respectively. Insulation and its associated parameters represents a complex subject, and the present treatment will be confined to direct applications to cable design rather than a discussion of the physical and mathematical theories related to dielectrics in general. Before proceeding to practical insulations a brief summary of the essential properties will be given.

2.2 BREAKDOWN IN SOLID DIELECTRICS

There are a number of possible mechanisms for electrical breakdown in a solid dielectric:[1,2,3,4,5,8]

(a) Disruptive

In this, the high-energy fields cause molecular disintegration of the material associated with high energy release.

(b) Conductive

Ionic charge carriers move in an electric field, producing a leakage current. Additional ions are produced by collision of electrons and molecules so that the current increases with electric stress. Eventually the large ion and electron flow represents a conducting path and breakdown occurs. Several mathematical theories for this process have been developed.

(c) Thermal

Heat is evolved in a dielectric due to its electric resistance and (on a.c.) dielectric loss. The losses created result in an increase in temperature, often in a localized area, and hence there is a further increase in loss which produces yet higher temperatures, and so on.

(d) Electromechanical

Compressive electrostatic forces become such that mechanical followed by electrical failure occurs. Moisture moving under the influence of these forces in a solid dielectric may form a conducting bridge between electrodes or partially bridge the dielectric so causing excessive electric stresses in the remaining dielectric and hence breakdown.

(e) Electrochemical

Impurities such as waters, acids, and particles of foreign matter cause leakage currents resulting in reactive by-products which may be detrimental to the dielectric. If a dielectric contains a particle of higher relative permittivity (ϵ_r) or of higher electric conductivity, than the bulk material, the field inside the particle is lower than the bulk value. At the edges of the particle the high-voltage gradient may lead to local breakdown and carbonization by the resulting high temperature which decreases the resistivity. Even in a multi-layer dielectric the inclusion of a particle may gradually produce a conducting path and hence breakdown.

(f) Partial Corona Discharges[6,7]

Insulation often includes small inclusions or cavities (voids containing air or gas) of different ϵ_r to the bulk material thus creating higher electric stresses. The ions always present in the gas acquire velocity and kinetic energy due to the field and collide with un-ionized molecules. If the moving ions possess sufficient energy an electron is removed from the neutral molecule creating further ions, and a gaseous discharge is created. The high-speed electrons in striking the cavity walls cause erosion of the dielectric material and eventual breakdown. Breakdown may also be caused by chemical attack by the ozone created by the discharge or by surface tracking caused by the carbonization of the dielectric surfaces. A further aspect of this phenonmenon is found in the bubbles created in liquid insulation, in which discharges may occur causing eventual breakdown. A rapid increase in dielectric strength strength is observed as the thickness of a film of insulation (gas or solid) is reduced. In a gas this is due to the inability of the electrons to form collisions as the distances they have to travel approach the mean free path.

2.3 BREAKDOWN IN LIQUIDS[1,2,3,4,5]

Several theories have been proposed to account for breakdown phenomena in liquids. In uncontaminated liquids (if such exist in practice) the electric stress causes the formation of bubbles by the creation of heat or cavitation. These bubbles elongate and align themselves in the field direction and the discharges occurring eventually bridge the electrodes causing breakdown.

Krasucki[9] observed that in a very viscous liquid (hexachlordiphenyl) bubbles are initiated and grow from a pointed electrode, and that the rate of growth of the bubbles is consistent with all the electric energy developed being used in forming the vapour. Other theories involve the presence of particles, e.g. cellulose fibres, which under the action of the electrostatic forces move and eventually form a bridge between the electrodes, causing breakdown.

2.4 BREAKDOWN IN GAS[1,2,3,4,5]

Gases have low d.c. conductivities and low loss-tangents and are hence of interest as insulators. The current–voltage characteristic for a gas is illustrated in Figure 2.1. At the lower voltages there is ohmic conductivity due to electron emission from the electrodes, and also ions and electrons in the gas created by cosmic rays and light. The transition region immediately before breakdown is produced by the collisions with the gas molecules and the exciting electrons from the cathode. After this the avalanche or 'Townsend' phenomenon takes over with breakdown.

Breakdown occurs when

$$\gamma \exp(\alpha d) = 1$$

where d is the electrode separation and α and γ are ionization coefficients. For a uniform gap α and γ are functions of (V/Pd), where V is the breakdown voltage and P the pressure. Hence V is a function solely of (Pd) as was found experimentally by Paschen and the Paschen curves for air, hydrogen, and SO_2 are shown in Figure 2.2.

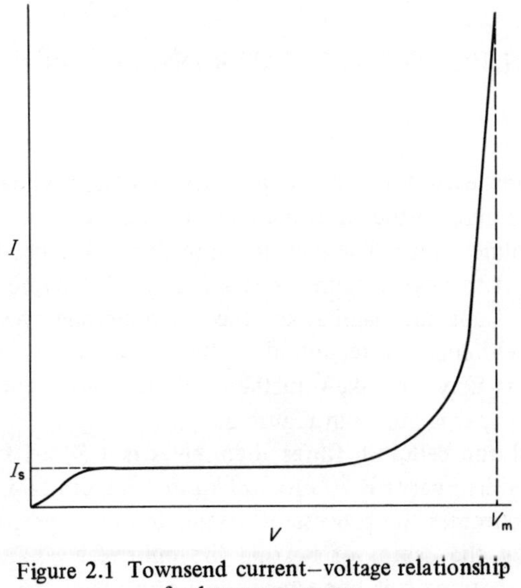

Figure 2.1 Townsend current–voltage relationship for low-pressure gas.

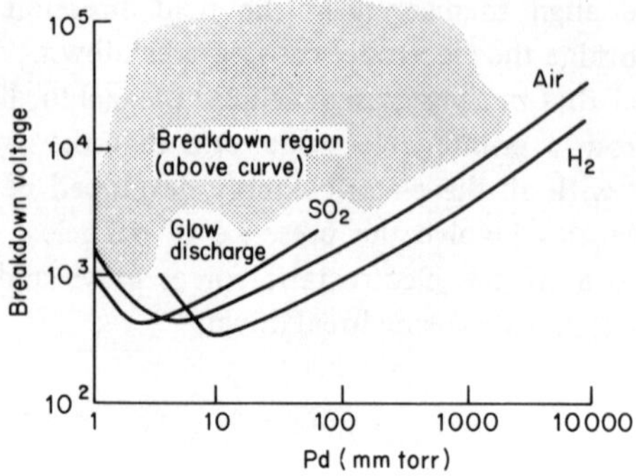

Figure 2.2 Breakdown voltage (pressure x spacing) for air and hydrogen (Paschen's Law).

The breakdown strength of a gas below atmospheric pressure (but above a moderate vacuum) is poor, and this is very relevant to cable insulation (see later section on cavities). The breakdown voltage is high at large values of (*Pd*) because the mean free path is short, and high at low values of (*Pd*) because of the small number of ionizable molecules between the electrodes. Inert gases, e.g. nitrogen, are not readily ionized. Compounds containing fluorine and chlorine are excellent in that they contain atoms which 'mop up' electrons and form negative ions which are large and relatively immobile and cannot contribute to the breakdown avalanche. Sulphur hexafluoride (SF_6) is a good example of these gases, which are known as electro-negative, and its properties are discussed further in the section on gas spacer cables.

2.5 OIL IMPREGNATED PAPER INSULATION[6,10,11,12]

Paper

The cellulose fibres used for cable paper tape are largely derived from wood, notably spruce and pine. In the Kraft (sulphate) process chips of wood are boiled in a superheated alkaline liquor. The suspension of fibres obtained is washed and then beaten by passing it through rotating rollers and finally matted on travelling gauze and dried. Fibres of cotton, manilla, and flax are sometimes used. A water washing process helps to eliminate water-soluble salts in the paper, reducing its power factor; this process may be taken further by the use of de-ionized water. The structure of Kraft paper is shown in Figure 2.3.

The density of the cellulose fibres themselves is 1.54×10^3 kg/m^3 at 20 °C, whereas that of finished paper is 700 to 900 kg/m^3. Impermeability is an important property, as it determines the process of drying and oil impregnation in the paper. It is controlled by the degree of heating the pulp receives in the paper-making process. Mechanical strength is important for the automatic tape-lapping process in which a lapping force of 22 N (5lb) is used. The tensile strength is 75 MPa

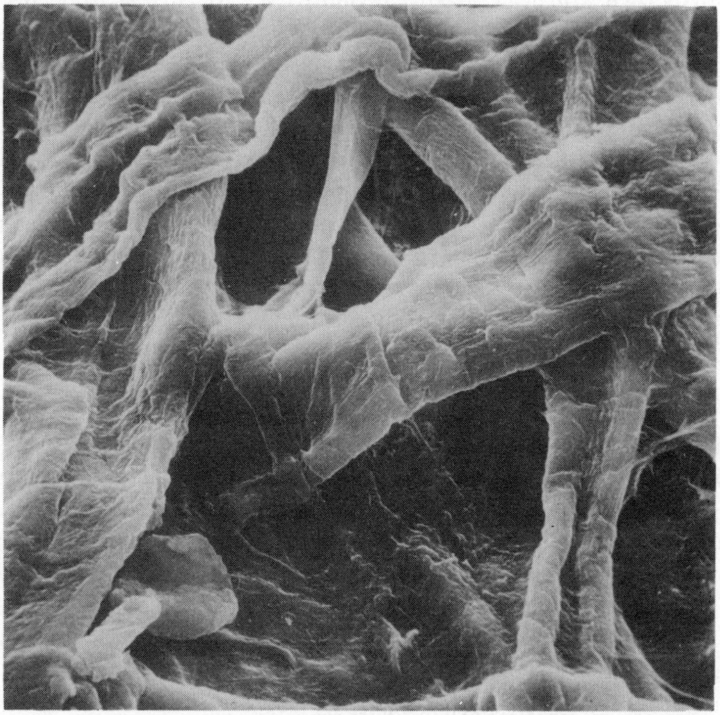

Figure 2.3 Structure of paper (magnification × 1200).

(11,000 lb/in^2) longitudinally and 34.5 MPa (5000 lb/in^2) transversely. Under normal atmospheric conditions paper absorbs moisture to about 8% of the paper dry weight. This will seriously reduce the electric breakdown strength if not corrected by suitable drying under vacuum before impregnation with oil. A smooth finish and higher density can be obtained by calendering, a process in which the paper is passed over hot rollers at the end of the manufacturing process.

Insulating Oils

These may be divided into two types, (*a*) very viscous oils mainly used in 'solid-type' cables and containing rosin or hydrocarbon polymers, and (*b*) low viscosity oils for 'oil-filled' systems, either high or low pressure. An oil commonly used in self-contained oil-filled systems is Dodecylbenzene (DDB) which posesses a viscosity of 12 centistokes at 20 °C and 5 centistokes at the rated cable temperature. Thin mineral oil has a viscosity of 25 centistokes at 20 °C reducing to 8.5 at 50 °C. On the other hand thick mineral oil has a viscosity of 3500 centistokes at 20 °C reducing to 48 at 95 °C. Typical physical parameters of paraffinic oil are as follows: viscosity (centistokes) 90 at 20 °C, 8.4 at 80 °C; density 863 kg/m^3; thermal conductivity 0.18 W/m-deg C, relative permittivity 2.20, tan δ at 20 °C, 0.1×10^{-4}, at 80 °C, 5×10^{-4}, and at 120 °C, 50×10^{-4}.

Pipe-type cables utilize refined mineral oil or synthetically produced oil from isobutylene. For forced-cooling, low-viscosity oils of either type are used. More recently alkylates of lower viscosity, linear decylbenzene (d.b.) and branch nonylbenzene (n.b.), have been introduced with the intention of increasing the distance between cable feeding points and for long submarine cables (up to 80 km). These alkylates absorb hydrogen evolved in the insulation over a much wider range of operating temperature than mineral oils. Also, water is produced by the ageing process in cellulose which accelerates thermal degradation, and this is again absorbed by alkylates.

Paper–Oil Insulation

Paper tapes[14] are wound helically around the cable conductor screen. A gap (butt gap) exists between adjacent turns to allow the tapes to move laterally with the bending of the cable without 'riding up' on one another. This latter process causes creases and tears and impaired electric strength. As the butt gap represents a weak location electrically, the gaps are staggered between layers by suitable registration of the tapes in the winding process. In Figure 2.4 50/50 and 65/35 registrations are shown. The radius to which a cable may be bent is of great practical importance and is related to the tape parameters as follows:

$$\frac{\text{radius of one complete helix}}{\text{bending radius of cable}} = \frac{\text{butt gap}}{\text{tape width}}$$

which in practice is in the order of 0.1. Before the main insulating paper tape is applied, the stranded conductor is taped with two layers of semiconducting (carbon

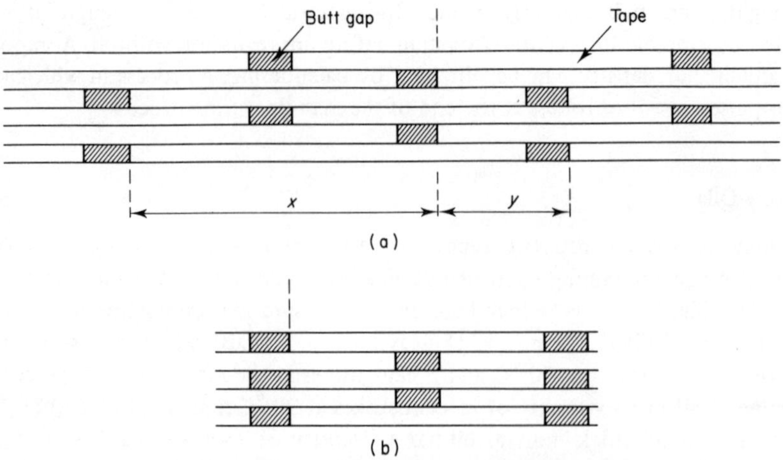

Figure 2.4 Tape registration for lapping (not to scale): (a) 65/35, (b) 50/50. Butt gaps usually in the order of 1 mm width. Registration = x/y

black) paper or metallized paper to form a screen. This evens out the local increase in electric stress due to the individual strands.

Solid-type (nondraining) insulation is used for a.c. voltages up to 35 kV and at higher voltages for direct current transmission, mainly underwater links. With this form of insulation, cavities or voids exist in the insulation because of the dissimilar thermal expansion coefficients of the paper, oil, and sheath. On heating, when on load, the sheath is stretched beyond its elastic limit and does not return to the original volume on cooling, thus forming cavities which fill with air at below atmospheric pressure. A limiting maximum stress of 4.0 kV/mm is often used for a.c. operation.

The above phenomena may be corrected by the measures which give rise to further types of cable, namely:

(*a*) Oil-filled, in which a light, low viscosity oil is used for impregnation and is kept under pressure either within the cable sheath itself or within a containing pipe.
(*b*) Gas-filled in which gas under pressure is maintained in the dielectric.
(*c*) Compression-cable, in which oil-impregnated insulation is kept under pressure by an externally applied gas pressure.

The oil-filled cable is of particular interest in the context of high-power transmission. The effect of registration on breakdown is shown in Figure 2.5.

Particles and other contaminants are invariably present in insulating oil. Under the high electric stress obtaining in the butt gaps (the relative permittivity of oil is approximately 2.0 compared with a value of 6.0 for paper) these particles are acted on by the electrostatic forces and tend to form a bridge across the butt gap causing a local breakdown in the gap. This produces heat and cracking of the oil resulting in gases, mainly hydrogen and methane (i.e. bubbles), being formed. Discharges take place in these bubbles which elongate in the direction of the field and eventually the discharges bridge the butt gap eroding the walls and eventually causing breakdown.

The following breakdown voltages are of interest: power frequency[15] 60 Hz, impulse waveform e.g. 1/50 μs for lightning, and much longer, e.g. 200/3000 μs, for switching surges and direct voltage. The voltage at which discharges commence in the butt gaps is very important in insulations prone to such action. In Figure 2.6 the improvements of a.c. and impulse breakdown strengths with increase in oil pressure (self-contained oil-filled cables) are shown.[16]

The impulse strength is often the limiting factor in the design of insulation: it can be improved by the use of higher density paper and higher viscosity oils.[17,18] Increased paper density results in higher permittivity and increased capacitance. The direct breakdown voltage is of the same order as the impulse value. It is seen in Figure 2.6 that for oil-filled systems an increase in oil pressure from say 2 atmospheres to 20 to 30 atmospheres increases the impulse strength by 10 to 15% and the a.c. strength by 70%.

The breakdown strength of the oil-impregnated tapes increases with decrease in individual tape thickness, as shown in Figure 2.7 for oil-filled cables. As the electric

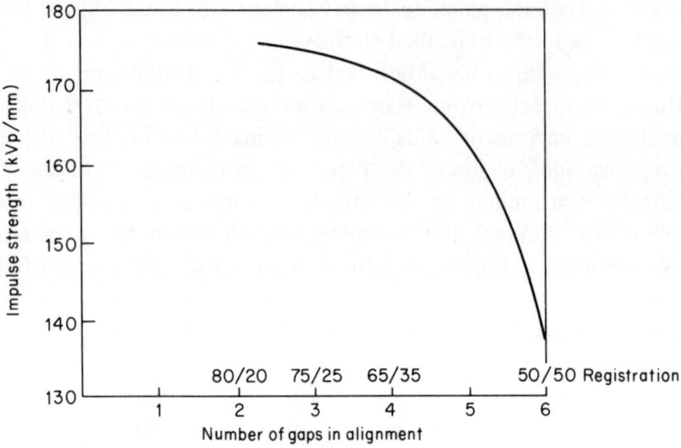

Figure 2.5 Relation between impulse strength and registration (results for paper/oil model tests).

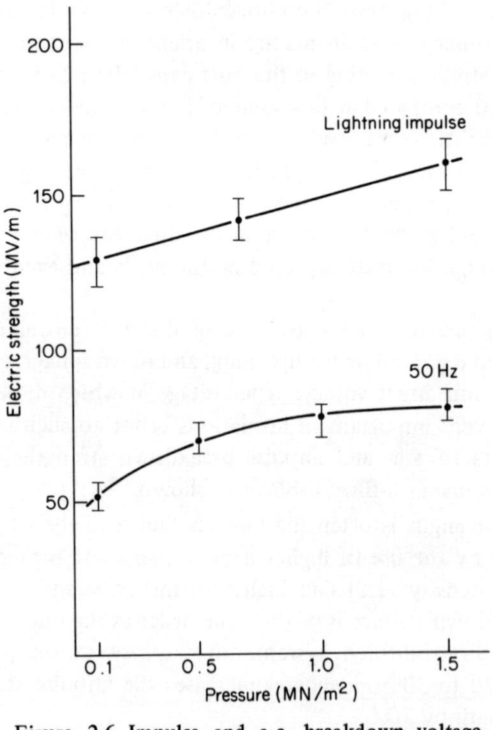

Figure 2.6 Impulse and a.c. breakdown voltage versus oil pressure for paper/oil insulation (oil-filled).[13] (Reproduced by permission of IEE.)

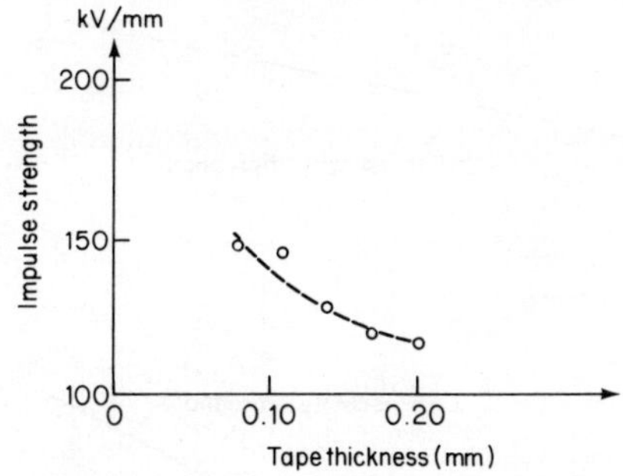

Figure 2.7 Impulse breakdown voltage versus sheet thickness – paper/oil insulation. (Total thickness 1 mm). From Lanfranconi, G. M., *I.E.E.E. Trans., PAS-93*, 1535–1543. (Reproduced by permission of IEEE.)

stress is high at the conductor and decreases towards the outer screen, economy and a reduced insulation thickness may be obtained by the use of tapes of varying thickness throughout the insulation. The relation between impulse strength and width of butt gap is shown in Figure 2.8 and between impulse strength and paper density in Figure 2.9.

Great care is taken to exclude moisture in the manufacturing process and in hand-splicing operations. Bubbles of gas are formed by electrolysis of the water absorbed in the fibres, and if the rate of gas generation exceeds the ability of the oil to absorb it into solution the bubbles grow and corona discharges occur, possibly leading to eventual breakdown.

Although it would seem attractive to reduce tape thickness (improved electrical

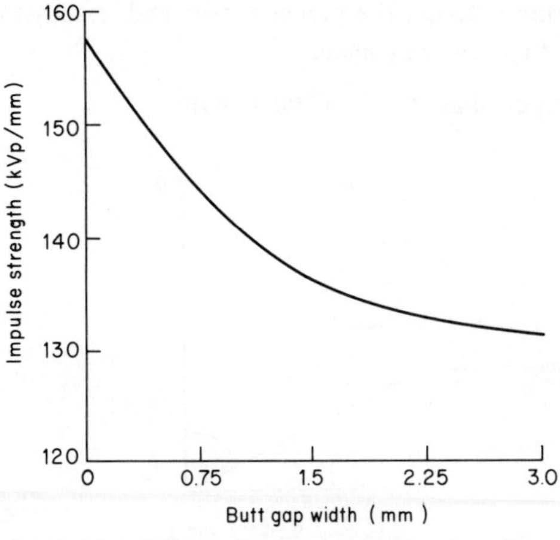

Figure 2.8 Effect of butt gap width on impulse strength – paper/oil.

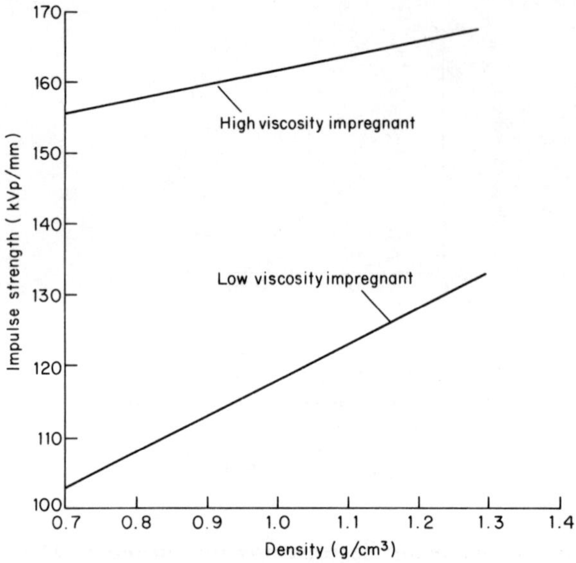

Figure 2.9 Relation between impulse strength and apparent paper density (model tests paper/oil insulation).

performance) the resulting reduction in transverse rigidity causes the paper to buckle rather than slide with cable bending. This is due to the interlayer friction which in turn depends on the lapping tension, i.e. high tension gives high friction and vice versa. If, however, the tension is low the tape will be slack and depart from its helical form. The taping tension is therefore constrained between upper and lower limits and these converge as the tape thickness decreases. Because of the moisture absorption in paper and the consequent dimensional change, lapping is carried out in an atmosphere of controlled, low humidity.

At higher operating voltages the power factor and resulting dielectric loss in the insulation become of great importance:

$$\text{Dielectric loss per phase} = V^2 \omega C \tan \delta \text{ W/m} \tag{2.1}$$

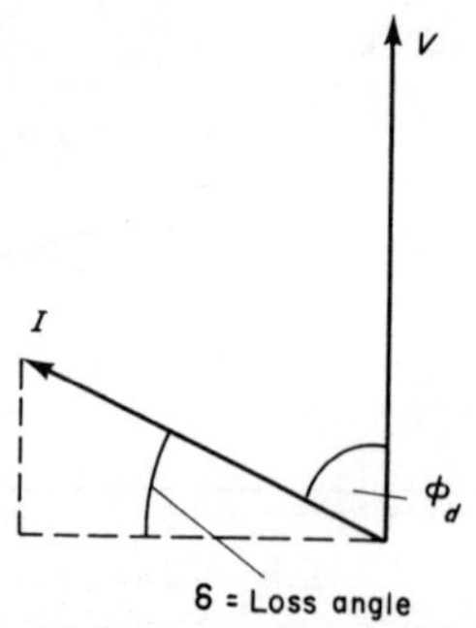

Figure 2.10 Loss tangent in dielectric. Loss = $VI \cos \phi_d = VI \sin \delta \doteqdot \omega CV^2 \tan \delta$.

where V is the line to neutral voltage, C is the insulation capacitance per metre, and $\tan \delta$ is the loss tangent (see Figure 2.10).

In LPOF cables the dielectric loss at 400 kV (50 Hz) is approximately half of the conductor ohmic loss. The variation of the dielectric loss-angle ($\tan \delta$) with temperature is important. In Figure 2.11 the $\tan \delta$–temperature curves for the paper/oil dielectric used in high-pressure oil-filled cables is shown. The dielectric loss increases with temperature, and a mechanism for *thermal instability* resulting in breakdown therefore exists. This is especially so in those parts of the insulation where high temperature and electric stress exist. The process is essentially a function of the temperature distribution and will be discussed in the section on thermal design. The effect of moisture content is shown in Figure 2.12.

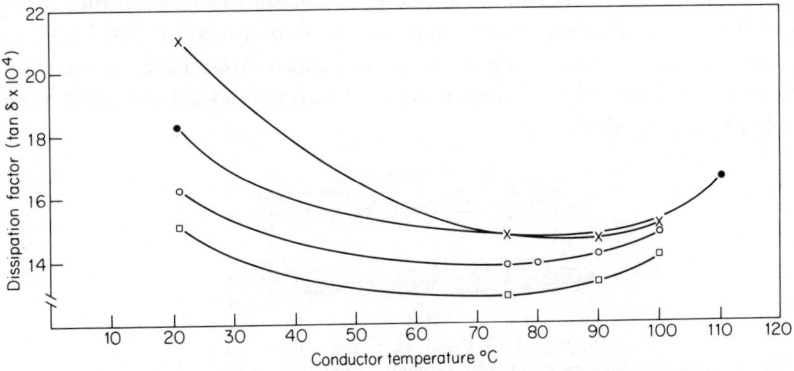

Figure 2.11 Tan δ–temperature relationship for paper/oil insulation (HPOF cable). 550 kV. Results for four test cables at Waltz Mill. (Reproduced by permission of IEEE.)

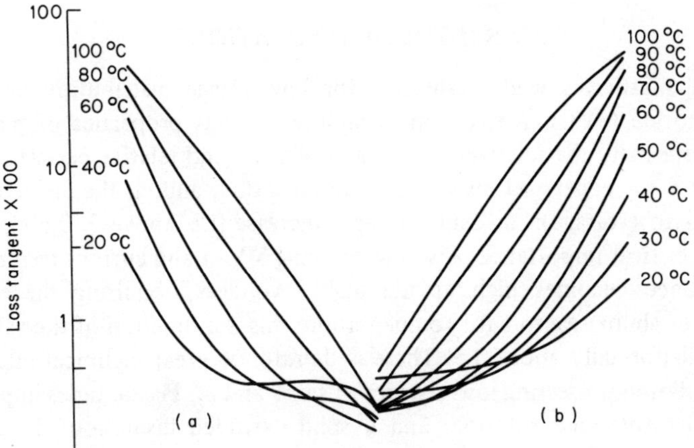

Figure 2.12 Effect of moisture content and temperature on the loss at 50 Hz of a dense paper: (a) unimpregnated; (b) oil impregnated. (Reproduced by permission of IEE.)

Screens

The screens of carbon-black paper tapes, used to 'even out' the stress variations around the conductor strands, improve the breakdown strength.[19,20] A further advantage is their ability to 'mop up' particles in the oil in the butt gaps adjacent to the conductor. Screens are also used to provide an equipotential surface on the outside surface of the insulation. The a.c. and impulse breakdown stresses are increased, but the dielectric loss also increases – although not to a serious extent. This increase in loss occurs in the oil spaces in contact with the carbon-black screen and can be minimized by the provision of only very thin gaps between the paper tapes adjacent to the screen or by the use of Duplex two-ply (carbon-plain) paper. A typical thickness for Duplex screens is 0.1 mm equally divided between the two plies. A typical arrangement of conductor and insulation screens is shown in Figure 2.13.[13] Using the Duplex screen, improvements in insulation for LPOF cables have been obtained, raising the impulse breakdown stress from in the order of 100 kV/mm to 105 kV/mm. Greater gains are seen in the a.c. breakdown strength, from 45 kV/mm to 60 kV/mm.

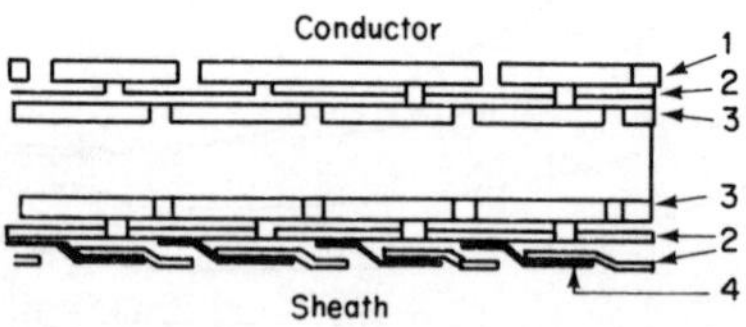

Figure 2.13 Conductor and insulation screens (LPOF cables): (1) plain carbon paper; (2) Duplex; (3) insulation; (4) AL tape.[13] (Reproduced by permission of IEE.)

2.6 SYNTHETIC INSULATIONS

The use of plastics is well established for low-voltage distribution cables, e.g. Polyvinylchloride (PVC). Apart from economy, various properties of plastics are attractive, especially the relatively low values of tan δ and relative permittivity. The value of tan δ for paper–oil insulation is about 0.002, and, as the dielectric loss is proportional to $(\text{voltage})^2$ as cable voltages increase (i.e. above 230 kV), this loss becomes of critical importance. Also the charging MVAr and current resulting from the capacitance becomes high at the higher voltages, requiring the expensive installation of shunt reactors for compensation; this is a function of the value of ϵ_r, for paper/oil normally about 3.5. There is therefore a great technical incentive to develop insulation possessing lower values of tan δ and ϵ_r. Plastic tapes impregnated with oil constitute one solution, and a solid extruded insulation of, say, polyethylene another. The problem of particle inclusion and cavities or voids has resulted in disappointing progress for the solid-wall insulation, and as yet line to line voltages above 138 kV have not been realized commercially on a large scale, although the French[21] have had test installations with such cables at 225 kV since 1969. The development of the plastic-tape oil insulation has also met with problems, mainly connected with the mechanical and chemical effects of the oil on the

Table 2.1 Insulation materials.

Material	Relative permittivity at 50 Hz	tan δ at 50 Hz	Thermal resistivity (K m/W)	Material/oil-impregnated-paper insulation cost ratio (1975, UK)	Tensile strength (N/mm^2)
Oil-impregnated paper for EHV cable	3.5	0.0020	5.00	1	–
High-density polythene	2.35	0.0001	3.5	2	15–40
Polypropylene	2.35	0.0003	4.78	2	30–40
Polytetrafluoroethylene	2.10	0.0001	6.62	20	15–35
Polyphenylene oxide	2.56	0.0004	5.15	16	–
Polysulphone	3.14	0.001	3.86	15	70
Polycarbonate	3.00	0.001	5.15	10	50–80
Polyethylene tarephthalate (Melinex)	2.96	0.002	5.95	8	150–300
Tenax paper	2.0	0.0002		50	–
Tyvek paper	2.0	0.0001		5	61
Polypropylene/paper laminate	2.7	0.0007	approx. 3	approx. 5	–

plastic. Also the ability of plastic tapes to be applied by automatic lapping machines (tensile strength) and properties when bending the cable (friction, etc.) are critical and rule out the use of some plastics which electrically would be attractive.[22,23,24,25]

Various properties of plastics of interest are given in Table 2.1. For tape–oil purposes the woven or cellulose-structured films are of interest, e.g. of polyethylene or polycarbonate, as they theoretically combine the advantages of paper (permeability, and oil in intimate contact with the material) with the lower loss and permittivity of the tape. The tape material showing most promise to date is a laminate tape of polypropylene film and paper (Low Loss Polypropylene/Paper or LLPP/Paper). Here the advantages of paper and plastic are combined. This tape has been used commercially for through-joint construction.

Synthetic Papers and Films Impregnated with Oil

Of the 50 materials which have been evaluated only two possess an impulse strength comparable with cellulose: there are polypropylene film/cellulose paper laminates

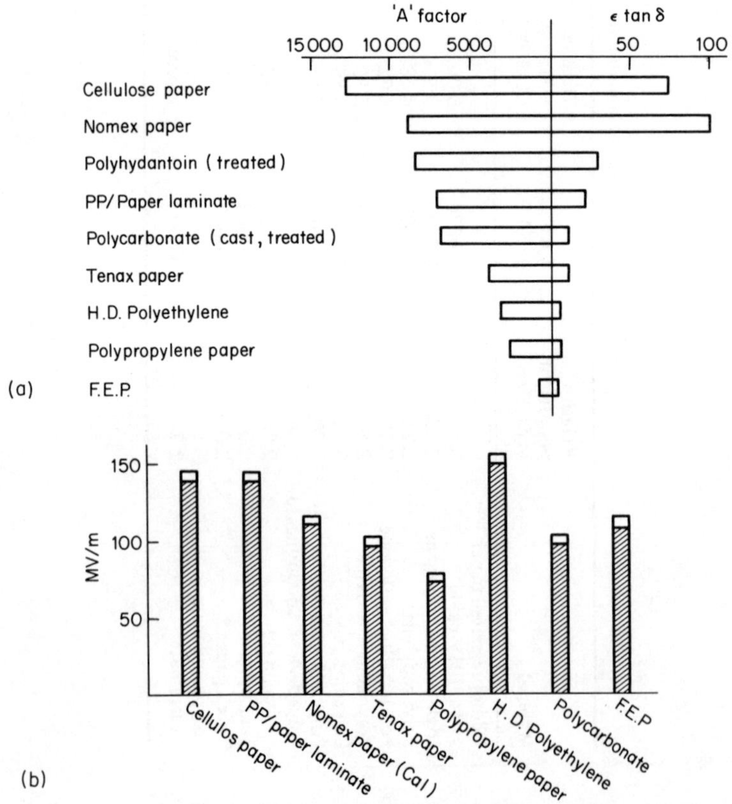

Figure 2.14 (a) Properties of synthetic tapes. (b) Impulse strengths (1/50 μs). (Impregnation with DDB sheet thickness 0.1 mm.) (Reproduced by permission of IEE.)

and high density polyethylene. The mechanical performance may be summarized by the use of a factor (A) given by[13] $(4E_1E_2G^2)^{1/4}/\eta$, where E_1 and E_2 are the elastic moduli in the longitudinal and transverse directions, G the shear elastic modulus, and η the static friction coefficient. A comparison of several materials is shown in Figure 2.14. It is seen that high A factors unfortunately corresponds with high $\epsilon \tan \delta$. A cable combining polyethylene tape and SF_6 gas under pressure has been investigated.[26]

Extruded Polyethylene

Although due attention has to be given to such breakdown mechanisms as thermal and electromechanical, by far the most acute problem is that of degradation due to treeing and partial (corona) discharges. In single-core solid extruded polyethylene cables up to 70 kV (line-neutral) the *mean* working stress (2.5 kV/mm) to date has been similar to that prevailing in solid (nondraining) paper–oil insulation.[27] The stress at the conductor is 40–80% higher according to the ratio of conductor to dielectric diameters. Cables using low density polyethylene for 225 kV (line to line) are in use in France with a stress of 8 to 9 kV/mm. Crosslinked polyethylene allows the use of higher temperatures (i.e. 90 °C) on the conductor because of its ability to remain solid at a higher temperature than ordinary polyethylene.

Three materials are now available for solid extruded cables for high voltage. These are:

High and low density polyethylene
Cross-linked polyethylene
Ethylene-propylene rubber (EPR).

2.7 PARTIAL (CORONA) DISCHARGES

In all dielectrics the possibility of small cavities and particle inclusions exists. In taped insulation the butt gaps formed between adjacent tapes form a volume of different permittivity to the surrounding paper/oil composite; in them higher stresses result with greater likelihood of discharge or breakdown. In extruded cables, e.g. polyethylene, it has been found difficult to extrude without cavities and inclusions existing in the insulation. If E_c and ϵ_c are the stress and relative permittivity in a cavity and E_d and ϵ_d are the stress and relative permittivity in the bulk of the insulation in series with the cavity then:

$$E_c = \frac{\epsilon_d}{\epsilon_c} E_d \quad \text{(for polyethylene } \epsilon_d = 2.3 \text{ and } \epsilon_c \text{ (air)} = 1\text{).}$$

Consider first a cavity (ϵ_c) and insulation in series (ϵ_d) having a common area A (m^2) as shown in Figure 2.15. The capacitances are as follows:

$$C_d = \frac{\epsilon_d A \epsilon_0}{(d - d_1)} \quad \text{and} \quad C_c = \frac{A\epsilon_0 \times 1}{d_1}, \text{ where } \epsilon_c = 1.$$

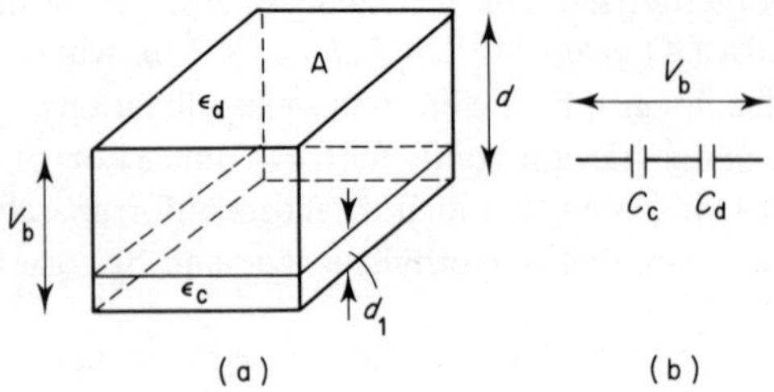

Figure 2.15 (a) Cavity and bulk insulation in series. (b) Equivalent circuit.

The voltage across the cavity,

$$V_c = \frac{C_d}{C_c + C_d} V_b$$

where V_b is the total voltage across the system. Hence,

$$V_c = \frac{V_b}{1 + (d/d_1 - 1)/\epsilon_d} \,. \tag{2.2}$$

In practice a cavity is often spherical in shape, and it has been shown that[7]

$$E_c = \frac{3\epsilon_d E}{(\epsilon_c + 2\epsilon_d)} = \frac{3E}{2} \quad \text{for } \epsilon_d > \epsilon_c \tag{2.3}$$

where E is the stress in the bulk of the insulation. Approximately, the breakdown strength of an air-filled spherical cavity of diameter d is equal to that of a parallel-plate capacitor with plates d apart. The general case of a cavity in a bulk of insulation is shown in Figures 2.16 and 2.17. Again, C_c = capacitance of the cavity and C_d that of the column of the same area; C_b = capacitance of the surrounding bulk insulation.

When the gas in the cavity breaks down, the voltage across C_c falls by ΔV_c (in about 10^{-7} s) and eventually the discharge extinguishes. The loss of charge in an

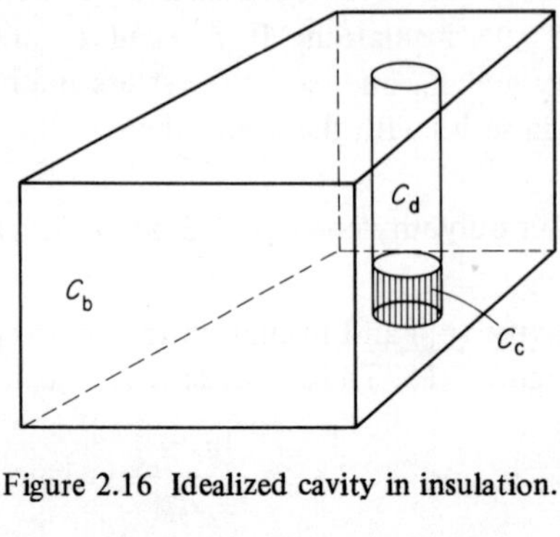

Figure 2.16 Idealized cavity in insulation.

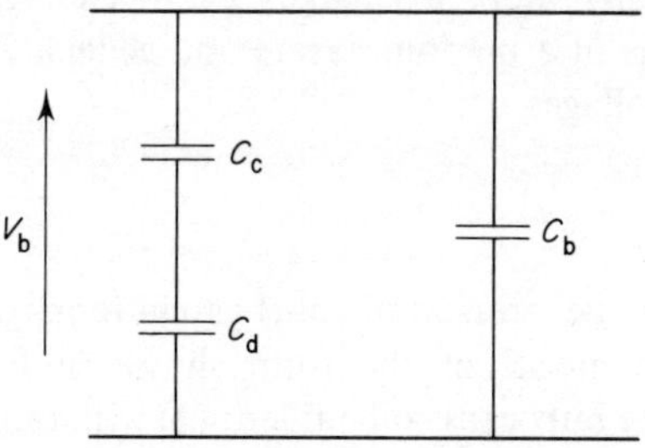

Figure 2.17 Equivalent circuit of cavity in insulation.

effective capacitance of $C_cC_d/(C_c + C_d)$ is

$$\Delta V_c \left(\frac{C_c C_d}{C_c + C_d}\right)$$

and this is made up by charge from C_b resulting in a reduction of ΔV_b in V_b.

$$V_c \left(\frac{C_c C_d}{C_c + C_d}\right) = C_b \Delta V_b$$

and as $C_d \ll C_c$ $(C \propto A/d)$

$$\Delta V_b = \frac{C_d}{C_b} \Delta V_c. \tag{2.4}$$

Internal discharges erode the cavity and the energy of the discharge raises cavity walls by hundreds of degrees Centigrade. Pits are formed and attain critical length and fine channels are propagated causing complete breakdown as the stress is very concentrated at the tip of fine channels. With only one discharge site in a cavity the waveform is as shown in Figure 2.18. Usually there are several sites with a

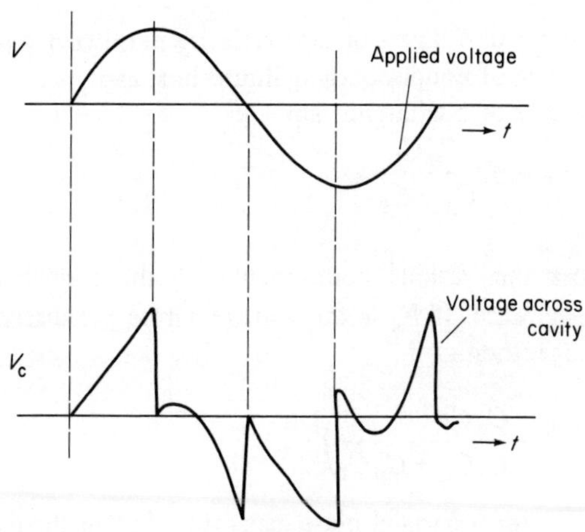

Figure 2.18 Typical waveshape of discharge voltage in a cavity.

waveform containing many more discharges per cycle. With direct voltages the incidence of discharges is of a random nature and generally the problem is not as acute as for alternating voltages.

Example 2.1

Obtain an expression for the stress in the butt gap in terms of the inception voltage for the taped insulation model of the form shown in Figure 2.19 (i.e. 65/35 registration), assuming the butt gaps to be filled: (*a*) with oil; (*b*) with gas.

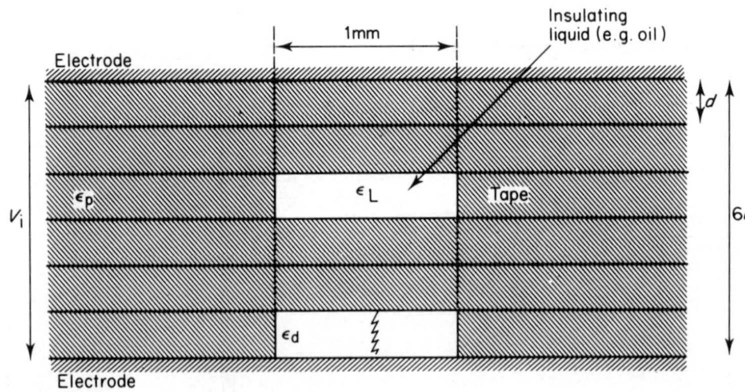

Figure 2.19 Cross-section of a taped dielectric. (65/35 registration, not to scale).

Solution

Let

C_p = capacitance of 4 layers of tape (relative permittivity ϵ_p)
C_L = capacitance of nondischarging liquid butt gap
C_c = capacitance of discharging gap

C_c be in series with $\dfrac{C_p C_L}{C_p + C_L}$.

The voltage across the various components of the dielectric will vary as the inverse of the capacitance. If V_c is the voltage across the discharging gap and V_i is the inception voltage, then:

$$V_c = V_i\left(\frac{C_p C_L}{C_c C_p + C_c C_L + C_p C_L}\right).$$

Now $C = \epsilon_0 \epsilon_r A/d$ for a parallel plate capacitor. But in the case under consideration A and d are constants for each layer. Therefore

$$V_c = V_i \left[\frac{(\epsilon_p \epsilon_L/4)}{(\epsilon_p \epsilon_d/4) + \epsilon_L \epsilon_d + (\epsilon_p \epsilon_L/4)} \right]$$

$$= V_i \left[\frac{\epsilon_p \epsilon_L}{\epsilon_p \epsilon_d + 4\epsilon_L \epsilon_d + \epsilon_p \epsilon_L} \right].$$

For paper $\epsilon_p = 6.0$, for oil $\epsilon_L = 2.00$, and for gas $\epsilon_g = 1.00$. If the discharging gap is full of oil then $\epsilon_d = \epsilon_L$. Therefore

$$V_c = 0.3 V_i$$

Stress in the butt gap = $0.3 V_i/d$. If the discharging gap is full of gas then $\epsilon_d = \epsilon_g$ and $V_c = 0.46 V_i$. The corresponding stress in the butt gap is $0.46 V_i/d$.

Measurement of Internal Partial (Corona) Discharges[7,28,29,30,31,32,33]

As already indicated, the problem of discharges is actute in solid extruded polyethylene insulation and in nondraining solid-type paper/oil insulation. In oil-filled systems the high oil pressure effectively prevents discharges. When considering the life of a cable it is necessary to quantify discharge activity and the definitions given in Table 2.2 are relevant. Large discharges result in more rapid breakdown than small ones for the same temperature and stress. However, tests on polyethylene cables have indicated that discharges of only a few pica coulombs (pC) with

Table 2.2

Term	Definition
Discharge inception voltage (V_i) kV r.m.s.	The lowest voltage at which discharges of a specified magnitude recur in successive cycles when an increasing alternating voltage is applied to insulation. V_i is quoted in r.m.s. volts, but is determined by the peak voltage.
Discharge extinction	The lowest voltage at which discharges of a specified magnitude (e.g. 5 pC) recur when an alternating voltage, which exceeded the discharge inception voltage, is reduced.
Discharge magnitude Q_b and energy	The loss of charge caused by a single discharge is measured at the terminals of a sample, i.e. $Q_b = C_b V_b$ where V_b is the voltage change across the sample of capacitance C_b. The energy dissipated by a single discharge $W \doteqdot 0.7 V_i Q$.
Discharge detector sensitivity	The magnitude of the smallest individual discharge which can be measured under particular test conditions. The sensitivity varies with the type of detector, the capacitance of the sample, and the presence of external interference.
Discharge detector resolution	The time interval between two discharges which can be separately distinguished by a discharge detector.

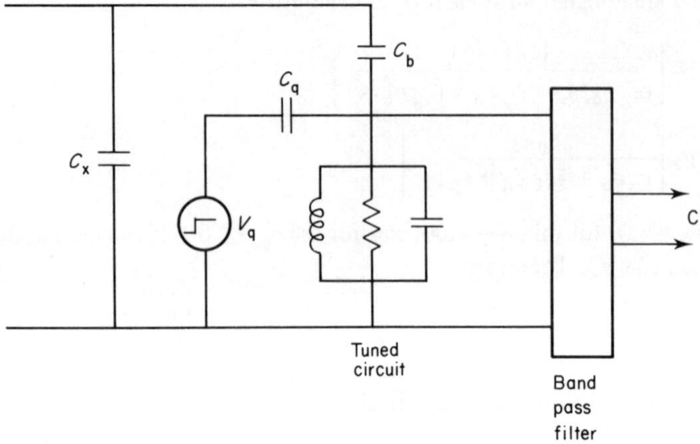

Figure 2.20 Circuit for discharge detector: C_x = test sample, C_q = calibration capacitance, C_b = blocking capacitor, V_q = calibration voltage. Discharge magnitude $q_x = e_q c_q(1 + C_x/C_b)$.

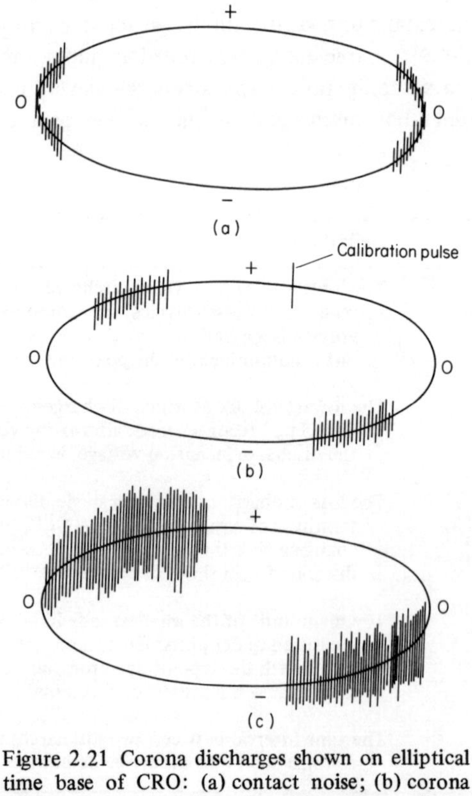

Figure 2.21 Corona discharges shown on elliptical time base of CRO: (a) contact noise; (b) corona partial discharges – just above inception voltage; (c) corona discharges – just before breakdown.

energies of about 10^{-8} Joules can cause accumulative deterioration ending in discharge through channels and breakdown. A quantitative knowledge of individual discharges is required.

The basic circuit of a common form of discharge detector[29] is shown in Figure 2.20. The test sample is C_x and the high voltage is blocked from the detector circuit by capacitor C_b. The discharges are normally displayed on an elliptic time base as shown in Figure 2.21. Various interchangeable input circuits may be used to match the capacitance of different test samples so that the resonance circuit is tuned to the midband frequency of the band-pass amplifier. The calibration pulse (V_q) is adjusted to give an excursion on the CRO screen equal in height to the discharge under study. When detection in long lengths of cable is required difficulties may arise due to travelling waves set up by the discharges, e.g. the superposition of incident and reflected waves. As discharges typically last between 10 and 100 ns travelling waves of length greater than 20 m are set up. Kreuger,[7] in work on PVC cables, found that discharges exceeding 1000 pC caused failure within hours for a stress of 6 kV/mm. Discharges in the range 30 to 200 pC caused failure in a month and those below 5 pC produced no noticeable effect.

2.8 TREEING IN EXTRUDED INSULATION

Research using transparent insulation has showed the formation of potential breakdown channels in the form of a tree, see Figure 2.22. These trees often appear to originate from discharging cavities or impurity particles.[34] Investigation has indicated that moisture can permeate extruded polyethylene causing treeing.[35] The

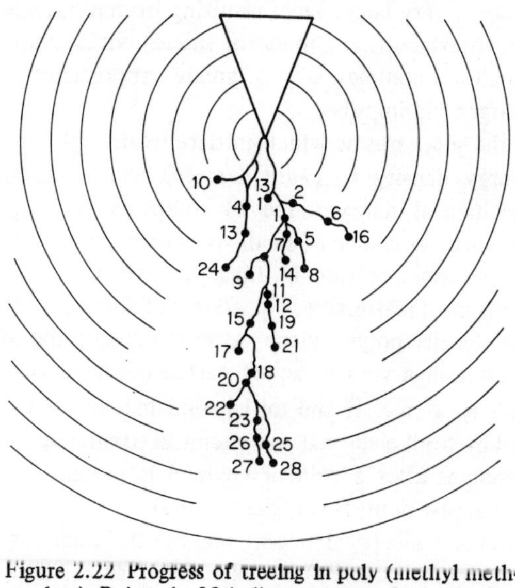

Figure 2.22 Progress of treeing in poly (methyl methacrylate). Points 1–28 indicate extent of growth after 1st, 2nd, – 28th discharge. (Reproduced by permission of IEE.)

following summary of treeing is based on the excellent review paper by Eichhorn.[36] Trees may be classfied as follows: electrical, water, and electrochemical. All are initiated at sites of high electric stress. In solid organic dielectrics treeing appears the most likely mechanism of long-term electrical failure as opposed to faster catastrophic failure. Reports that directly buried polyethylene cables which failed in service sometimes contained treeing caused a great renewal of interest in this phenomenon, although research has been in progress for many years.

A variety of tree formations have been observed and described as dendrite, fan, plume, delta, bush, broccoli, string, and bow. Trees consisting of hollow channels formed by the decomposition of material show up clearly in polyethylene, whereas water trees appear diffuse and temporary. Considerable force is required to cause the penetration into the material which starts at a surface imperfection or stress concentration. Water trees do not seem to consist of permanent hollow channels but rather fine paths along which moisture moves under the voltage gradient.

Electrical and water trees may grow from an electrode/insulation interface or from internal voids or contaminants. The term electrochemical treeing is applied to water treeing when the water contains soluble ions which move under the electric field. This tree often takes the 'bush' or 'broccoli' form. Analysis of some cables has indicated the presence of Cu_2O in the trees, implying that copper ions can move through polyethylene under some conditions. Because of this copper tape conductor shields are no longer used. Water and electrochemical trees can grow at stress levels much lower than those needed for electrical trees.

Trees which commence their growth at surfaces with an unlimited supply of air or water can grow completely through the dielectric and bridge the electrodes; these are called vented trees. Trees which start at an internal void or inclusion (nonvented) rarely grow very large. Trees resulting from moderate a.c. stresses at an electrode/insulation interface (i.e. conductor shield) which contains imperfections cause problems which are commercially of greater importance than trees resulting from internal discharges and impulses.

Possibilities for the mechanisms which initiate treeing (all requiring high stresses to provide the energy density to cause damage) are: localized heating, thermal decomposition, mechanical damage due to compression, fatigue cracking from repeated polarity changes, presence of small voids or inclusions, the development of 'hot electrons', and electron injection. Of these, electron injection causing ionization of the material seems most reasonable. Tip stresses of about 1.2 MV/mm (30 kV/mil) are required for low density polyethylene. After the absorption of sufficient energy the dielectric will contain a void in which partial discharge occurs, propagating a channel and leading to a tree. If the initial void has access to air or water, tree growth can proceed by both electrical and chemical (oxidative) attack. Tree growth due to partial discharges after a void or channel is present is an erosion process propagating at the sharpest point (i.e. highest stress).

Attempts to eliminate electrical treeing include the replacement of conducting tapes round the conductor (screens) by extruded screens, thus eliminating loose fibre ends which often protrude into the insulation layer and act as needle points. Provided tight interfacial bonds are obtained, this action is very successful. The other main approach has been the use of additives in the insulation to absorb injected electrons before they can react with the insulation. The additives or stabilizing compounds used are mainly of aromatic organics substituted with electron acceptors, i.e. are very effective in deep electron traps. N-N′-diphenyl-p-phenylenediamine, which is used as an anti-oxidant in polyethylene, has been found to be effective as a tree inhibitor, as has polysulfone The use of peroxides in polyethylene for cross-linking also gives an improvement, but this has been shown to be of a temporary nature.

Most efforts to minimize electrical treeing have concentrated on the use of very clean resins and the elimination of contamination. It has been claimed that the use of certain fillers resists tree propagation, e.g. powdered clay, carbon-black, and powdered quartz. Some workers, however, report no improvement with such fillers. Elimination of tree growth from internal voids has been attempted by the use of oil or SF_6 to fill the voids, as has the incorporation into the polymer of a semiconductive organic liquid which forms semiconducting coating on all free surfaces and prevents voltage gradients and discharges within voids. Microvoids caused by chemical crosslinking are difficult to eliminate in manufacture; however, the elimination of steam as the heat-transfer medium in crosslinking is being used to achieve void reduction. Several methods have been used to achieve this, including the use of oil or nitrogen with infra-red radiation curing, ultrasonic generation of heat, and the use of very long dies (up to 20 m) in which the extrusion, heating, and crosslinking are achieved in separate stages.

The presence of an electric field increases the penetration of moisture into organic materials with deleterious results. Polyethylene has a lower moisture permeability than any other plastic except PVC. Both water and a.c. stresses are required for water trees to grow. Increasing the concentration of occluded water decreases the tree inception voltage for water trees while increasing it for electrical trees. Moisture collects at points of high localized electric stress, resulting in water treeing. It appears that increased temperature leads to accelerated water tree growth, and thermal cycling appears to be very influential. No unifying explanation for the formation of water trees has yet been offered. Proposals include electrophoresis and dielectrophoresis, small partial discharges in channels almost filled with water resulting in cracking due to vaporization and expansion of the water, localized thermal degradation, and fatigue failure due to repeated compressive forces in high-stress regions.

Methods of minimizing water treeing follow the same general lines as for electrical treeing. An obvious method would be to exclude moisture, and the use of a metal sheath to provide an impermeable barrier to moisture would be an effective but expensive solution in power cables. Despite difficulties in detection due to opacity, treeing has been shown to occur in ethylene propylene (EPR) and ethylidene nobornene (EPDM).

2.9 STRESSES IN COAXIAL CABLES

In a.c. cables three stresses have to be considered: impulse, a.c., and inception for partial discharges. Oil-filled systems work in a region which is discharge-free and the design is based on the impulse stress, with the a.c. stress perhaps becoming critical

System voltage kV	Impulse level MV	Switching level MV
138	0.65	–
345	1.175	0.95
500	1.425/1.550	1.175
765	2.1	1.675
1200	2.4	1.95

at very high voltages. The above impulse (1/50 μs waveform) and switching (200/3000 μ waveform is typical) withstand levels are relevant to cables. To date the highest working a.c. maximum stress for oil-filled cable insulation is 15 kV/mm with the prospect of further increase with increase in oil pressure. The following treatment applies to 'capacitance-determined' stress distributions, i.e. a.c. and surge. For the concentric configuration shown in Figure 2.23 let the charge per metre length be Q coulombs. The stress at radius x, E_x, is given by

$$E_x = \frac{Q}{\epsilon\, 2\pi x}$$

where $\epsilon = \epsilon_0 \epsilon_r$. The voltage between conductor and outer dielectric screen or sheath,

$$V = \int_r^R E_x \, \mathrm{d}x = \frac{Q}{2\pi\epsilon} \ln \frac{R}{r}.$$

and as

$$Q = CV$$

$$C = \frac{2\pi\epsilon}{\ln(R/r)} \quad \text{F/m}$$

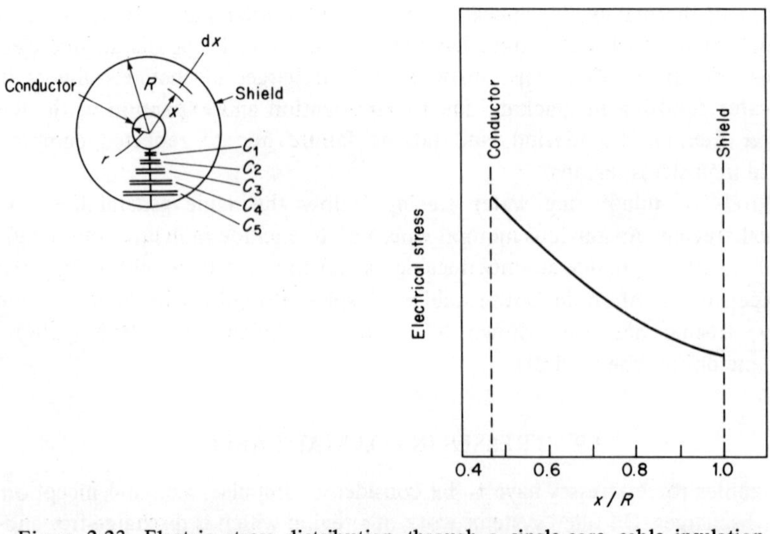

Figure 2.23. Electric stress distribution through a single-core cable insulation with uniform radial electric stress distribution.

then

$$E_x = \frac{V}{x \ln(R/r)} \quad \text{V/m} \tag{2.5}$$

and

$$E_r = \frac{V}{r \ln(R/r)} \quad \text{V/m.} \tag{2.6}$$

The stress which is a maximum at the surface of the conductor is further increased by the effects of conductor stranding by an amount dependent on the strand diameter but in the range 15–25%. This increase is overcome by the use of taped screens of metallized paper and carbon-black paper tapes, or in the case of extruded cables an extruded screen of semiconducting (e.g. carbon-loaded) plastic. The minimum value of E_r is found from $dE_r/dr = 0$, and occurs when $\ln(R/r) = 1$, i.e. $R = 2.718r$, when $E_r = V/r$. This optimum relationship is often overridden by other considerations for conductor radius.

Because of the variation of stress through the dielectric, economy may be obtained in taped dielectrics by the use of thinner tapes near the conductor and thicker tapes as the outer shield is approached. Because of limitations for bending the outer diameter of the dielectric is limited to 150 mm in oil-filled self-contained systems. The determination of stress in three-core cables, i.e. three conductors in one sheath, is more complex. However, at the voltages of interest in this work only single-core cables are likely to be used.

Great care must be taken when reviewing working and breakdown stresses in the literature. It is American practice to quote the *average value* of stress in the insulation whereas most other countries quote the *maximum* stress. In taped dielectrics it is the maximum stress that determines the dielectric performance and design and this would seem to be the preferable quantity to use. However, extruded plastic cables are designed on the basis of average stress by some authorities. In the case of average stress (E_a)

$$V = E_a(R - r)$$

where

$$E_a = \frac{V}{R - r} = \frac{V}{R_m \ln(R/r)}$$

where R_m is defined as

$$\frac{R - r}{\ln(R/r)}.$$

2.10 INSULATION LIFE

A number of factors affect the life of an insulation system:

(*a*) Temperature, which changes electrical properties such as tan δ and also mechanical and chemical properties.

(*b*) Mechanical, due to differential expansions between the insulation and the surrounding sheath and also the conductor.
(*c*) Mechanical, due to forces set up on the conductor during short-circuit conditions.
(*d*) Presence of partial (corona) discharges.
(*e*) Oxidation, in which volatile gases are driven off leaving the insulation with a higher carbon constituency.
(*f*) Treeing

Experience over many years on samples and real cables has indicated that the life of a cable at constant temperature is governed by the empirical equation, tE^n = Constant. This law is tested by maintaining constant stress on the dielectric and measuring time to failure. Life under service conditions is obtained by extrapolating the straight line resulting from the plot of log E against log t. This assumes that the law which has been shown to be valid over relatively short test times holds for a value of t of say 30 years or more.

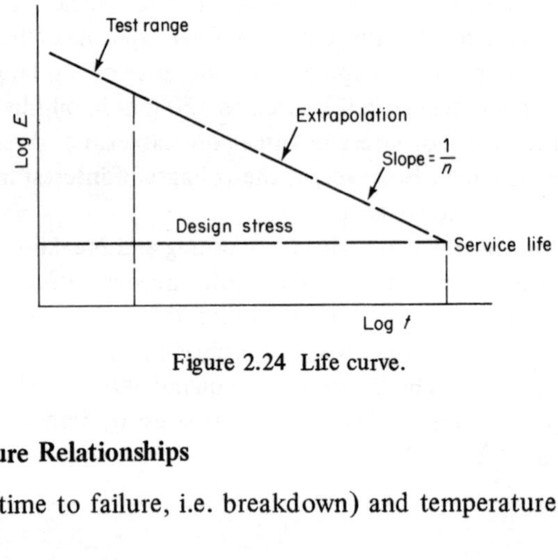

Figure 2.24 Life curve.

Life–Temperature Relationships

Insulation life (time to failure, i.e. breakdown) and temperature are related by the equation

$$\ln L = A + \frac{B}{T} \tag{2.7}$$

where

L = life in hours
T = absolute temperature (K)
A and B are constants for a material.

This is a form of the Arrhenius Equation for chemical reaction rates.

Other forms of this equation are

$$L = L_{T_0} \exp\left(\frac{B}{T} - \frac{B}{T_0}\right)$$

where L_{T_0} is the life corresponding to an absolute temperature T_0 (K) or

$L = Ae^{-m\theta'}$ where A and m are constants and θ' the temperature (°C).

These laws are responsive *only to temperature and the insulation type*. It is of interest to examine life over typical temperature transients occurring in cables.

Consider the rises in insulation temperature following a change in conductor current assumed linear or exponential. Let the temperature and time at the commencement of the assessment of deterioration to the insulation be θ'_0 and t_0 (deg C and hours); similar quantities but at the end of the period are θ'_f and t_f.

θ'_m is the maximum steady-state temperature (°C)

$\frac{1}{\alpha} = \tau$ = thermal time constant of insulation (hours)

θ' = temperature at any time t (°C)

D = dimensionless quantity by means of which cumulative damage to insulation is assessed.

Failure occurs when $D = 1 = \Sigma D$, i.e. the sum of the increments of damage which have occurred to the insulation over its life.

Damage dD occurs over time dt and when d$t = L$, d$D = 1$, and hence

$$\frac{dD}{dt} = \frac{1}{L}$$

which is a function only of temperature, i.e. $f(\theta')$. Consider the linear temperature rise at time t:

$$\theta' = \theta'_0 + \left(\frac{\theta'_m - \theta'_0}{t_m}\right) t$$

where θ'_0 = initial temperature, θ'_m = final temperature, t_m = time to θ'_m, and

$$\frac{d\theta'}{dt} = \frac{\theta'_m - \theta'_0}{t_m} .$$

By definition,

$$\Delta D = \int_{\theta'_0}^{\theta'_m} \frac{1}{L} \frac{t_m}{(\theta'_m - \theta'_0)} d\theta' = \int_{\theta'_0}^{\theta'_m} \frac{1}{L} dt$$

and, as $L = Ae^{-m\theta'}$,

$$\Delta D = \left[\frac{t_m}{(\theta'_m - \theta'_0)A} \frac{e^{m\theta'}}{m}\right]_{\theta'_0}^{\theta'_m} . \tag{2.8}$$

Next consider an exponential rise described by an equation of the form,

$$\theta = \theta_m(1 - e^{-\alpha t})$$

where θ, θ_m are the temperature rises above reference at time t and at steady state respectively, and

α = (thermal time constant)$^{-1}$.

It should be noted that it is the highest or hot-spot temperature which is of vital importance and not the average temperature. At the hot-spot insulation, failure will initially occur. From the temperature-rise equation:

$$\frac{d\theta}{dt} = \alpha\theta_m e^{-\alpha t}$$

$$= \alpha\theta_m \left(1 - \frac{\theta}{\theta_m}\right) = \alpha(\theta_m - \theta).$$

If actual temperatures are used instead of rises,

$$\frac{d\theta'}{dt} = (\theta'_m - \theta')$$

assuming that the coolant or ambient temperatures are constant. The damage increment between times t_0 and t_f is ΔD, where

$$\Delta D = \int_{t_0}^{t_f} \frac{1}{L}\, dt$$

$$= \int_{\theta'_0}^{\theta'_f} \frac{1}{L} \frac{d\theta'}{\alpha(\theta'_m - \theta')}.$$

Insulation life L, if considered a function of temperature only, is governed by the law $L = Ae^{-m\theta'}$. Hence

$$\Delta D = \frac{1}{\alpha} \int_{\theta'_0}^{\theta'_f} \frac{d\theta'}{Ae^{-m\theta'}(\theta'_m - \theta')}.$$

The solution is

$$\Delta D = \frac{e^{\theta'_m m}}{\alpha A} [(-\mathrm{Ei}(-x_f)) - (-\mathrm{Ei}(-x_0))] \tag{2.9}$$

where

$$x = m(\theta'_m - \theta') \quad \text{and} \quad -\mathrm{Ei}(-x) = \int_x^\infty \frac{e^{-y}}{y}\, dy$$

$$x_f = m(\theta'_m - \theta'_f) \quad x_0 = m(\theta'_m - \theta'_0).$$

Values of the exponential function $-\mathrm{Ei}(-x)$ are available in mathematical tables.

Similar reasoning may be applied for cooling. The temperature fall-time equation of the form $\theta = \theta_m e^{-\alpha_1 t}$ form which

$$\frac{d\theta'}{dt} = -\alpha_1\theta_m e^{-\alpha_1 t}$$

$$\frac{d\theta'}{dt} = -\alpha_1(\theta'_m - \theta').$$

Hence

$$\Delta D = -\frac{1}{\alpha_1}\int_{\theta'_0}^{\theta'_f} \frac{d\theta'}{Ae^{-m\theta'}(\theta'_m - \theta')}$$

$$= \frac{1}{\alpha_1}\int_{\theta'_f}^{\theta'_0} \frac{d\theta'}{Ae^{-m\theta'}(\theta'_m - \theta')} .$$

It should be noted here that $\theta'_0 > \theta'_f$. The damage increment incurred during the period of cooling is therefore of comparable importance to that during the period of heating.

Life of Cellulose Paper Insulation[37,38,39]

Cellulose paper suffers a loss of mechanical strength due to thermal ageing, certain gases, e.g. CO and CO_2, being evolved. At present it is accepted that the operating temperature of oil-filled cable insulation should not exceed 85 °C which should under normal conditions give a life expectancy of about 40 years. It is of interest, however, for economic reasons to consider the use of increased temperatures, e.g. 100 °C, for prescribed short periods. The following parameters are of interest for cellulose paper tapes after periods at given temperatures: friction coefficient, folding strength, burst strength, tensile strength, extension to break, tear strength, and degree of polymerization. Only four gases are produced by purely thermal processes, i.e. CO_2, CO, methane, and hydrogen in the volume ratios of 2000 : 600 : 2 : 1 respectively.

Gas-in-oil analyses in future may be able therefore to assess the physical state of the insulation with a knowledge of the previous thermal history. However at the present time this is not at all certain. The criterion of cable life could be taken as either: (*a*) the time to reach 10% of the original value of any of the mechanical or chemical characteristics; or (*b*) the time taken to saturate the impregnant with CO (the least soluble gas).

Prediction of Insulation Life – use of Statistical Methods[40,41,42,43]

The prediction of life from the results of model tests at higher operating stresses is of great interest. Generally the breakdown stress of extruded plastic insulation presents a larger scatter than taped insulation (e.g. paper/oil) and the use of statistical methods is essential. Weibull's distribution is applicable where the occurrence of an event in any part of an object means failure of the object as a whole. In the cable context the breakdown is related to the weakest point.

If log (percentage of failure) is plotted against log (stress at breakdown) then a straight line results for insulations. The probability of failure,

$$P(E,t) = 1 - \exp(-kE^b t^a) \tag{2.10}$$

where

$P(E,t)$ is the percentage of failures at stress E in time t

a is a factor determining the distribution
b is the slope of Weibull line
k is a constant.

The breakdown voltage of cables insulated with polyethylene is a function of the contaminants, voids, and 'skips' in the conductor shield (causing protrusions into the insulation). The stress at a protrusion depends on its size and shape; the ratio of the maximum stress at the tip of the protrusion to the stress that would exist at that point without the protrusion varies from roughly 80 for long and narrow shield protrusions to about 6 for almost hemispherical ones. Apart from protrusions and contaminants discharges occurring in voids or cavities can initiate breakdown. The influence of the diameter of a spherical void on the partial discharge inception voltage[44] may be calculated by Paschen's Law.

Voltage breakdown results from ionization in the voids or insulation, and if Q_c is the critical cumulative charge over time t to breakdown, then

$$Q_c = A(E_m - E_i)^n t$$

where

A is a constant depending on the imperfection
E_m is the maximum voltage stress at the imperfection
E_i is the discharge inception stress at the imperfection
t is the time to breakdown
n is a constant.

When

$$E_m \gg E_i,$$

$$E_m^n t = \frac{Q_c}{A}$$

and as E_m/E_a is roughly constant (E_a is the average stress in the insulation which *approximates* to the value at the point without the imperfection) then $E_m^n t$ is a measure of the imperfection level or content of the insulation. The exponent n defines the slope of the life-expectancy curve and can be modified by changing the ionization conditions within the intermolecular spaces.

Because the imperfections occur in a random manner a statistical distribution of breakdown voltages with time is usually obtained when short cable samples (a few metres in length) are employed to predict cable life. The Weibull statistical distribution is applied as follows:

$$P(E, t) = 1 - \exp -\left[\left(\frac{t}{t_0}\right)^a \left(\frac{E}{E_0}\right)^b \frac{lr^2}{l_0 r_0^2}\right] \quad . \tag{2.11}$$

If it is wished to compare model tests with the performance of an operating cable for the same probability of breakdown, then

$$\frac{E}{E_0} = \left[\frac{l_0}{l}\left(\frac{r_0}{r}\right)^2\right]^{1/b} \quad \text{or} \quad \left[\frac{\text{volume}_0}{\text{volume}}\right]^{1/b}$$

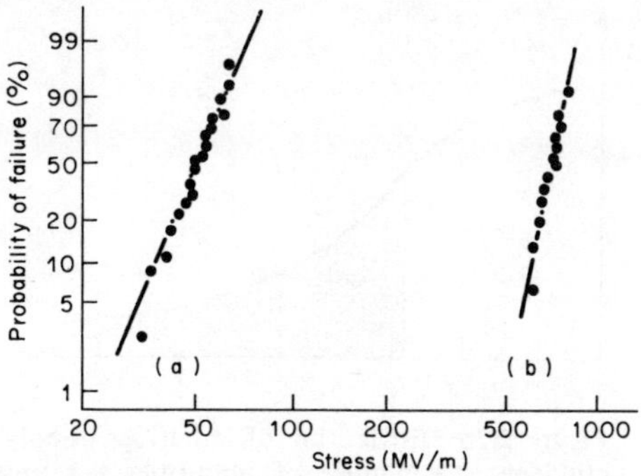

Figure 2.25 Weibull Plots – probability of breakdown of cross-linked polyethylene cable insulation. (a) Steam-cured cables, (b) small samples prepared in laboratory. (Reproduced by permission of IEE.)

where

E and E_0 are the operating cable and test model stresses
l and l_0 are the operating cable and test model lengths
r and r_0 are the operating cable and test model radii

and suffix 0 refers to the model test-cable sample. A Weibull plot for polyethylene samples is shown in Figure 2.25. The value of the slope (b) is 7. l_0, r_0, t_0, and E_0^b are often a set of parameters chosen such that P is 63.2%. a is called the time exponent and b the stress exponent.

Two tests are performed to establish the values of a and b and the characteristic stress E_0(i.e. for $P = 63.2\%$). These are the step test to give a and E_0, and the life test to give b. In the step test a stress E_1 is applied for a time t_0 to a group of N samples and the number of failures n_1 counted. This is repeated at the increased stress E_2 and the new number of failures n_2 counted. This is repeated until all the samples have failed. If l and r are considered constant, then from the Weibull equation,

$$\ln\left\{\ln\left(\frac{1}{1-P}\right)\right\} = a\ln\left(\frac{t}{t_0}\right) + b\ln(E) - b\ln(E_0).$$

If $\ln\{\ln[1/(1-P)]\}$ is plotted against E then the slope gives b and the intercept E_0.

From the Weibull equation, with again l and r constant, if P is constant then

$$\left(\frac{t}{t_0}\right)^a \left(\frac{E}{E_0}\right)^b = \text{Const.}$$

or

$$t^a E^b = \text{Const.}$$

and

$$tE^{b/a} = tE^n = \text{Const.}$$

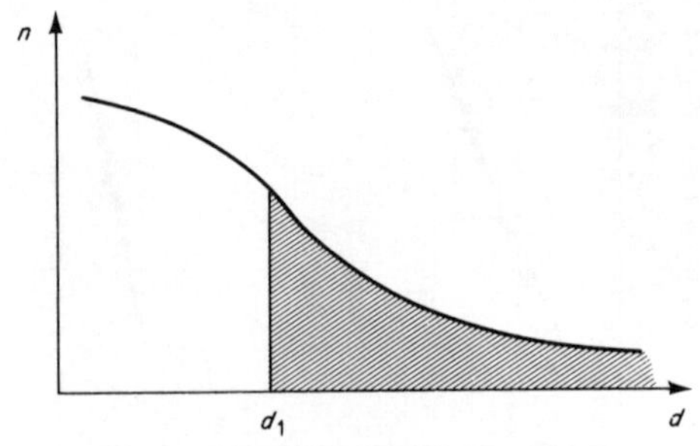

Figure 2.26 Distribution of impurities in polyethylene: n = number of impurities per unit volume; d = maximum dimensions of impurity. (Reproduced by permission of CIGRE from Report 21–07, 1972.)

This is the traditional life relationship and if ln (t) is plotted against ln(E) the slope is n, the life index, and hence parameter a is found.

Thus model results can be interpreted to full practical cable lengths. With cables in service the interest is in failure rates per 1600 km (1000 miles) per year whereas the statistical testing possibly involves only something like 0.3 km of cable per year, i.e. 0.02% of the total. The extrapolation of model tests to predict the performance of full-length cables must therefore be treated with caution. However, the model tests and associated techniques are of great value when comparing insulations, manufacturing methods, or innovations.

A typical form of impurity distribution in polyethylene is shown in Figure 2.26. There are three distinct regions in the life of a cable corresponding to different ageing processes, and these are illustrated in Figure 2.27. At high stresses the size of the impurities (mainly conducting in nature) has a great influence on breakdown; this is region (I) in Figure 2.27. Below these stresses the role of the impurities

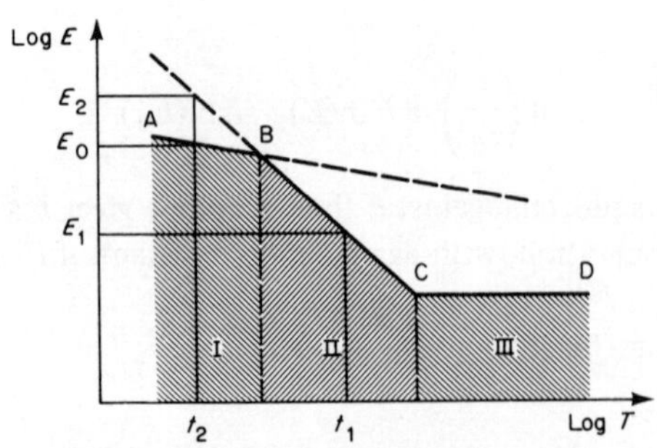

Figure 2.27 Theoretical life curve for a polyethylene insulated cable: AB = impurity level; CD = inception of coron (partial) discharges. (Reproduced by permission of CIGRE from Report 21–07, 1972.)

becomes small and the process of partial (corona) discharges takes over, giving region (II). Finally, in region (III) the stress is below the partial discharge level and there is no effective deterioration of the insulation.

An expression which has been proposed to represent the relationships implied in Figure 2.27 is:

$$(E - E_i)^n (t + t_0) = \text{Constant}$$

where E_i is discharge inception stress and t_0 a time interval. If $E \gg E_i$ and $t_0 \ll t$ the equation is obtained which relates to the partial discharge region (II). As t approachs zero, E approaches the intrinsic electric stress of the insulation. This equation gives a reasonable qualitative representation of the ageing process.

REFERENCES

1. Sillars, R. W., 'Electrical Insulating Materials and their Applications', Peter Peregrinus, London, 1973.
2. Adamczewski, I., 'Ionization, Conductivity and Breakdown in Dielectric Liquids', Taylor and Francis, London, 1969.
3. Saums, H. L. and Pendleton W. W., 'Materials for Electrical Insulation and Dielectric Functions', Hayden, New Jersey, 1973.
4. Kuffel, E. and Abdullah, M., 'High Voltage Engineering', Pergamon, Oxford, 1969.
5. Harrop, P. J., 'Dielectrics', Butterworths, London, 1972.
6. Robinson, D. M., 'Dielectric Phenomena in High Voltage Cables', Chapman and Hall, London, 1936.
7. Kreuger, F. H., 'Discharge Detection in High Voltage Equipment', Heywoods, London, 1964.
8. Whithead, S., 'Dielectric Breakdown of Solids', Oxford University Press, 1951.
9. Krasucki, Z., 'Breakdown of liquid Dielectrics', *Proc. Royal Society*, A, **282**, 393–404 (1966).
10. Casey, J. P., 'Pulp and Paper', Vol. III, Interscience, New York, 1961, Second Edition,.
11. Coste, H., 'Fundamentals of Papermaking Fibres', British Paper and Board Makers Assoc. (Inc), London p. 301.
12. Barnes, C. C., 'Electric Cables', Pitman, London, 1964.
13. Miranda, F. J. and Gazzana-Priaroggea, P., 'Self Contained Oil Filled Cables, A review of progress', *Proc. I.E.E.*, **123**, 229–238 (1976).
14. Gazzana-Priaroggia, P. *et al*, 'A Brief Review of the Theory of Paper Lapping of a Single-Core High-Voltage Cable', *Proc. I.E.E.*, 1960, Monograph No. 390S (108C page 25).
15. Waldron, R. C., 'Oil-paper insulated EHV cables – 60 Hz breakdown', *I.E.E.E. Trans., PAS 67*, 53–60.
16. Lanfranconi, G. M., Maschio, G., and Occhini, E., 'Self Contained oil-filled cables for high power transmission in the 750–1200 kV range.' *I.E.E.E. PAS-93*, 1974, 1535–1543.
17. Sie, T. H., 'Oil-paper insulation: Impulse breakdown voltages', *I.E.E.E. Trans., PAS 69*, 862–868.
18. Gazzana-Priaroggia, P. and Palandri, G., 'Research on the Electric breakdown of Fully Impregnated Paper Insulation for High-Voltage Cables', *Trans. A.I.E.E.*, **74**, 1343–1365 (1955).
19. Gazzana-Priaroggia, P. *et al*, 'Dielectric Phenomena Related to the use of

Carbon Black Screening in High Voltage Fully-Impregnated paper cables', *I.E.E. Trans. PAS-83*, 137–152 (1963).
20. Toriyama, Y. *et al.*, 'Effect and Physical Behaviour of Carbon-Black Papers used as a Conductor Screen', *Proc. I.E.E.*, **116**, 566–574 (1969).
21. Jocteur, R. *et al*, 'Extruded Polyethelene High Voltage Cables', *CIGRE*, Paper No. 21–07, 1972.
22. Vermer, J. and Boone, W., 'Characteristic Properties of Pol-Impregnated Laminated Synthetic Insulation for Extra-High Voltage Cables', *Proc. I.E.E.*, **114**, 981–988 (1967).
23. Edwards, D. R. *et al.*, 'Polymer and Polymer/Paper Laminated Tapes for ehv Oil-Filled Cables', *CIGRE*, Paper No. 15–05, 1972.
24. Gibbons, J. A. M. and Stannet, A. W., 'Synthetic Insulation to Replace Oil-Impregnated Paper in Mains Transmission Cables', *I.E.E. Electronics & Power*, 492–495 (May 1st 1975).
25. Eich, E. D., 'Laminated Synthetic Insulation for Extra-High Voltage Cable', *CIGRE*, Paper No. 203, 1966.
26. Gibbons, J. C. M., Howard, P. R., Skipper, D. J., Body, R. S., and Hawley, W. G., 'Gas-pressurised Polythene Tape Dielectric for Extra-High-Voltage Cables', *CIGRE*, Paper No. 201, Paris, 1964.
27. Little, A. D., Inc., for the Electric Research Council (New York), *Underground Power Transmission.*
28. Mason, J. H., 'Discharge Detection and Measurement', *Proc. I.E.E.*, **112**, 1407, (1965).
29. Mole, G., 'Measurement of the Magnitude of Internal Corona in Cables', *I.E.E.E. Trans., PAS 89*, **2**, 204, (1970).
30. Mildner, R. C., 'Continuous Discharge Testing of Extruded Cables', *Proc. I.E.E.*, **109A** Suppl., 139 (1962).
31. Gooding, F. H. and Slade, H. P., 'Corona Level Scanning of High Voltage Cable', *Trans. Amer. Inst. Elect. Engrs.*, 1957, **76**, Part III, 999 (1957).
32. Endacott, J. D. and Smith, E. L., 'Control of quality by Discharge Detection Scanning During the Manufacture of High-Voltage Plastic-Insulated Power Cables', *Proc. I.E.E.*, **109A**, Suppl. 3, 119 (1962).
33. Kreuger, F. H., 'Scanning of Screened Plastic Cores for Discharges', *Proc. I.E.E.*, **109A**, Suppl. 3, 129 (1962).
34. Katz, C. and Bernstein, B. S., 'Electrochemical Treeing at Contaminants In Polyethelene and Cross Linked Polyethelene Insulation', *Proceeding of 1973 Conference on Electrical Insulation and Dielectric Phenomena, National Academy of Sciences, Washington*, pp. 307–315.
35. Eichhorn, R. M., 'Effect of Moisture on Needle Testing of Polyethelene', *Proceedings of 1973 Conference on Electrical Insulation and Dielectric Phenomena, National Academy of Sciences, Washington*, pp. 289–297.
36. Eichhorn, R. M., 'Treeing in Solid Extruded Electrical Insulation', *I.E.E.E., Trans. Electrical Insulation, EI-12*, 1976, pp. 2–18.
37. Lawson, W. G. *et al.*, 'Thermal Ageing of Cellulose Paper Insulation', *I.E.E.E. Trans. on Electrical Insulation, EI-12*, 1977. pp. 61–66.
38. Kagaya, S. *et al*, 'Ageing of Oil-Filled Cable Dielectric', *I.E.E.E. Trans., PAS-89*, 1420–1428 (1970).
39. Grazzana-Priaroggia, P. *et al.*, 'The influence of Ageing on the characteristics of Oil-Filled Cable Dielectric', *Proc. I.E.E.*, **108.**, A, 1–13 (1961).
40. Kanko, R. and Sugiyama, K., 'Statistical Considerations on Impulse Breakdown Characteristics of cross linked Polyethylene Insulated Cables', *I.E.E.E. Trans., PAS-94*, 367–377 (1975).
41. Oudin, J. M. *et al.*, 'Reliability and Development Towards High-Voltage of Synthetic Insulated Cables', *CIGRE*, No. 21–10 (1968).

42. Kreuger, F. H. and Lelie, M. C., 'The Weibull distribution for evaluation of cable dielectrics', *CIGRE*, M71–7, SC21 (1971).
43. Occhini, E., 'A statistical approach to the discussion of the dielectric strength in electric cables', *I.E.E.E. Trans.*, *PAS-90*, 2671–2682 (1971).
44. Bahder, G. *et al.*, 'Criteria for determining performance in service of cross linked polyethylene insulated power cables', *I.E.E. Trans, PAS 95*, 1552–1560 (1976).

PROBLEMS

Problem 2.1

A large parallel plane air gap is 20 mm wide and a voltage of 60 kV (r.m.s.) is applied across it. Calculate the maximum stress. A slab of cellulose insulation of $\epsilon_r = 4$ and thickness 2 mm is then inserted into the gap next to one of the plates. Calculate the new stress in the air and comment on your result. (Answer: 32.5 k/Vcm.)

Problem 2.2

A single-core self-contained cable comprises two dielectric materials. For the dielectric next to the conductor the relative permittivity is 5 and the radial thickness t mm, next to the outer screen the relative permittivity is 2.5. If the conductor radius is 10 mm and the outer screen 50 mm determine the value of t such that across each dielectric exists an equal voltage. (Answer: $t = 19$ mm.)

Problem 2.3

A single-core self-contained cable comprises three different concentric insulations of relative permittivities from the conductor outwards of 5.4, 3.6, and 2.0. If the stress at the conductor screen and at the screens between the three insulations are to be equal and 10 kV/mm, calculate the radial thicknesses of the annular insulations. The conductor and outer screen radii are 5 mm and 22.5 mm. (Answers: $\epsilon_{r1} = 5.4$, $t_1 = 2.5$ mm; $\epsilon_{r2} = 3.6$, $t_2 = 6$ mm; $\epsilon_{r3} = 2.0$, $t_3 = 9$ mm.)

Problem 2.4

The insulation round a conductor of radius 20 mm is to be capacitance graded by the use of two types of tape of relative permittivity 3 and 5. The voltage between the conductor and outer screen is 200 kV and the maximum working a.c. stress is to be 15 kV/mm. The design impulse stress is 90 kV/mm and the BIL 1350 kV. Design the cable noting that for minimum thickness of insulation the maximum stress in each wall of tape must be the same. Calculate the ratio of volume of insulation if only one tape type is used. (Answer: $\epsilon_r = 5$ tape next to conductor and radius to interface 33.3 mm. Radius to outer screen 32.5 mm. Using one type of tape, outer radius 42.4 mm.)

Problem 2.5

Determine the radius of the conductor (r) for a single core of an oil-filled cable system to give the minimum stress for a given voltage and outer screen radius (R). If V is 289 kV and maximum stress is 12 kV/mm (r.m.s.) determine the optimum value of r and R (Answers: 25 and 65 mm).

Problem 2.6

The paper/oil insulation of a single-conductor cable contains a void (a small pocket of air or gas). If the relative permittivity of the insulation is 3.5 calculate the stress across the void in terms of the stress in the adjacent insulation.Comment on the practical implications of your result. (Answer: 3.5.)

Problem 2.7

An oil-impregnated taped single-core cable has a tape registration of 65/35. Determine the stress in the butt gap in terms of the average applied stress E (E is assumed constant over the insulation thickness): (a) When the tape material is cellulose paper; (b) when the tape material is polyethylene. Repeat the calculations when the butt gaps are completely filled with gas. (Answers: $0.3E$, $0.18E$, $0.46E$, $0.26E$.)

Problem 2.8

The probability of survival at a time (t) from a cable energization of mean stress E is given by

$$P(t, E) = \exp\left[-\left(\frac{t}{t_0}\right)^a \left(\frac{E}{E_0(2)}\right)^b\right]$$

$$= \exp\left[-\left(\frac{t}{t_0}\right)^a \frac{E}{E_0(1)}^b \left(\frac{L_0(2)}{L_0(1)}\right)\right]$$

where $L_0(1)$ and $L_0(2)$ are lengths of the test models and actual cable respectively, t_0 is the time to breakdown of model cable of mean stress $E_0(1)$, and E and t refer to the actual cable (stress and time). A large number of tests on specimen cable lengths of length 100 m and radius 3 mm give the following parameters: $a = 0.5$, $b = 12.5$, $t_0 = 27$ hours for $E_0 = 20$ kV/mm. Calculate the maximum allowable mean working stress of a 20 km length of cable of the same material and of the same radius as the test lengths if the cable is to have 90% probability of survival after 40 years of constant energization (Answer: 7.5 kV/mm).

3

Thermal Performance and Current Rating

Electric transmission is normally the province of the electrical engineer, who will be at home with the concepts of the previous chapter. The general area of thermal performance, embracing as it does heat transfer in complex equipment, represents less familiar territory to most readers of this book. It has already been emphasized that the basic limitation on the power-carrying capacity of conventional underground cables is that of current-carrying capability or ampacity, which is determined by temperature limitations. The dependence of insulation life on temperature is considered in some detail in Chapter 2. Implicit in these considerations is the method of installation of the cables, i.e. whether directly buried, in ducts, under water, above ground, in tunnels and so on. The cable environment presents the most difficulty when predicting the maximum temperature in the insulation because of the many variables present. The ambient temperature, depending as it does on the seasons and weather conditions, influences the permissible current. In Europe this is advantageous as the peak loads occur in the winter when a larger temperature rise is possible. In North America, however, the situation is often reversed, with peak loads (due to air-conditioning) occurring in the summer with a restricted temperature rise.

Power-system loads are usually cyclic to some degree, and this fact may be used advantageously in order to obtain the highest possible rating in an underground circuit. Although the steady-state rating is used when deciding on a cable type and size for a given transmission load, utilization is desirable of the slow transient temperature response of buried systems to reduce the conductor size of cables in the usual situation of two circuits in parallel. This is illustrated in Figure 3.1 in which each of the parallel circuits is rated at the full-load current (I) and normally each will carry ($I/2$). With one cable out of circuit the other will carry its rated current. By using the temperature–time characteristics shown it is apparent that if circuit rearrangements can be effected in a period of a few hours each cable could be of lower rating, with resulting economic advantage. However, repairs to underground circuits have traditionally taken rather long periods, very much longer than for overhead lines.

Practical measurement of the highest temperature in cable insulation is very

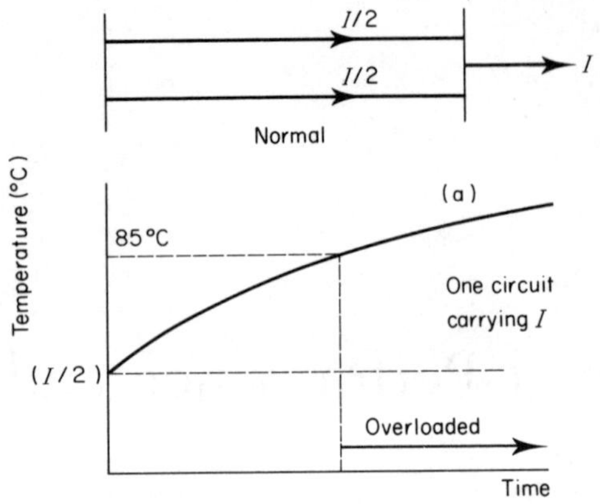

Figure 3.1 Exploitation of long thermal time-constants in buried cable systems comprising two circuits in parallel. (a) Cable rated at $2I/3$ but carrying I.

difficult for commercial installations at working voltage, and monitoring of the sheath temperature is often used. The accurate prediction of thermal conditions in a cable system is therefore of prime importance. Indeed, considering the capital costs at the voltage and power levels under study, quite small differences in predicted rating represent considerable sums of money. It is essential therefore that the process of calculation and modelling be as accurate and realistic as possible. This is often difficult because of the uncertainties of the cable environment, and over the last decade or so considerable effort has been devoted internationally to this subject. It is hoped to present the basis of this work in the following sections.

3.1 CALCULATION OF TEMPERATURE AND HEAT FLOW – BASIC IDEAS

The heat-flow rate q (W) through a thermal resistance R_T (deg C/W) and the temperature rise θ (deg C) across it are related by the equation $\theta = R_T q$. When electrical engineers initially undertook thermal calculations, so great was their attachment to Ohm's Law that their unit of thermal resistance was known as the 'thermal Ohm'.

Also,

$$R_T = g \frac{l}{A}$$

for a solid of constant cross-section A (m^2), where g = thermal resistivity (deg C-m/W) and l = length (m). Values of thermal resistivity (g) for various materials met in cable engineering are given in Table 3.1.[1]

The ability of a material to store heat is referred to as its thermal capacity (C), and C (J/deg C) = Mass (kg) x Density ρ (kg/m^3) x Specific heat c_p (J/kg-deg C). The quantity of heat stored, Q (J) = $C\theta$. This is analogous to storage of electricity

Table 3.1 Thermal resistivities

Material	Thermal resistivity g deg C-m/W
Paper insulation in solid-type cables	6.0
Paper insulation in oil-filled cables	5.0
Paper insulation in cables with external gas pressure	5.5
Polyethylene	4.0
Polyvinyl chloride	7.0
Compounded jute and fibrous materials	6.0

Values taken from I.E.C. Report No. 287 (1969). (Reproduced by permission of the International Electrotechnical Commission, to whom the copyright belongs)

in a capacitor. Heat is dissipated by radiation and convection (free or forced) from the cooling surfaces.

The simplest form of thermal transient results from the application of a step function of heating q (due to I^2R, magnetic, or dielectric losses) to a body of thermal capacity C which dissipates heat into a cooling medium via a heat-transfer coefficient h (W/m^2-deg C). The heat transfer coefficient may be replaced by a thermal resistance, $R_T = 1/hA$, *where A* is the area of body in contact with the coolant. It may be shown that the relationship between temperature-rise and time is

$$\theta = \theta_m(1 - e^{-t/\tau})$$

where θ_m is the steady-state temperature rise ($\theta_m = qR_T$) and τ is the thermal time-constant ($\tau = RC$). The graphical interpretation of the time-constant is shown in Figure 3.2. The equivalent thermal network is shown in Figure 3.3. The response of the simple R–C network to a time-varying load is illustrated in Figure 3.4, in which τ is the termal time constant and θ_{m1}, θ_{m2}, and θ_{m3} are steady-state temperature corresponding to the various current loadings.

Most items of power equipment possess temperature rise–time characteristics having more than one exponential term. Often when carrying out heat runs the temperature will appear to no longer rise with the time scale being used, usually a few hours. It is quite possible, however, for the temperature to continue rising

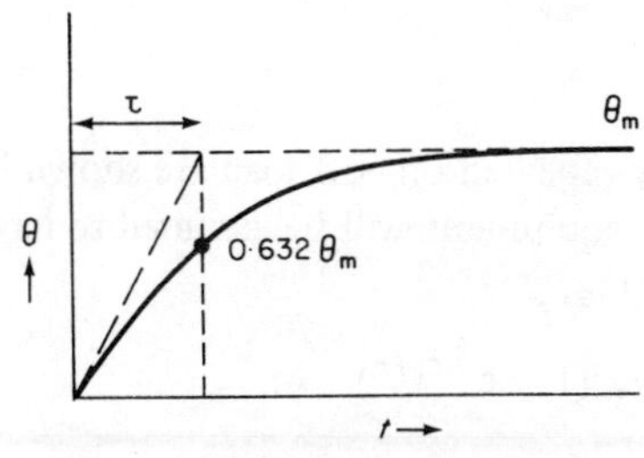

Figure 3.2 Temperature–time relation for application of step-function of heat-basic thermal model.

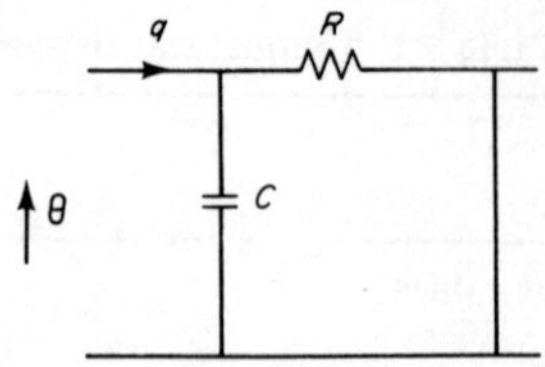

Figure 3.3 Basic thermal network representation.

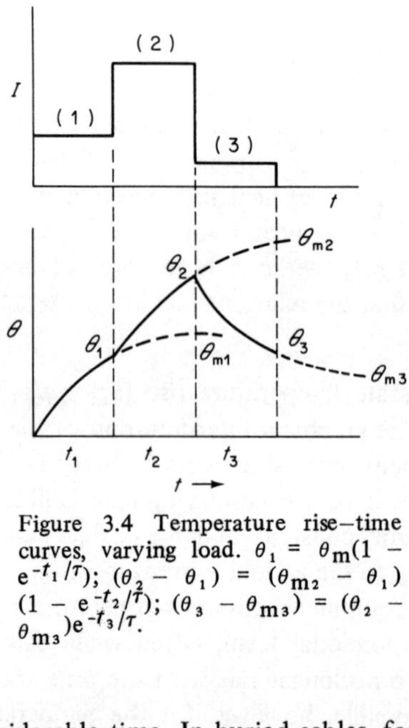

Figure 3.4 Temperature rise–time curves, varying load. $\theta_1 = \theta_m(1 - e^{-t_1/\tau})$; $(\theta_2 - \theta_1) = (\theta_{m2} - \theta_1)(1 - e^{-t_2/\tau})$; $(\theta_3 - \theta_{m3}) = (\theta_2 - \theta_{m3})e^{-t_3/\tau}$.

appreciably over a considerable time. In buried cables, for example, many hundred hours elapse before steady conditons are approached.

Cyclic Loading

The effects of a regularly repeated on–off load are shown in Figure 3.5. In order to simplify the analysis the equipment will be assumed to have a single time-constant. From Figure 3.5; on heating,

$$\theta_2 - \theta_1 = (\theta_m - \theta_1)(1 - e^{-t_1/\tau})$$

on cooling,

$$\theta_1 = \theta_2 e^{-T_1/\tau}.$$

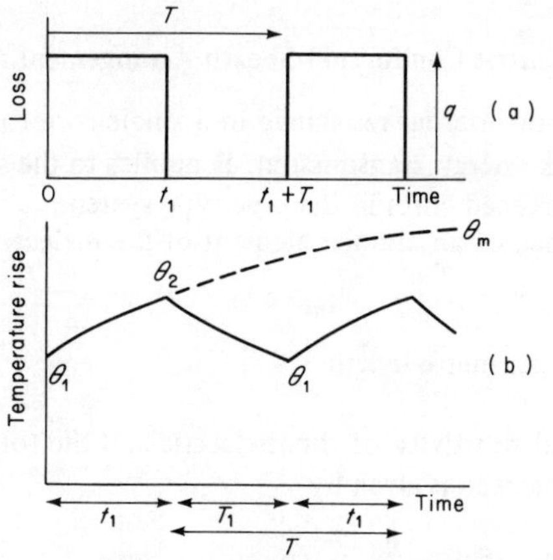

Figure 3.5 (a) On–off load cycle: $T = t_1 + T_i$; (b) corresponding temperature curves. θ_m = steady-state temperature rise with q dissipated continuously.

From these,

$$\theta_2 - \theta_2 e^{-T/\tau} = (\theta_m - \theta_2 e^{-T_1/\tau})(1 - e^{-t_1/\tau})$$

from which,

$$\theta_2 = \theta_m \left(\frac{1 - e^{-t_1/\tau}}{1 - e^{-T/\tau}} \right). \tag{3.1}$$

In all the above equations the losses have been assumed constant. In practice an I^2R loss increases with increase in temperature due to the temperature coefficient of resistance. The losses will normally consist of a constant plus a variable portion, i.e. $q_0 + a\theta$.

The above represents the simplest form of representation and in practice the temperature–time equation possesses several exponential terms and composite time-constants as follows:

$$\theta = A_1(1 - e^{-t/\tau_1}) + A_2(1 - e^{-t/\tau_2}) + A_3(1 - e^{-t/\tau_3}) + \ldots$$

This equation is typical of the conductor transient in a buried self-contained cable. Sometimes the equivalent thermal network idea will not suffice and a full thermal field solution is required. This situation occurs when considering temperatures in joints and terminations when the heat-flow is three-dimensional.

3.2 THERMAL RESISTANCES

In this section the methods of calculation of the thermal resistances existing in cable systems will be given. Undue complexity and refinement will be avoided and readers are referred to the list of references for very refined treatments.

(i) Dielectric in a Concentric Conductor–Sheath Arrangement (Figure 3.6)

This represents the major internal resistance in a single-core cable, the type almost exclusively used in bulk energy transmission. It applies to the self-contained cables and to the individual screened cores in the pipe-type system.

The thermal resistance of an annular element of the dielectric of radius x and of thickness dx is,

$$g_d \frac{dx}{2\pi x} \text{ deg C/W per metre length}$$

where g_d is the thermal resistivity of the dielectric and the total thermal resistance conductor to sheath or screen is given by

$$\int_r^R g_d \, dx/2\pi x \text{ (deg C/W)} = \frac{g_d}{2\pi} \ln (R/r) \text{ (deg C/W)}. \tag{3.2}$$

This formula may also be used for the sheath, the bedding layer, and the protective covering.

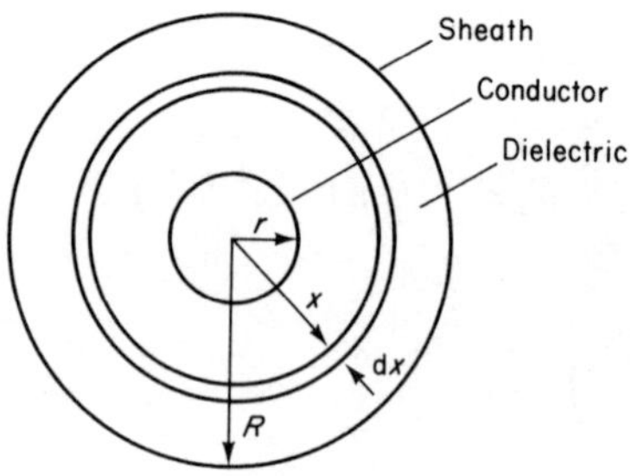

Figure 3.6 Thermal resistance of single-core cable.

(ii) Thermal Resistance of the Cable Environment

The cables are assumed to be directly buried with the trench refilled with the original soil, i.e. they are situated in a semi-infinite homogeneous medium. The ground surface is assumed to be an isothermal plane. The thermal resistivity (g) of soil varies widely depending on the nature of the soil (clay, sand, etc.), and its mositure content. Values for g frequently taken are 1.2 °C-m/W and 0.90 °C-m/W. The value used in different countries varies somewhat but is in the above region. One policy when specifying a cable is to carry out a thermal resistivity survey of the proposed route. This will highlight the point of highest g, the value to be used in the calculation of the current rating. The mean value of g taken over the route is of little use as it is the hottest section of the cable that will fail, little heat being conducted longitudinally from the sections of highest temperature.

The external thermal resistance of a single cable or pipe may be found by considering the image of the line heat-source in the thermal field as shown in Figure 3.7. If q watts is the total loss in one metre of the cable, the temperature difference

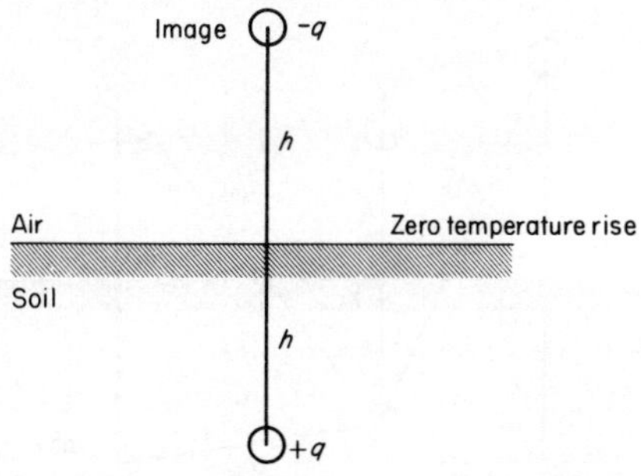

Figure 3.7 Image method for determining thermal resistance due to surrounding soil – one cable (radius r).

between the external surface of the cable and the ground surface is

$$\frac{qg}{2\pi} \ln \frac{2h}{r}$$

where h is normally of the order of 1 m. The external thermal resistance is therefore

$$\frac{\theta}{q} = \frac{g}{2\pi} \ln \left(\frac{2h}{r}\right) \text{ deg C/W per m length.} \tag{3.3}$$

With several, equally loaded, similar cables in a group which do not touch each other, the method of images is again used. Consider Figure 3.8. The temperature rise of a cable in the group is the sum of its own independent rise and that due to each of the other cables. The temperature rise at the surface of the l cable due to a loss q watts per m in cable k is

$$\theta_{kl} = \frac{g}{2\pi} q \ln \frac{d'_{lk}}{d_{lk}} .$$

Also, the temperature rise of l due to a loss q in m is

$$\frac{qg}{2\pi} \ln \left(\frac{d'_{lm}}{d_{lm}}\right) .$$

Hence the temperature rise of l due to both k and m is

$$\frac{qg}{2\pi} \ln \left(\frac{d'_{lm}\, d'_{lk}}{d_{lm}\, d_{lk}}\right) .$$

The total temperature rise of cable l therefore will be

$$\frac{qg}{2\pi} \ln \left(\frac{2h}{r}\right) + \frac{qg}{2\pi} \ln \left(\frac{d'_{lm}\, d'_{lk}}{d_{lm}\, d_{lk}}\right) .$$

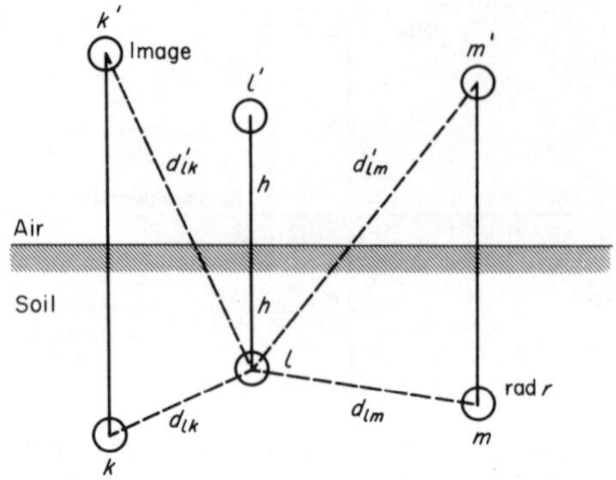

Figure 3.8 Image method for determining external thermal resistance for a group of buried cables.

For the popular configuration of three cables laid in a horizontal formation with a spacing a between adjacent cable centres and a buried depth of h,

$$d_{kl} = a;\ d'_{kl} = [(2h)^2 + a^2]^{1/2}$$

$$d_{lm} = a;\ d'_{lm} = [(2h)^2 + a^2]^{1/2}.$$

Hence,

$$\theta_l = \frac{qg}{2\pi}\left[\ln\frac{2h}{r} + \ln\left(\frac{4h^2 + a^2}{a^2}\right)\right]. \tag{3.4}$$

The effective thermal resistance of the centre cable of the group is (θ_l/q) deg C/W.

The image method may be applied to a multi-source, multi-sink situation (e.g. cooling water pipes) but great problems ensue from the reflection of the images, which therefore involves large amounts of computation. In general, for several cables laid at various depths, the effective thermal resistance of a reference cable (normally the hottest and most central) is given by

$$\frac{g}{2\pi}\ln\left[\left(\frac{h}{r}\right) + \left\{\left(\frac{h}{r}\right)^2 - 1\right\}^{1/2}\left\{\left(\frac{d'_{ab}}{d_{ab}}\right)\left(\frac{d'_{ac}}{d_{ac}}\right)\left(\frac{d'_{ad}}{d_{ad}}\right)\text{ etc.}\right]$$

where d_{ab} = distance from reference cable a to neighbouring cable b and d'_{ab} = distance of cable a from image of b and so on for a with respect to cables c,d, etc. When the cables are touching in trefoil formation the effective thermal resistance of any one of the cables is given by[1]

$$0.366\,g\left[\log_{10}\left(2\frac{h}{r}\right) + 2\log_{10}\left(\frac{h}{r}\right)\right] \text{ deg C/W per cm length}$$

where h is measured to the centre of the trefoil group. The equation holds for the apex at both the top and bottom. The units of g are deg C-cm/W.

In pipe-type cables the cable to screen thermal resistance is readily calculated, as is the resistance from the pipe to ambient. The effective thermal resistance from cable screens to pipe is more complex, however, involving natural convection in the oil space. Similarly, the corresponding resistance in air ducts must be evaluated. For values of the equivalent diameter (D_e) of the insulated cores between 75 to 125 mm, the effective thermal resistance, cable screen to pipe wall, is given by[1]

$$\frac{100A}{1 + (B + C\theta_m)D_e} \text{ deg C/W per cm.} \tag{3.5}$$

For three cores in a pipe-type cable, $D_e = 2.15 \times$ outside diameter of core (cm); for 2 cores $D_e = 1.65$ core outside diameter (cm); θ_m is the mean temperature (°C) of the filling medium (e.g. oil). Values of A, B, C, and D are given in Table 3.2. An assumed value is initially used for θ_m and an iteration process used. A more fundamental approach may be made by use of formulae for natural convection. As (3.5) is empirical it should be used with caution for dimensions outside the range suggested.

Table 3.2 Constants A, B, and C

Installation	A	B	C
In metallic conduit	5.2	1.4	0.011
In fibre duct in air	5.2	0.83	0.006
In fibre duct in concrete	5.2	0.91	0.010
In asbestos cement duct in air	5.2	1.2	0.006
In asbestos cement duct in concrete	5.2	1.1	0.011
Oil pressure pipe-type cable	0.26	0.0	0.0026

Values taken from I.E.C. Report No. 287 (1969). (Reproduced by permission of the International Electrotechnical Commission, to whom the copyright belongs)

3.3 CHARACTERISTICS OF SOIL[2,3,4,5,6,7,8,9]

For self-contained and pipe-type systems the thermal resistance of the soil accounts for a significant component of the conductor–ambient resistance. Hence a precise knowledge of soil thermal characteristics is vital. An important characteristic of some soils is the increase in the thermal resistivity when dried out. The heat from the cable causes the moisture in the soil to migrate, leaving air in the interstices previously occupied by water and the thermal resistivity increases due to the high resistivity of the air. This in turn increases the cable temperature which increases the dielectric loss and to a lesser extent the copper loss, thus causing increased moisture migration. A possible result is *thermal instability* and the electrical failure of the insulation.

Soil is classified into several types, e.g. chalk, peat, and, made-up, and clay. The moisture content and hence thermal resistivity is affected by the presence of an impermeable cover such as tarmac and concrete which tends to prevent evaporation. The term '*Backfill*' is now reserved for the material immediately surrounding an underground cable. Special backfills are such that even when completely dry their resistivity does not exceed agreed values. Hence the use of special backfill is often to ensure stability of environment rather than consciously to reduce the thermal resistance value. Initial work on the drying-out process suggested that the cable or pipe surface should not exceed 50 °C in summer.

There are two main classes of backfill:[8]

(*a*) Selected sands (dried out resistivity $\ngtr$ 2.7 deg C-m/W). These tend to be the most cohesive. A high density is required, and this implies a high degree of compaction on installation.
(*b*) Stabilized backfills, i.e. composite materials selected to give a dried-out resistivity $\ngtr$ 1.2 deg C-m/W.

It is unlikely that a dried out value of resistivity as low as 1.2 deg-m/W is obtainable from natural sands, and manmade mixes are used as follows:

Cement-bound sand in proportion 1 : 14 by volume.
Gravel or shingle/sand in proportion 1 : 1 by weight, particle size of shingle not to exceed 10 mm and preferably rounded in shape.
Bitumen-bound sand in proportion 1 : 9 by volume.

A proposed British policy[8] with regard to soils and backfills is as follows. In winter a resistivity of 1.05 deg C-m/W is ascribed to all soils and backfills (including troughs – see later). In summer a resistivity of 1.20 deg C-m/W is ascribed to soils and backfills which are not fully dried-out. For situations in which full drying-out exists the resistivity for soil is taken as 3.0 deg C-m/W, for selected-sand backfills 2.7 deg C-m/W, and for stabilized backfills 1.20 deg C-m/W. In light of the above criteria it is recommended that soil (route) resistivity surveys are not required except where there is suspicion that in winter resistivities greater than 1.05 deg C-m/W exist. The establishment of any such code must be conservative in nature because of a lack of precise knowledge of soil performance in practice. Such knowledge is difficult to obtain because of the following practical reasons:

(*a*) Few cables are fully loaded and so experience onerous enough to give useful data on long-term effects is scarce.
(*b*) Few installations possess temperature monitoring.

Trench dimensions with special backfills are illustrated in Figure 3.9 for self-contained systems. The effective thermal resistance of an installation involving special backfill and local soil cannot be calculated using equation 3.4, and a full field plot is required. However, a method of determining the effective thermal resistance when the cables are surrounded by a rectangular section of material of different thermal resistivity to the soil is given in Example 3.2. In this particular case it refers to a duct set in concrete but can be applied to any other material.

The possibility of controlling the drying out of soil, and hence its thermal resistivity, by the installation of automatically controlled water sprinklers along the route has been investigated. Care must be taken when adding water in such a fashion not to undermine the wall of soil around the cables or pipe.

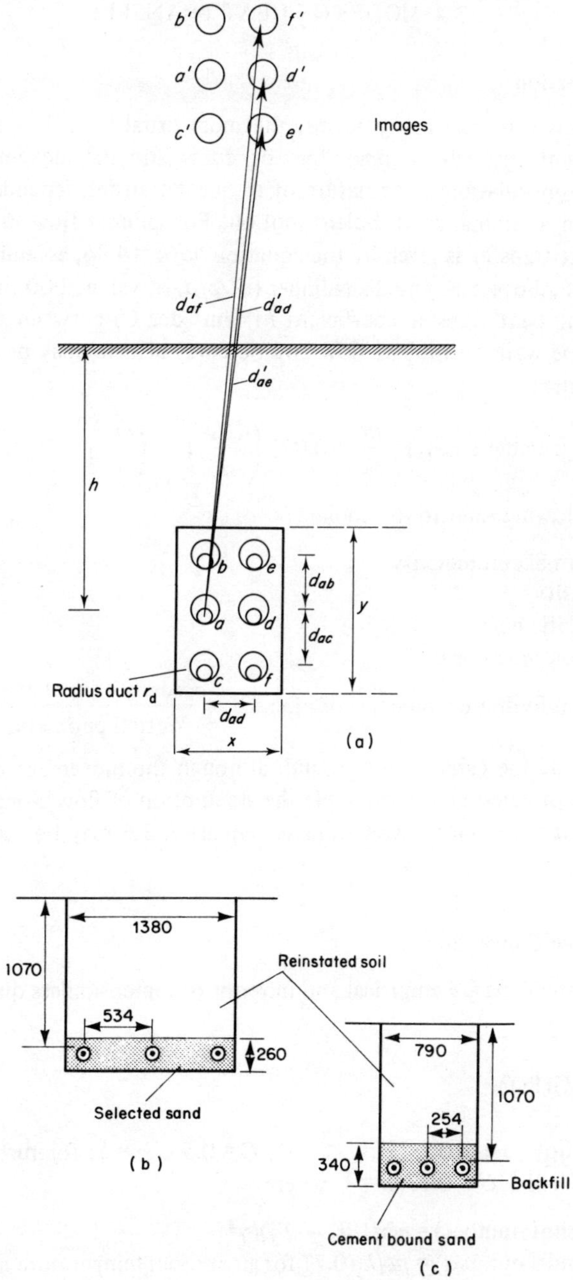

Figure 3.9 (a) Typical duct installation with images of cables also shown. Ducts installed in concrete. (b) Section of LPOF direct buried cable installations. Cables, 275 kV, 1935 mm², 760 MVA – rural route. (c) As (b) for urban route. Dimensions in mm.

3.4 MODES OF HEAT TRANSFER

Forced Convection

In practice this can take many forms, the most usual in cable systems being the flow of coolant through cooling pipes or ducts and the movement of wind in above-ground installations. The nature of the heat transfer depends upon whether the fluid is in a laminar or turbulent motion. For laminar flow in smooth heated tubes the heat transfer is given by the equation $hd/k = 4.36$, assuming entry effects are negligible. Above a Reynold's Number $(dU\rho/\mu)$ of value 2000 the flow becomes turbulent. The heat transfer coefficient h(W/m^2 deg C) between the coolant and containing-pipe walls (turbulent flow and no entry effects) may be calculated from the expression,

$$\text{Nusselt number } (N_{\text{NU}}) = \frac{hd}{k} = 0.023 \left(\frac{dU\rho}{\mu}\right)^{0.8} \left(\frac{c\mu}{k}\right)^{0.4} \tag{3.6}$$

where the following refer to the coolant:

k = thermal conductivity
ρ = density
c = specific heat
μ = absolute viscosity
d = mean hydraulic diameter of pipe, i.e. $4\,\dfrac{\text{Cross-sectional area}}{\text{Wetted perimeter}}$.

For portions of the cable above ground, although the movement of wind may be considered as directed across the cable the assumption of flow along the cable gives a more pessimistic value of heat transfer (equation 3.6 may be used) and may be preferable.

Natural or Free Convection

Again the expressions are empirical and in terms of dimensionless quantities.

Generally,

$$\frac{hd}{k} = C(\text{GrPr})^n. \tag{3.7}$$

For laminar flow, i.e. $10^4 < \text{GrPr} < 10^9$, $C = 0.53$, $n = \frac{1}{4}$; for turbulent flow, i.e. $10^9 < \text{GrPr} < 10^{12}$, $C = 0.13$, $n = \frac{1}{3}$, where,

Gr(Grashof number) = $a\beta d^3 (T_s - T_a)/\gamma^2$
Pr(Prandtl number) = $\mu c/k$ (0.71 for air at room temperature and atmospheric pressure)
γ = kinematic viscosity (m^2/s)
a = acceleration due to gravity (9.81 m/s^2)
β = volume expansion coefficient ($1/K$)
$(T_s - T_a)$ = surface minus ambient (or coolant) temperature (deg C).

For laminar flow the heat transfer resistance per unit length is

$$\frac{1}{h(\text{Area})} = \frac{1}{k\,0.53\,(\text{GrPr})^{1/4}\pi}$$

i.e. proportional to

$$\frac{1}{(T_s - T_a)^{1/4}}$$

and the heat transferred = $(T_s - T_a)/R$, i.e. proportional to $(T_s - T_a)^{5/4}$.

Radiation

This mode of heat transfer is important when considering the ratings of cables installed in tunnel and shafts and for above-ground installations. Radiation is a maximum when the cable surface approaches the 'black body' condition when the surface emissivity is unity. Usually the finishes accorded to cables have an emissivity approximating to 0.9. The formula for the heat transferred by radiation is as follows:

$$q_r = A_n K_r \sigma({}_aT_s^4 - {}_aT_w^4) \text{ per m length} \tag{3.8}$$

where ${}_aT_s$ and ${}_aT_w$ are the absolute temperatures of the cable surface and surrounding wall (K). The heat transfer coefficient h at temperature ${}_aT_s$ is

$$\frac{q_r}{A_n({}_aT_s - {}_aT_w)} = K_r\sigma({}_aT_s^3 + {}_aT_s^2\,{}_aT_w + {}_aT_s\,{}_aT_w^2 + {}_aT_w^3). \tag{3.9}$$

For a single cable or pipe, A_n = radiating area = πd per unit length,

$$K_r = \text{effective emissivity factor} = \frac{1}{1/\epsilon_s + \beta(1/\epsilon_w - 1)}$$

where

$$\beta = \frac{n\pi d}{\pi d_w}\left(\frac{1}{1 - (\pi n d/d_w)(\epsilon_s/\epsilon_w)(n-1)(1-\epsilon_w)}\right).$$

Typically, $\epsilon_s \simeq \epsilon_w \simeq 0.9$ and $K_r = 0.9$,

ϵ_s and ϵ_w = emissivities of cable/pipe surface and surface of containing structure, e.g. tunnel wall, and d and d_w are the corresponding diameters

n = number of cables

σ = Stefan Boltzman coefficient = 5.7×10^{-8} W/m^{2}K^4.

When several cables or pipes are in close physical proximity the mutual radiating area between their surfaces must be evaluated. A simple method for cables is given in Reference 10.

3.5 LOSSES

Heat originates from the various losses in a cable system. These are, conductor loss (including skin and proximity effects), sheath loss due to eddy and circulating currents (the latter may be largely eliminated by cross-bonding), eddy and hysteresis loss in steel armouring and in steel pipes and ducts, and finally dielectric loss. A further source of heat in above-ground parts of an installation emanates from solar radiation (q_s),

$$q_s = \alpha ds \tag{3.10}$$

where

d = diameter (m)
α = solar absorption coefficient (maximum value of 1.0 for black surface; 0.3 is typical for a shiny surface)
s = intensity of solar radiation (W/m^2); this varies geographically, a typical value being 800 W/m^2.

Further refinements may be introduced to account for the angle of the sun and the reader is referred to the literature on overhead-line thermal performance.

A.C. Losses in Armouring – Single-core Cables[11,12]

The loss in the sheath due to induced eddy currents (sheath open circuited) resulting from the circumferential flux of the conductor may be obtained approximately by the use of 'thin sheet' formulae. The skin depth in sheaths (approximately 7 cm in lead) is much greater than the thickness and their a.c. resistances are approximated to by the d.c. values. The greatest loss is in the armouring: because of the spiralling of the wires there are components of loss resulting from axial flux along the wires and circumferential flux cutting the armour at right angles. There are therefore two components of the eddy loss in the armouring; these are derived below. Also there is the hysteresis loss. Typically the loss in the sheath and armour of a submarine cable is roughly 70% of the conductor loss.

(a) Calculation of circumferential loss

The value of the peak flux density in the circumferential direction is obtained by the use of the approximate representation of the armour wires shown in Figure 3.10.

$$H_0 = \frac{\sqrt{2}I}{2\pi R}$$

where R is the mean radius of the centre line of the wires and I is the conductor current.

The magnetic reluctance

$$s = \frac{nl_2}{\mu_0 A} = \frac{H_0(l_1 + l_2)n}{\Phi_0}$$

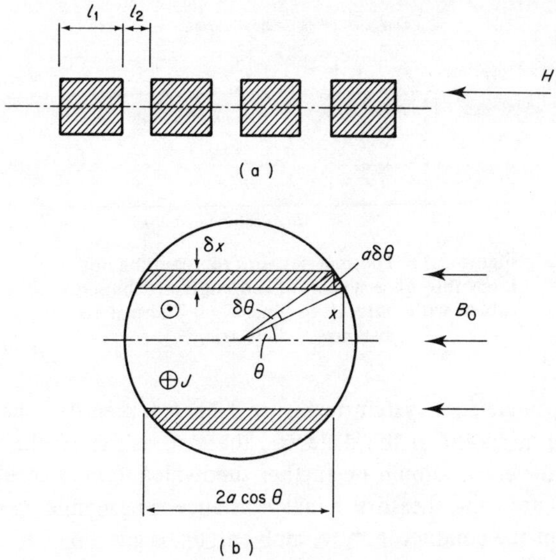

Figure 3.10 Losses in steel wire armour – magnetic fields and induced currents. (a) Armouring wires; (b) Analysis of one wire. J = induced eddy current, B_0 = flux density.

where n is the number of wires. From Figure 3.10 it may be shown that the eddy-current loss in one wire is $\pi\omega^2\sigma B_0^2 a^4/8$ W/m where $\sigma = 10^7$ Ω-m for steel.

(b) Calculation of axial loss in the armour

$$H_0 = \frac{\sqrt{2}I}{l_T}$$

where l_T is the length of a wire in travelling one complete spiral.

The eddy current loss per wire may be shown to be approximately equal to

$$\frac{\omega^2 B_0^2 \pi\sigma a^4}{16}\ \text{W/m}.$$

(c) Total hysteresis loss

This is given by

$$(\text{Vol. of one wire in m}^3)\ (\text{No. of wires})\ B_0^{1.6} \times 2512\ \text{W}$$

where B_0 is the flux density (Wb/m²).

3.6 STEADY-STATE CURRENT RATING

The evaluation of temperature rise is faciliated by the use of equivalent lumped-constant thermal circuits. This is shown for a cable of a directly buried,

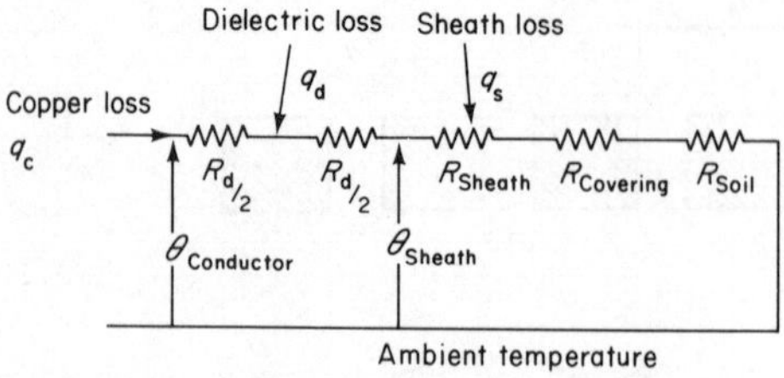

Figure 3.11 Thermal network representing hottest cable of a system of self-contained buried cables with natural cooling. R_d = Thermal resistance of dielectric.

naturally cooled, single-core system in Figure 3.11. It is seen that the dielectric loss is injected at the midpoint of the dielectric thermal resistance; this is an approximation, and the dielectric should be further subdivided if more accurate representation is desired. Often the sheath thermal resistance is negligible, in which case the temperature rise of the conductor above ambient, θ_c, is given by

$$\theta_c = q_c(R_d + R_{cov} + R_{soil}) + q_d(\tfrac{1}{2}R_d + R_{cov} + R_{soil}) + q_s(R_{cov} + R_{soil}).$$

θ_c is the conductor operating temperature, minus ambient, and as q_c is I^2R_{ac}, the current rating (I) is obtained. Often the expected load on a circuit will be of a cyclic nature with time. If this is well defined a load loss factor will be used to modify the conductor current and hence the conductor, sheath, and armour losses. The dielectric loss is not affected. A cable on full load will have a load factor of 100% and a load loss factor of 100%. In Britain the load loss factor is related to the load factor (LF) by the following:

$$\text{Load Loss Factor} = 0.2(\text{LF}) + 0.8(\text{LF})^2.$$

Often in bulk transmission circuits a LF of 75% is assumed, resulting in a loss factor of 57%.

The value of tan δ, and hence the dielectric loss, varies with temperature, and accurate values of conductor temperature will involve adjustment for the change in

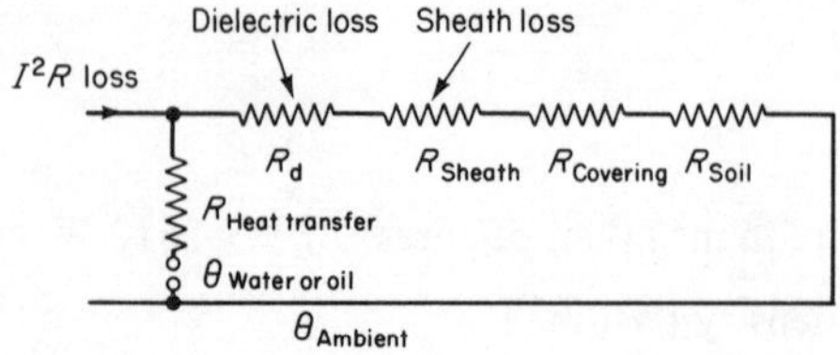

Figure 3.12 Thermal network for hottest cable of self-contained system incorporating heat transfer to moving fluid in conductor duct (internal cooling). $R_{heat\ transfer}$ = heat transfer resistance from duct wall to fluid.

tan δ. As in steady-state calculations the conductor temperature is usually specified; this adjustment is not a problem but must be borne in mind in situations where the temperatures are not specified. If heat is removed at the conductor by the flow of a coolant in the conductor duct of a self-contained cable the steady state network of Figure 3.11 is modified as shown in Figure 3.12 by the inclusion of a heat transfer resistance. A widely used paper for the calculation of steady-state ratings is Reference 14. Other basic references are numbers 13, 15, 16, and 17, and much of this information is summarized in Reference 1.

Joints and Terminations

When calculations are made for joints and terminations, at least two-dimensional heat flow must be assumed. In such cases numerical methods are used in conjuction with digital computers. The cable, joint, or termination is represented by finite difference networks. Simplified radial heat-flow networks for each longitudinal section into which the joint is subdivided are interconnected at nodal points by longitudinal thermal resistances representing heat-flow paths along conductor and sheath/spinning. The number of longitudinal interconnected sections can be altered to vary the fineness of representation, which becomes coarse in the remote cable where axial variations of temperature with length approach zero.

Longitudinal thermal resistances between axial sections of the dielectric and soil are not included since their values are very high compared with those of the conductor and sheath. The dielectric at each longitudinal position is divided into a number of annular rings in which the dielectric losses are injected. Conductor and sheath losses for each section are also injected at the corresponding node points. Programs exist[20] which construct the finite difference mesh for the joint and termination. Simultaneous equations are set up relating the heat balance for every node in the mesh as follows:

$$T_i = \frac{Q_i + \sum_j T_j/R_{ij}}{\sum_j 1/R_{ij}} \tag{3.11}$$

where

Q_i is the heat loss at node i
T_i, T_j the temperature at nodes i and j
R_{ij} the thermal resistance between nodes i and j.

The equations can be solved by an iterative method or by direct matrix inversion. The iterative method allows changes in temperature-dependent losses to be made readily during the computation. Temperature distributions for normally buried 275 kV, 1935 mm^2, straight and stop joints obtained using this method[22,23,24,25,26] are shown in Figure 3.13.

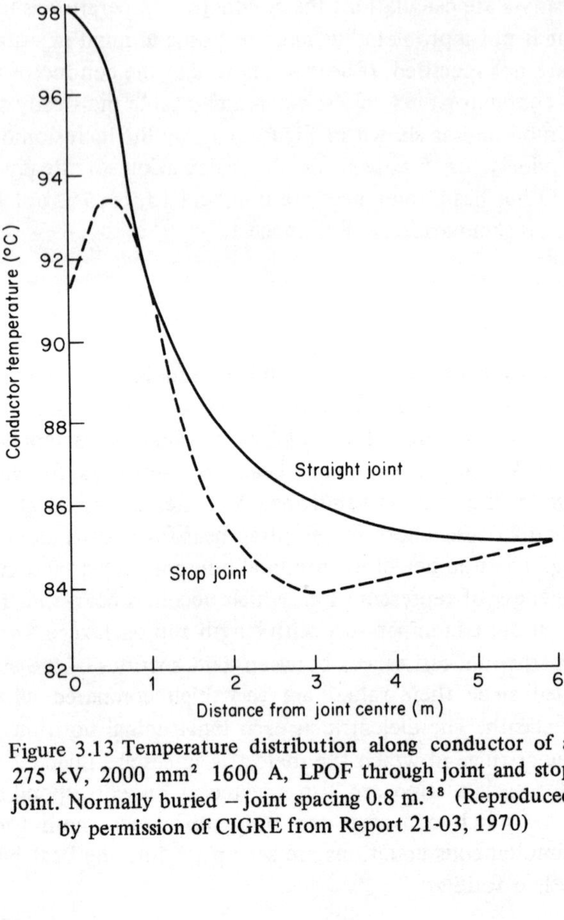

Figure 3.13 Temperature distribution along conductor of a 275 kV, 2000 mm² 1600 A, LPOF through joint and stop joint. Normally buried – joint spacing 0.8 m.[38] (Reproduced by permission of CIGRE from Report 21-03, 1970)

Example 3.1

Two 3-phase 138 kV, 60 Hz circuits comprising 6 self-contained, extruded, cross-linked, polyethylene cables are installed in a buried concrete duct as shown in Figure 3.9*a*. Calculate the MVA rating of the two circuits assuming a maximum insulation temperature of 80 °C and an ambient temperature of 5 °C. The cable details are as follows:

Area of concentric stranded conductor	= 760 mm²
A.C. Resistance of conductor	= 32.3 μΩ/m
Diameter over conductor	= 36 mm
Diameter over insulation	= 76 mm
Diameter over outside covering	= 90 mm (PVC *g* = 7 (deg C-m/W)
Soil *g*	= 1.2 deg C-m/W
Polyethylene *g*	= 4.0 (deg C-m/W)
Fibre duct *g*	= 4.80 deg C-m/W (thickness 3.3 mm)
Concrete, *g*	= 0.85 deg C-m/W
Inside diameter of each duct	= 0.13 m

x	= 0.5 m (see Figure 3.9a)
y	= 0.7 m (see Figure 3.9a)
Axial spacing between centre lines of ducts (vertical and horizontal)	= 0.25 m
Depth to centre of middle duct (h)	= 1.1 m.

Solution

Assume the dielectric loss to be negligible. Thermal resistance of dielectric per metre

$$R_d = \frac{4.0}{2\pi} \ln\left(\frac{76}{36}\right) = 0.47 \text{ deg C/W.}$$

Thermal resistance of protective covering

$$R_{cov} = \frac{7.0}{2\pi} \ln\left(\frac{90}{76}\right) = 0.19 \text{ deg C/W per metre.}$$

Thermal resistance from cable surface to duct wall

$$R_{c-d} = \frac{A}{1 + (B + C\theta_m)D_e} \text{ per m}$$

where θ_m = mean temperature in air gap, assumed to be 40 °C, D_e = external diameter of cable (cm) = 9 cm. From Table 3.2, A = 5.2, B = 0.91, C = 0.01. Hence

$$R_{c-d} = \frac{5.2}{1 + (0.91 + 0.01 \times 40)9} \text{ deg C/W per m length}$$

$$= 0.41 \text{ deg C/W per metre length.}$$

Thermal resistance from ducts to ambient:

(a) Assume first the ducts are in an homogeneous environment of

g = 0.85 deg C-m/W.

Thermal resistance of middle cable (including effects of mutual heating) to ambient is

$$R_{a-s} = \frac{g_c}{2\pi} \ln\left[\left(\frac{h}{R_d} + \sqrt{\left\{\left(\frac{h}{R_d}\right)^2 - 1\right\}}\right)\left(\frac{d'_{ab}}{d_{ab}}\right)\left(\frac{d'_{ac}}{d_{ac}}\right)\left(\frac{d'_{af}}{d_{af}}\right)\left(\frac{d'_{ad}}{d_{ad}}\right)\left(\frac{d'_{ae}}{d_{ae}}\right)\right]$$

where (from Figure 3.9a) h = 1.1 m, R_d = 0.13,

$$\frac{d'_{ab}}{d_{ab}} = \frac{2.45}{0.25} \quad \frac{d'_{ac}}{d_{ac}} = \frac{1.95}{0.25} \quad \frac{d'_{af}}{d_{af}} = \frac{2.46}{0.3535} \quad \frac{d'_{ad}}{d_{ad}} = \frac{2.21}{0.25} \quad \frac{d'_{ae}}{d_{ae}} = \frac{1.86}{0.3535}.$$

Hence,

$$R_{a-s} = \frac{0.85}{2\pi} \ln[(8.46 + \sqrt{(71.58 - 1)}^{1/2}(24742.4)]$$

$$= 1.75 \text{ deg C/W per m length.}$$

(b) Correction[1] *to* R_{a-s} *to account for concrete to be algebraically added to* R_{a-s} *is*

$$\frac{N(g_s - g_c)}{2\pi} \ln\left[\left(\frac{L_d}{R_{ed}}\right) + \left[\left(\frac{L_d}{R_{ed}}\right)^2 - 1\right]^{1/2}\right]$$

where N = number of loaded cables = 6, R_{ed} = equivalent radius of concrete bank (cm) which is given by

$$\log_{10} R_{ed} = \frac{1}{2}\frac{x}{y}\left(\frac{4}{\pi} - \frac{x}{y}\right)\log_{10}\left(1 + \frac{y^2}{x^2}\right) + \log_{10}\left(\frac{x}{2}\right)$$

where x is in cm and L_d = depth to centre of bank = h = 1.1 m. Hence correction is

$$\frac{6(1.2 - 0.85)}{2\pi} \ln\left[\frac{110}{31.6} + \left(\left(\frac{110}{31.6}\right)^2 - 1\right)^{1/2}\right] = 0.64 \text{ deg C/W per m length.}$$

Therefore,

$$R_{a-s} = 1.75 + 0.64 = 2.39 \text{ deg C/W.}$$

Dielectric loss is

$$2\pi 60 \left(\frac{138000}{\sqrt{3}}\right)^2 C \times 0.0005 \text{ W/m} = 0.2 \text{ W/m per cable}$$

where

$$C = \frac{2\pi\ 8.85 \times 10^{-12} \times 2.3}{\ln(76/36)} = 0.000171\ \mu\text{F/m.}$$

To calculate current rating per cable (I) use a chain of thermal resistances with losses injected. Hence

$$I^2 R_{ac}(3.46) + 0.2(2.895) = (80 - 5) = 75$$

and

$$I = 816 \text{ A.}$$

Total MVA of system = $\sqrt{3} \times 138000 \times 816 \times 2 = 390$ MVA.

Example 3.2

In this example the traditional empirical approach to the rating of a pipe-type cable will be given. A solution[21] has been reported using a computer, although this is also semi-empirical.

Determine the rating of a 500 kv, 60 Hz, HOPF pipe-type cable. Data:

Diameter over conductor, 41.4 mm
Thickness of insulation, 34 mm
Diameter of pipe, 300 mm
Depth of pipe centre, 1 m
Soil g, 0.9 deg C-m/W
Ambient, 25 °C
Conductor R_{ac} = 0.044 Ω/ph/km
Dielectric g = 5.0 deg C-m/W; $\tan \delta = 0.002$, $\epsilon_r = 3.5$
Pipe loss = 0.28 x conductor loss (assumed)
Operating temperature = 80 °C.

Solution (Figures 3.14 and 3.15)

From Reference 1 (see also Figures 3.14 and 3.15).

Thermal resistance cable surface to pipe (R_{c-p}) is

$$\frac{100\,A}{1 + (B + C\theta_m)D_e}.$$

For this system,[1] $A = 0.26$, $B = 0$, $C = 0.0026$, $D_e = 2.15(41.4 + 68) = 235.2$ mm = 23.52 cm. Assume mean oil temperature to be 55 °C, then

$$R_{c-p} = \frac{100 \times 0.26}{1 + (0.0026 \times 55)(23.52)} = 5.95 \text{ deg C/W per cm length}$$

$$R_{\text{dielectric}} = \frac{500}{2\pi} \ln \left(\frac{235.2}{41.4}\right) = 77.32 \text{ deg C/W per cm length}$$

$$R_{\text{soil}} = \frac{90}{2\pi} \ln \frac{2 \times 100}{15.0} = 36.9 \text{ deg C/W per cm.}$$

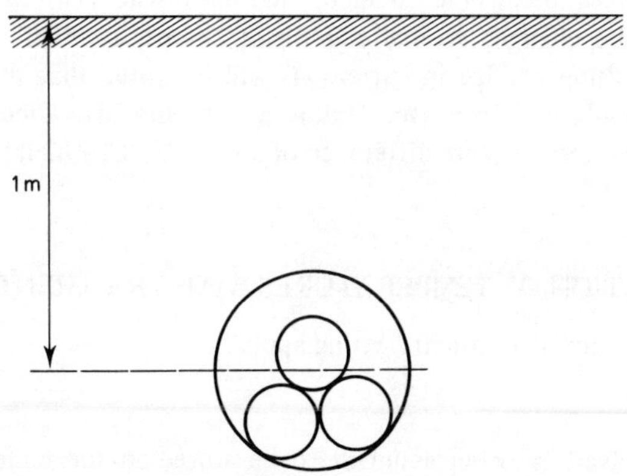

Figure 3.14 Example 3.2. Pipe-type HPOF cable.

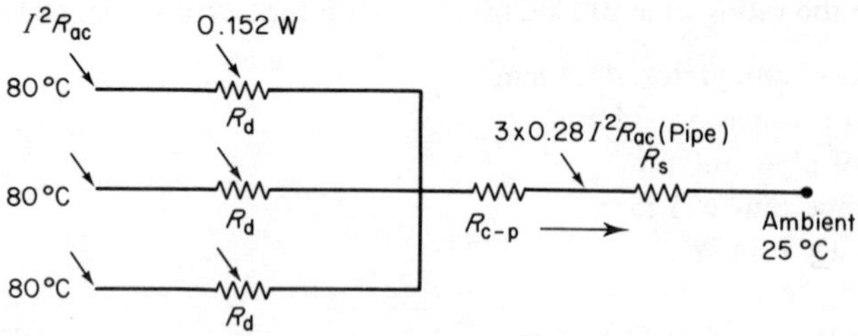

Figure 3.15. Example 3.2. Equivalent steady-state thermal network.

Conductor loss $= I^2\ 0.044 \times 10^{-5}$ W/cm

Pipe loss $= 3 \times 0.28 I^2 \times 0.044 \times 10^{-5}$ W/cm

$$\text{Capacitance per core} = \frac{2\pi\ 8.85 \times 10^{-12} \times 3.5}{\ln(109.4/41.4)}\ \text{F/m}$$

$$= 200\ \mu\mu\text{F/m}$$

Dielectric loss = 15.21 W/m = 0.152 W/cm per core

Temperature difference, conductor to ambient = 55 deg C

$$= 0.44 \times 10^{-6} I^2 (77.32) + 3 \times 0.44 \times 10^{-6} I^2 (5.96 + 36.9) + 0.152\left(\frac{77.32}{2}\right)$$

$$+ 3 \times 0.152(5.96 + 36.9) + 3 \times 0.28 \times 0.44 I^2 \times 10^{-6} \times 36.9.$$

Hence, $I = 530$ A.

With this value of current the mean oil temperature is 64 °C. As the initial guess for θ_m was 55 °C it is worthwhile recalculating I with θ_m taken as, say, 62 °C.

The new value of R_{c-p} is

$$\frac{100 \times 0.26}{1 + (0.0026 \times 62)(23.52)}\ \text{deg C/W per cm} = 5.43\ \text{deg C/W}.$$

In the context of the whole chain of thermal resistances this is small and the current will not be changed to any significant extent. It is left to the reader to ascertain the resulting change in current. It will be noted that the current magnitude is not particularly impressive. This is due to the high dielectric loss which alone gives rise to a temperature difference of about 25 °C, a high proportion of the available 55 °C.

3.7 CALCULATION OF TEMPERATURES WITH TRANSIENT OPERATION

The following regimes of transient heating apply:

(a) Short circuit

All the heat evolved is often assumed to be stored in the conductor or sheath depending on the nature of the fault.

(b) Cyclic loading

Calculations are often based on an idealized daily load curve of full-load current for 8 hours and no-load for 16 hours. This results in a lower temperature rise than when permanently on full load which is allowed for in practice by the use of Cyclic Rating Factors combined with steady-state calculations.

(c) Short-time emergency loading

An overload may be sustained for perhaps several hours without excessive temperatures due to the long thermal time-constant of cables if naturally cooled. Such an overload may be required by system operational requirements such as the outage of one circuit of a parallel combination.

(d) Change in cooling conditions

E.g. loss of coolant supply in artificially cooled systems.

In high-capacity cables the short-circuit current (I_{sc}) is probably restricted by the mechanical forces created rather than by temperature rises. The change in conductor electrical resistivity must be accounted for when calculating the short-circuit temperature rise. The following formula is typical:

$$\left(\frac{I_{sc}^2}{A^2}\right) t_{sc} = \left(\frac{m + 20}{\rho_{20}}\right) c_p \rho \ln\left(\frac{m + T_n}{m + T_0}\right) \tag{3.12}$$

where

A = conductor area (m^2), t_{sc} = short circuit period (s)
m = 234 for copper, 228 for aluminium, and 230 for lead
c_p = specific heat of conductor (kJ/kgK)
T_n = temperature at end of short circuit (°C)
T_0 = initial temperature (°C)
ρ_{20} = electrical resistivity of conductor (20 °C) (Ω-m)
ρ = conductor density (kg/m^3).

The maximum conductor temperature or short circuit is often taken as 120 °C whilst the following other limits apply; copper wire or tape screen 350 °C, lead sheath 200 °C, aluminum sheath 200 °C. If heat-flow into the insulation is to be included[27] a numerical method using a computer is best employed, of the type described in a later section.

The temperature rise during a short circuit depends upon the mean effective current of the fault which, over the fault period, generates the same heat as the actual current, which will normally have a d.c. component of magnitude dependent on circuit conditions. If an iterative numerical method is used the time step per iteration can be made small enough during the fault period to give the values of instantaneous heating including the effects of asymmetry. On cooling (after the fault) a much longer time-step may be used.

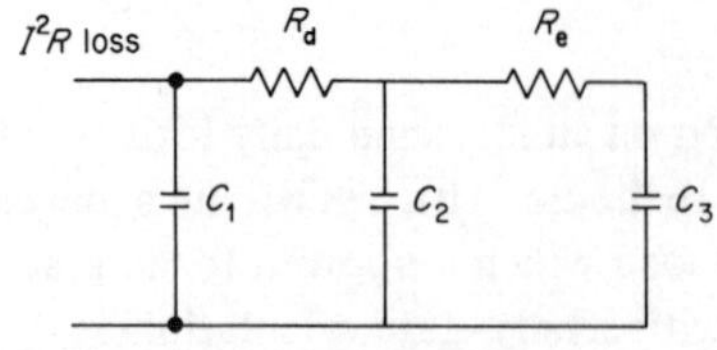

Figure 3.16 Simple model for transient thermal response of buried cable: C_1 = thermal capacitance of conductor + (p x capacitance of insulation), C_2 = thermal capacitance of sheath + $(1 - p)$ x capacitance of insulation), C_3 = effective capacitance of soil, R_d = thermal resistance of insulation, R_e = effective thermal resistance of soil, $p = 1/(2 \ln(R/r)) - 1/(R^2/r^2 - 1)$ where R = outside radius of cable and r = conductor radius.

A lumped-constant thermal network may be used for transient calculations[25] by the connection of appropriate thermal capacities between the nodes of the steady-state network and the ambient or reference line. For a self-contained cable, if uniformly radial heat-flow is assumed, i.e. the sheath is isothermal, the thermal network is the R–C ladder type. The number of nodes depends on the number of annular cylinders into which the dielectric is divided. The conductor and sheath temperatures may be obtained from the simple method[28] shown in Figure 3.16.

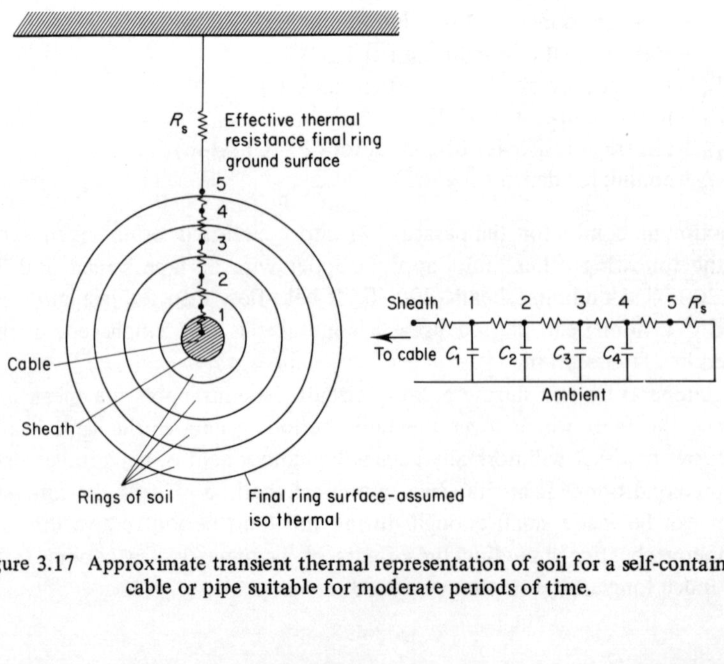

Figure 3.17 Approximate transient thermal representation of soil for a self-contained cable or pipe suitable for moderate periods of time.

For directly buried cables (self-contained) or pipes, the soil may be represented approximately by annular rings concentric with the cable as shown in Figure 3.17. The accuracy of this somewhat crude model will depend upon the transient time of interest. If relatively short (e.g. less than 100 hours) accuracy will be reasonable, becoming worse with increasing times. In this method radially uniform flow of heat is assumed.

For an accurate simulation,[29] the temperature rise at the cable sheath or pipe surface due to the soil alone is described by the equation:

$$\theta(t) = \frac{q(t) g_{\text{soil}}}{4\pi} \left[\left\{ -\text{Ei}\left(-\frac{d^2}{16\alpha t}\right) + \text{Ei}\left(-\frac{h^2}{\alpha t}\right) \right\} + \sum_{i=1}^{i=N} \left\{ -\text{Ei}\left(-\frac{r_{1,i}^2}{4\alpha t}\right) + \text{Ei}\left(-\frac{r_{2,i}^2}{4\alpha t}\right) \right\} \right] \tag{3.13}$$

where i = number of cable, $r_{1,i}$, $r_{2,i}$ = distances of cable i and its image from reference cable, h = depth, d = cable diameter, α = thermal diffusivity = $1/c_p \rho g$, and Ei is the exponential integral which is tabulated in various reference books.

If $q(t)$, the value of the heat transmitted from the cable to the soil, is known at the instant t, then the temperature of the cable surface is known and the cable thermal network may be solved. However, $q(t)$ varies from its value at $t = 0$ to q (steady state) at $t = \infty$. A reasonable approach[30] for obtaining $q(t)$ is to use

$$\frac{q(t)}{q(t)_{\text{steady-state}}} \quad \text{(cable surface–node short circuited)}$$

$$= \frac{\text{Temperature across cable at time } t}{\text{Steady-state temperature rise across cable}}. \tag{3.14}$$

The calculation of the temperatures in a thermal ladder network with several R–C sections is tedious if done analytically. Electrical analogues[24,31] have been used to obtain the thermal transients but numerical methods on digital computers are preferable,[25,32,33,34,35] The outline of such a method follows.

Consider the thermal network shown in Figure 3.18 and the relationships between nodes a,b, and c. The total heat input to node a is

$$K_{ba}(T_b - T_a) + K_{ca}(T_c - T_a) + q_{ga} \tag{3.15}$$

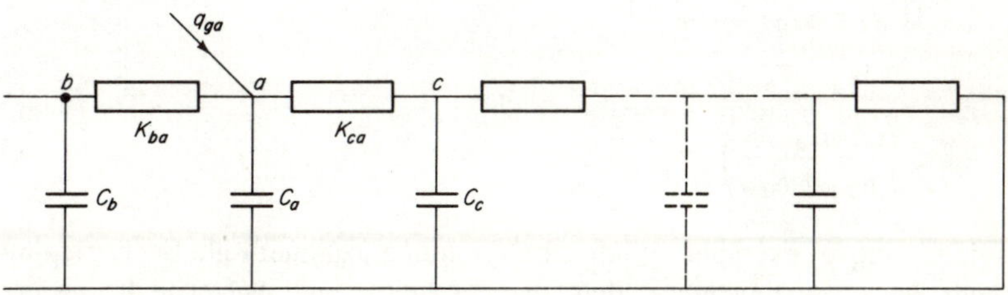

Figure 3.18 Ladder network for computation of temperatures.

where K_{ba} = thermal conductance b to a and q_{ga} = heat generated in the volume represented by node a, e.g. ohmic or dielectric loss. Over a finite time interval, Δt, the average rate of heat flow to region a is given by,

$$q_a = C_a(T_a^1 - T_a)/\Delta t \text{ (W)} \tag{3.16}$$

where

C_a = thermal capacity of region a (J/deg C)
T_a^1 = temperature at a after Δt
T_a = temperature at a before Δt.

Hence

$$q_a = \sum_{n \neq a} K_{na}(T_n - T_a) + q_{ga} \tag{3.17}$$

where n = subscript for region or node adjacent to a. Thus

$$C_a(T_a^1 - T_a)/\Delta t = \sum_{n \neq a} K_{na}(T_n - T_a) + q_{ga}. \tag{3.18}$$

For nodes b and c adjacent to a,

$$\frac{K_{ba}\Delta t}{C_a} T_b + \frac{K_{ca}\Delta t}{C_a} T_c + \frac{q_{ga}}{C_a} \Delta t - \frac{(K_{ba} + K_{ca})\Delta t}{C_a} T_a = T_a^1 - T_a$$

and

$$T_a^1 = \frac{K_{ba}\Delta t}{C_a} T_b + \frac{K_{ca}\Delta t}{C_a} T_c + \left[1 - \Delta t\left(\frac{K_{ca} + K_{ba}}{C_a}\right)\right] T_a + \frac{q_{ga}\Delta t}{C_a}.$$

Labelling the coefficients of T_b and T_c, P and Q respectively,

$$T_a^1 = PT_b + QT_c + (1 - P - Q)T_a + \frac{q_{ga}\Delta t}{C_a}. \tag{3.19}$$

This equation is successively applied to all the nodes in the network commencing with a known or estimated steady-state temperature distribution, e.g. all temperature rises zero for a cable originally non-energized. A suitable time step Δt is chosen and then the equation for T_n at each node solved using the temperature rises above ambient existing at the nodes, e.g. T_{n-1}, T_n, and T_{n+1}. If the coefficient of T_a is negative, mathematical instability occurs and the time step must be chosen such that $1 - P - Q > 0$ for each node, i.e.

$$1 - (K_{ba} + K_{ca})\Delta t/C_a > 1$$

or

$$\Delta t < \left(\frac{C_a}{K_{ba} + K_{ca}}\right). \tag{3.20}$$

This is a simple example of an *explicit* numerical method. Such step-by-step methods have the advantage that the conductor and dielectric losses may be adjusted for temperature at each step when their relationships are non-linear. A flow chart for the method is shown in Figure 3.19.

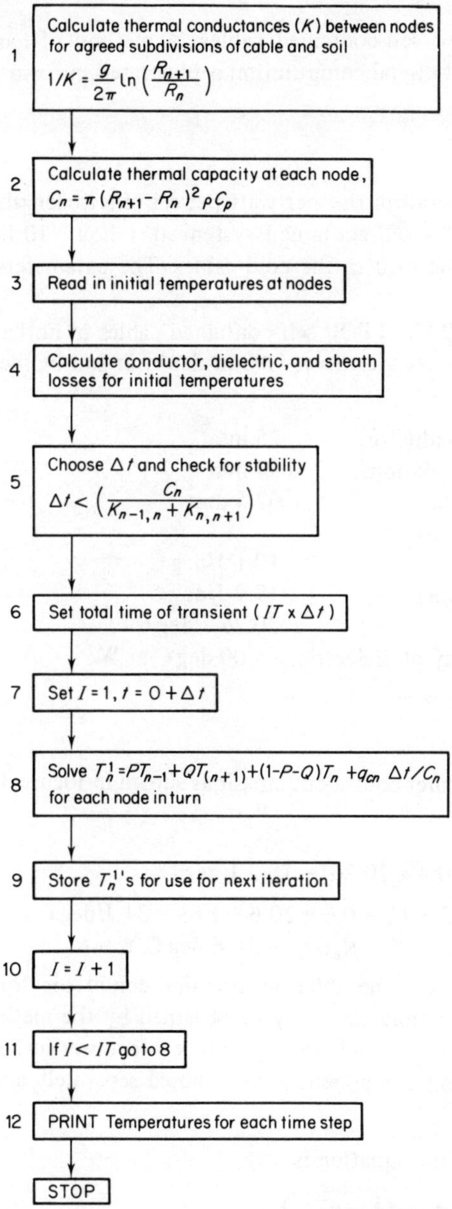

Figure 3.19 Flow diagram for explicit method for computation of transient temperatures applied to radial heat flow.

The above ladder-network model for the cable core assuming radial heat-flow through the dielectric may be applied to each core of a pipe-type cable. An accurate modelling of the oil in the pipe is much more difficult and usually the oil is represented by a single, lumped, thermal capacity. For short transient times the effect of the soil may be neglected without serious error or included in the same way as suggested for self-contained cables. A method of handling the pipe-type arrangement without digital computation is illustrated in Example 15.4.

Example 15.3

Determine the temperature rise per watt per centimetre length of the conductor of the hottest cable of a self-contained system at 1 hour, 10 hours, and 100 hours after switching on the load to the cold cables. The parameters of the cable system are as follows.

Three 138 kV, 60 Hz, LPOF self-contained cables in horizontal formation with 200 mm spacing between centres; 1.00 m depth in soil of thermal resistivity 0.90 deg C-m/W.

Diameter over conductor,	25 mm
Diameter over insulation,	47 mm
Diameter over cable,	62.5 mm
Thermal capacity of:	
Conductor,	10.1 J/deg C
Sheath and covering,	17.9 J/deg C
Dielectric,	20.26 J/deg C
Thermal resistivity of dielectric,	5.00 deg C-m/W.

Solution

Use the Van Wormer equivalent circuit as shown in Figure 3.16

$$p = 0.4$$

$$C_1 = 10.1 + (0.4 \times 20.26) = 18.2 \text{ J/deg C}$$

$$C_2 = (1 - p)C_i + C_s = 0.6 \times 20.6 + 17.9 = 31 \text{ J/deg C}$$

$$R_1 = 53.6 \text{ deg C/W} \quad R_{\text{cover}} = 16.6 \text{ deg C/W} = R_2.$$

The temperature rise–time relation for the conductor for a step-function of conductor heating at time zero may be obtained by the method by Neher.[23] This initially ignores the soil and assumes the cable outer-surface to be at a fixed temperature. The soil component is determined separately and then superposition applied.

The temperature–time equation is

$$R_a(1 - e^{-at}) + R_b(1 - e^{-bt})$$

where R_a and R_b are determined from a, b, R_1, and R_2. a and b are found by evaluating constants M and N as follows.

$$M = \tfrac{1}{2}C_1(R_1 + R_2) + (C_2R_2) = 890 \text{ s}$$

$$N = C_1R_1C_2R_2 = 0.4 \times 10^6 \text{ (s}^2\text{)}$$

$$\left.\begin{matrix} a \\ b \end{matrix}\right\} = [M \pm (M^2 - N)^{1/2}]/N.$$

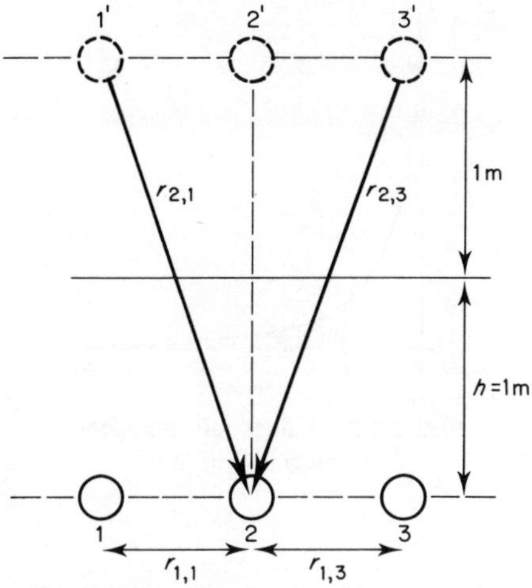

Figure 3.20 Disposition of cables in Example 15.3.

Hence, $a = 0.003$ and $b = 0.0007$. Also,

$$R_a = \frac{1}{(a-b)}\left[\frac{1}{C_1} - b(R_1 + R_2)\right] = 2.66$$

$$R_b = R_1 + R_2 - R_a = 67.59.$$

Temperature rise–time relation per watt input per cm length is

$$R_a(1 - e^{-at}) + R_b(1 - e^{-bt}) = 2.66(1 - e^{-0.003t}) + 67.59(1 - e^{-0.0007t}).$$

Response of the soil

Using the method suggested by Morello,[30] which is effectively the same as given by Neher,[23] the temperature rise of the hottest cable of the group (assuming equal loading) is given by equation (3.13), i.e.

$$q(t)\frac{g_s}{4\pi}\left[-\mathrm{Ei}\left(\frac{-d^2}{16\alpha t}\right) + \mathrm{Ei}\left(-\frac{h^2}{\alpha t}\right) + \sum_{i=1,i\neq 2}^{i=N} -\mathrm{Ei}\left(-\frac{r_{1,i}^2}{4\alpha t}\right) + \mathrm{Ei}\left(-\frac{r_{2,i}^2}{4\alpha t}\right)\right].$$

The distances are indicated in Figure 3.20: $d = 6.25$ cm, $h = 100$ cm, $r_{1,1} = 20$ cm $= r_{1,3}$, $r_{2,1} = 201$ cm $= r_{2,3}$,

$$\alpha = \frac{1}{c_p \rho g_s} = 0.0047\ \mathrm{cm^2/s}\quad g_s = 90\ \mathrm{deg\ C\text{-}cm/W}$$

$$q_t = q_{t=\infty}\left(\frac{\text{temperature across cable at time } t}{\text{steady-state temperature across cable}}\right).$$

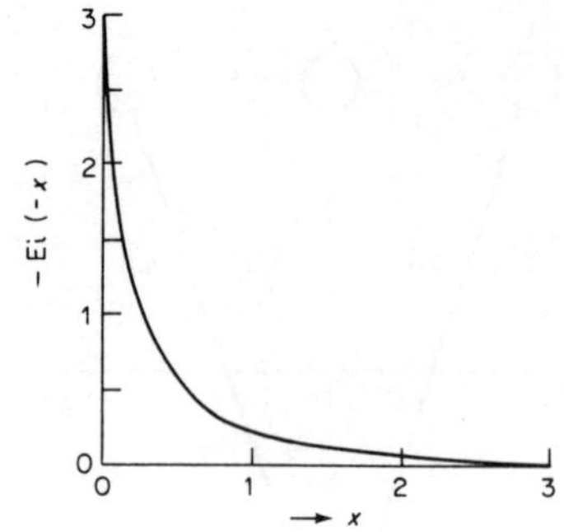

Figure 3.21 Values of exponential integral $-\text{Ei}(-x)$.

(a) For t = 3600 seconds

Using either Mathematical Tables for $-\text{Ei}(-x)$ or the graph shown in Figure 3.21

$$q(1\text{ hour}) = 1\left(\frac{62.2 + 2.66}{70.2}\right) = 0.925$$

$$\theta_s = \frac{q(t)}{4\pi}\,90 - \text{Ei}\left(-\frac{6.25}{16 \times 0.0047t}\right) + \text{Ei}\left(-\frac{100^2}{0.0047t}\right) - \text{Ei}\left(-\frac{20^2}{4 \times 0.0047t}\right)$$

$$+ \text{Ei}\left(-\frac{201^2}{4 \times 0.0047t}\right) - \text{Ei}\left(\frac{-20^2}{4 \times 0.0047t}\right) + \text{Ei}\left(\frac{-201^2}{4 \times 0.0047t}\right)$$

$= 19.85$ deg C.

Hence the conductor temperatures are as follows:

$$\theta_c(1\text{ hr}) = 19.85 + 64.86 = 84.7\text{ deg C}$$

$$\theta_c(10\text{ hrs}) = 28.36 + 70.2 = 98.6\text{ deg C}$$

$$\theta_c(100\text{ hrs}) = 57.3 + 70.2 = 127.5\text{ deg C.}$$

All these temperature rises relate to a conductor heat input of 1 W per cm with the dielectric loss assumed negligible.

3.8 ARTIFICIAL COOLING OF SELF-CONTAINED CABLES[20,36,37,38,39,40,41,42,43]

To overcome the severe restriction on current capability with direct-buried cables, artifical cooling may be used. The options available are as follows:

(*a*) *External pipe cooling*. The passage of water through plastic pipes laid beside the cables is illustrated in Figure 3.22. The normal thermal resistance presented to the cable losses is shunted by the thermal resistance into the water. The water increases in temperature along the route so that there is a limit to the individual lengths which

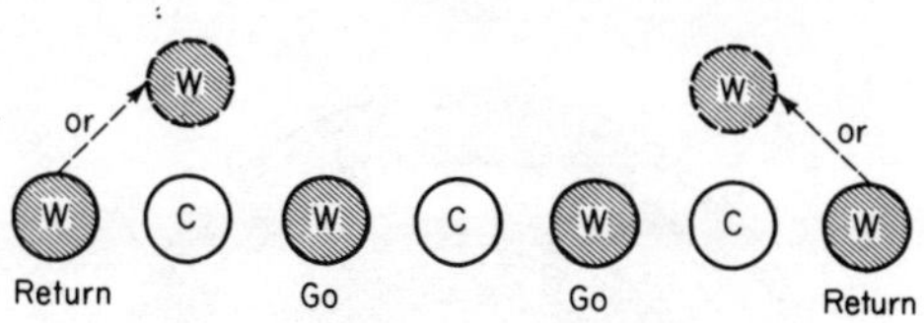

Figure 3.22 External cooling by water pipes. Go and Return water flows indicated. W = water pipe, C = cable.

can be cooled in this manner. The effective thermal resistance comprises the following: the soil between the cable and pipes, the pipe wall, the heat-transfer resistance into the water, and the thermal resistance of the cable itself. This system is relatively simple and retains the mechanical advantages of the direct buried cable. Longer cooling lengths are obtained by the use of larger-diameter cooling pipes, e.g. 150 mm diameter. These would be flexible and of reinforced plastic in order to withstand the soil when not containing water under pressure.

(*b*) *Surface cooling*. A more intensive water-cooled system than the use of external pipes involves the placing of each cable in a rigid plastic pipe of approximately 250 mm diameter and pumping water through the pipe. It is more expensive than method (*a*) but with a 2000 mm^2 (3 in^2) conductor, cable currents of over 3200 A may be achieved. This method is known as surface, integral, or sheath cooling[44,45,46] (Figure 3.23). In the integral scheme problems may occur due to the movement of the cable in the pipe caused by electro-mechanical forces. Because of this and of cost, the external water-pipe scheme is largely preferred and the surface-cooled arrangement only used if the ratings required cannot be obtained otherwise. A further problem arises from the high temperatures in the centres of joints because of the higher dielectric thermal resistance. In naturally-cooled schemes this is not normally a problem because of increased axial spacing in the joint bay. With the cable conductor at 85 °C the joint will be considerably higher unless preventive measures are taken.

(*c*) *Internal cooling*. That is, circulation of a coolant in a conductor duct provided in each cable core. The coolant may be:

(i) insulating oil which forms part of the insulation system in paper–oil systems[47,48,49]
(ii) water,[41] which has a much higher capacity than oil for absorbing heat. However, the water must be contained in a watertight tube inside the conductor duct (as shown in Figure 3.24. It has possibilities for cables with solid extruded insulation and has been used for generator connections at relatively low voltage.

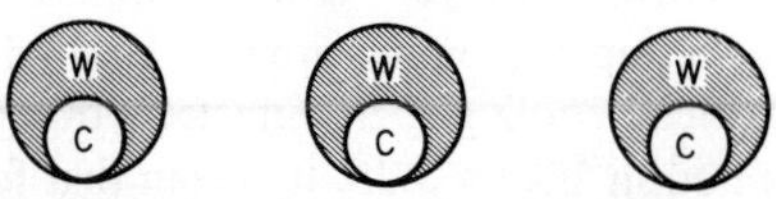

Figure 3.23 Surface or integral cooling – cables installed in pipes.

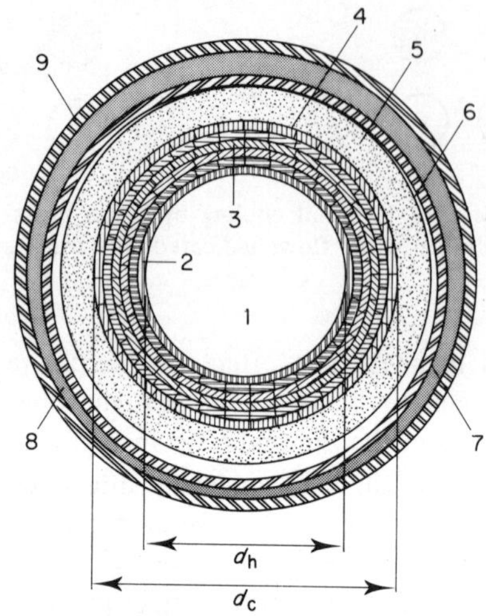

Figure 3.24 Cross-section of a 110 kV cable with internal water cooling 1. Water channel, channel diameter d_h; 2. watertight pipe; 3. stranded conductor, aluminium, diameter d_c; 4. semiconducting papers; 5. insulation; 6. shielding tapes; 7. corrugated aluminium sheath; 8. anti-corrosion protection; 9. PVC jacket.[41] (Reproduced by permission of CIGRE from Report 21-09, 1974)

The coolant voltage must be decreased to ground potential before entering pumps, etc. In a water-cooled scheme special terminations are used inside of which the coolant flows through a helical channel providing a reasonable electrical resistance between working voltage to ground. The electrical resistivity of water tends to decrease with length of service, and experience indicates that a resistivity of 200 kΩ-cm is reasonable and that regenerating plant is required to raise the water resistivity to this level when it has decreased to, say, 20 kΩ-cm. A high resistivity is essential to keep the I^2R loss developed by the water column to a reasonable level. The internal cooling system has the great advantage of removing heat directly from the major source. On the other hand the volumetric flow rates possible are limited by the conductor duct-size and hence the temperature rise of the coolant per given length is large.

It is possible to use fluorocarbon liquids as coolants in the conductor duct, e.g. Freon 12. The liquid absorbs heat, vaporizes, and absorbs heat at constant temperature and is passed to a cooler. This method is still in the research stage and the viability of a practical scheme has yet to be established. An advantage is that natural convective flow is set up and pumps are not required.

The necessity for artificial cooling if transmitted powers are to be increased at higher voltages is illustrated in Figure 3.25. It is seen that for each mode of cooling

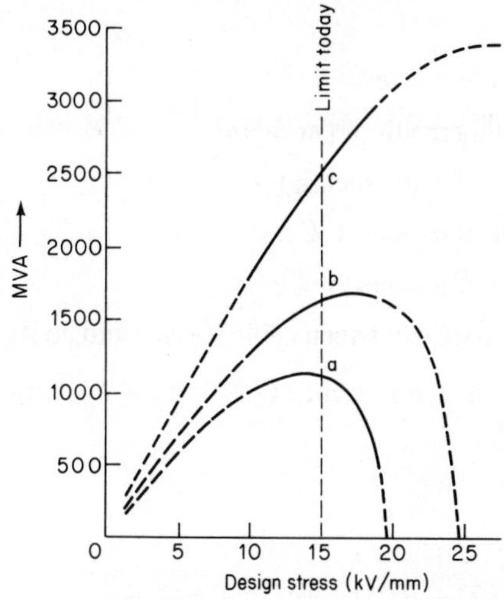

Figure 3.25 Relation between transmittable MVA and maximum working stress on cable dielectric (buried oil-filled 1935 mm² conductor cable). (a) Natural cooling (normally buried), (b) use of selective backfill (normally buried), (c) artificial cooling.

there exists a maximum power at a certain dielectric stress after which, owing to increased dielectric losses, the power falls. With normal burial in soil of thermal resistivity 1.20 deg C-m/W the limit is reached at a stress of 15 kV/mm. It is therefore evident that *increase in voltage without recourse to intensive artificial cooling is of little avail* when considering very high capacity circuits.

Thermal Analysis

Consider first the thermal conditions in a pipe or duct containing a line heat-source of q (W/m) and through which a coolant flows at a rate of W (m^3/s). Assume *all* the heat to be absorbed by the fluid (liquid or gas) and that the changes in fluid volume due to temperature are negligible. The heat absorbed in a length dx is $q\mathrm{d}x = WC\mathrm{d}\theta$, where C is the volumetric specific heat (J/m^3K). Over the total cooled length L,

$$qL = WC(\theta_L - \theta_0) \tag{3.21}$$

where θ_L and θ_0 are the exit and inlet temperatures respectively. Usually heat will be lost to the ambient via the soil as well as to the coolant, involving a complex scheme of thermal resistances. In artificially cooled schemes the reference temperature is that of the coolant and the temperature rise of this is a limiting parameter.

Consider a pipe, tunnel, or duct traversed by a coolant at a flow rate of W (m^3/s) entering at θ_0 °C. The cables emit a loss in the steady state of q watts/m and the thermal resistances from the cable surface to ambient is R_1 deg C/W per metre length. Let

q_s = heat to soil per metre

q_p = heat to cooling medium per metre

T_s = temperature of cable sheath (°C)

T_p = temperature of coolant (°C)

T_a = temperature of ambient (°C)

R_2 = thermal resistance between cable surface and coolant (deg C/W per m).

At any distance (x) from the entrance ($x = 0$, $T_p = T_{p0}$), the following equations apply (see Figure 3.26),

$$T_p = T_s - q_p R_2$$

and

$$q_p\, dx = q\, dx - \left(\frac{T_s - T_a}{R_1}\right) dx = CW\, dT_p.$$

Hence,

$$q_p = q - \left(\frac{T_s - T_a}{R_1}\right) = CW \frac{dT_p}{dx} \tag{3.22a}$$

$$\frac{dT_p}{dx} = \frac{dT_s}{dx} - R_2 \frac{dq_p}{dx}$$

$$q_p = q - q_s = q - \frac{T_s}{R_1} \quad \text{and let } \lambda_1 = WC(R_1 + R_2)$$

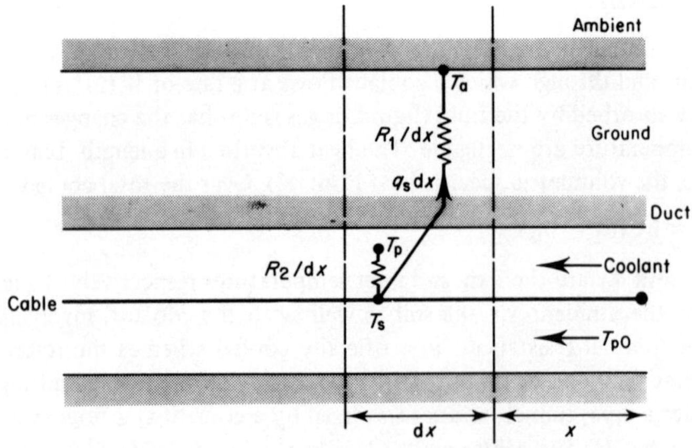

Figure 3.26 Coolant flow in duct, pipe or tunnel – calculation of temperature rise.

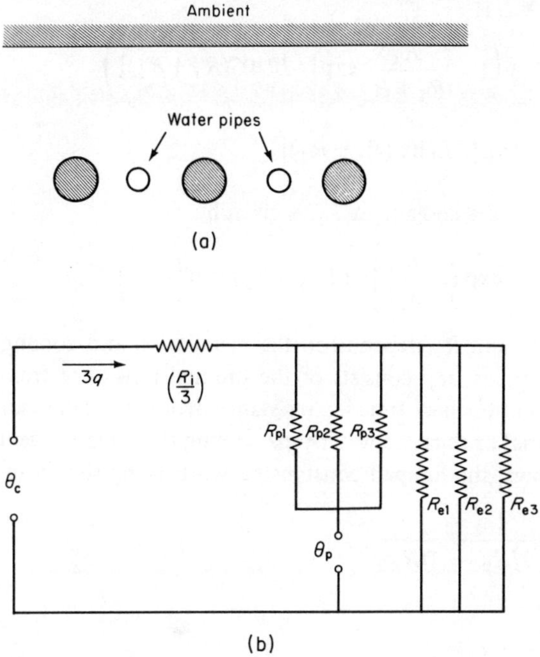

Figure 3.27 External cooling of self-contained cables by separate water pipes. (a) Physical arrangement, (b) Equivalent thermal network. R_{p_1}, R_{p_2}, R_{p_3} – thermal resistances between cable surface and pipes. R_{e_1}, R_{e_2}, R_{e_3} – thermal resistances between cable surfaces and ambient including mutual heating. R_i – thermal resistance between conductor and cable surface. Suffix 2 refers to centre cable. θ_c – conductor temperature; θ_p – pipe water temperature; $3q$ – total loss.

and so

$$\frac{\mathrm{d}q_p}{\mathrm{d}x} = -\frac{1}{R_1}\frac{\mathrm{d}T_s}{\mathrm{d}x}.$$

Therefore,

$$\frac{\mathrm{d}T_p}{\mathrm{d}x} = \frac{\mathrm{d}T_s}{\mathrm{d}x} + \frac{R_2}{R_1}\frac{\mathrm{d}T_s}{\mathrm{d}x}$$

$$= \left(1 + \frac{R_2}{R_1}\right)\frac{\mathrm{d}T_s}{\mathrm{d}x}. \qquad (3.22b)$$

From (3.22a) and (3.22b)

$$\frac{\mathrm{d}T_s}{\mathrm{d}x} + \frac{T_s}{WC(R_1 + R_2)} = \frac{qR_1 + T_a}{WC(R_1 + R_2)}.$$

The solution when $x = L$ is

$$T_{sL} = (qR_1 + T_a)\left(1 - \frac{R_1}{R_1 + R_2}\exp[-L/WC(R_1 + R_2)]\right)$$

$$+ \frac{T_{po}R_1}{R_1 + R}\exp[-L/WC(R_1 + R_2)]. \tag{3.23}$$

The temperature rise of the coolant, ΔT_p, is given by

$$\Delta T_p = qR_1\left[1 - \exp\left(-\frac{L}{\lambda_1}\right)\right] + (T_{po} - T_a)\exp\left(-\frac{L}{\lambda_1}\right). \tag{3.24}$$

The evaluation of R_1 and R_2 depends on the installation and cooling system. For cables contained in pipes, R_1 consists of the thermal resistance from pipe to soil surface plus the radiation heat-transfer resistance from the cable surfaces to the pipe wall. For external or separate water-pipe cooling the effective value of R_1 may be obtained as follows, the lumped constant network being shown in Figure 3.27:

$$R_p = \frac{1}{1/R_{p1} + 1/R_{p2} + 1/R_{p3}}$$

and

$$R_e = \frac{1}{1/R_{e1} + 1/R_{e2} + 1/R_{e3}}.$$

The temperature rises are expressed above ambient (or soil at cable depth), i.e. $\Delta\theta_c$ for conductor, $\Delta\theta_p$ for water, $\Delta\theta_0$ water inlet, and $\Delta\theta_{cl}$ conductor at distance L. From Figure 3.28,

$$\Delta\theta_{cl} = 3q\left[\frac{R_i}{3} + R_e\left\{1 - \frac{R_e}{R_e + R_p}\exp\left(-\frac{L}{\lambda}\right)\right\}\right] + \Delta\theta_{po}\frac{R_e}{R_e + R_p}\exp\left(-\frac{L}{\lambda}\right) \tag{3.25}$$

where

$$\lambda = WC(R_e + R_p).$$

From 3.25,

$$q = \left(\frac{\Delta\theta_{cl}}{R_i}\right)\frac{1 - (\Delta\theta_{po}/\Delta\theta_{cl})[R_e/(R_e + R_p)]\exp(-L/\lambda)}{1 + (R_e/\frac{1}{3}R_i)[1 - R_e/(R_e + R_p)]\exp(-L/\lambda)}. \tag{3.26}$$

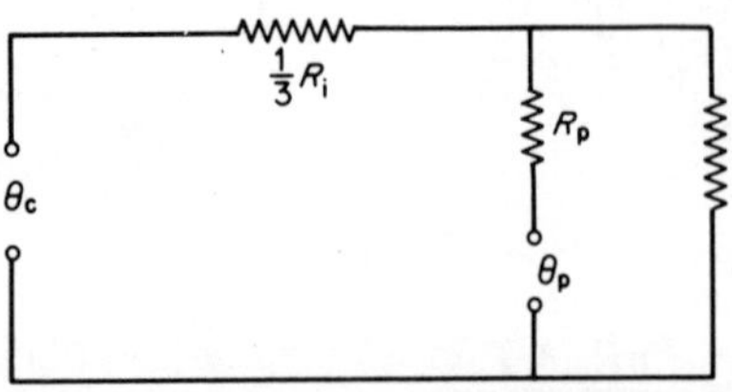

Figure 3.28 Simpler reduced form of network in Figure 3.27.

Table 3.3 Thermal resistance of soil between cable sheaths and water pipes (from CIGRE, 1960, Paper 21-5)

Arrangement	Outer cables to pipe °C/W per cm	Inner cable to pipes °C/W per cm
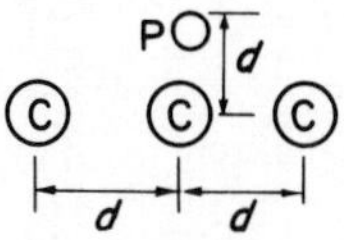	$g_s \log_{10} \dfrac{2.83d^4}{r_j r_{po}^3}$	$g_s \log_{10} \dfrac{2d^4}{r_j r_{po}^3}$
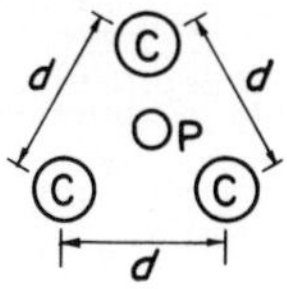	$g_s \log_{10} \dfrac{0.345d^4}{r_j r_{po}^3}$	$g_s \log_{10} \dfrac{0.5d^4}{r_j r_{po}^3}$
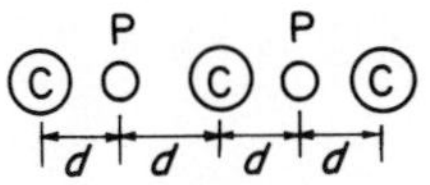	$g_s \log_{10} \dfrac{0.122d^2\sqrt{d}}{r_j r_{po}\sqrt{r_{po}}}$	$g_s \log_{10} \dfrac{0.0468d^2\sqrt{d}}{r_j r_{po}\sqrt{r_{po}}}$
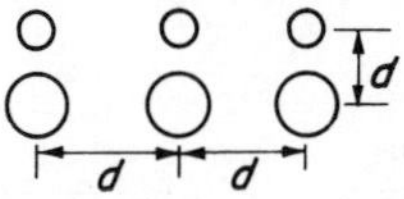	$g_s \log_{10} \dfrac{1.875d^2\sqrt{d}}{r_j r_{po}\sqrt{r_{po}}}$	$g_s \log_{10} \dfrac{3.17d^2\sqrt{d}}{r_j r_{po}\sqrt{r_{po}}}$

g_s = thermal resistivity of soil deg C-cm/W
r_j = outside radius of cable (cm)
r_{po} = outside radius of water pipe (cm)

Values for R_{e1}, etc. may be calculated by the normal soil-resistance formulae. The thermal resistance between the outer surfaces of the cables and the coolant (R_{p1}, etc.) are obtained by an image method. The soil resistances are summarized in Table 3.3. To these values 0.5 R_w must be added, where R_w is the thermal resistance of the pipe wall per cm plus the pipe-water heat transfer resistance, if this is not negligible. For a given water pressure, R_w is roughly constant for any pipe diameter; for 517 x 10^3 N/m^2 (75 psi), R_w = 0.25 deg C/W per m.

3.9 ARTIFICIAL COOLING OF PIPE-TYPE CABLES[50,51,52,53]

With pipe-type cables forced cooling may be affected by movement of the insulating oil past the three insulated cores. A general view of the system is shown in Figure 3.29 along with the equivalent thermal network. At any distance x from the pump the following heat-balance equations obtain;[50]

$$T_{c(x)} - T_{o(x)} = (T_c - T_s) + (T_s - T_o) = R_d(q_c + \tfrac{1}{2}q_d) + R_{fs}(q_c + q_d + q_s) \tag{3.27}$$

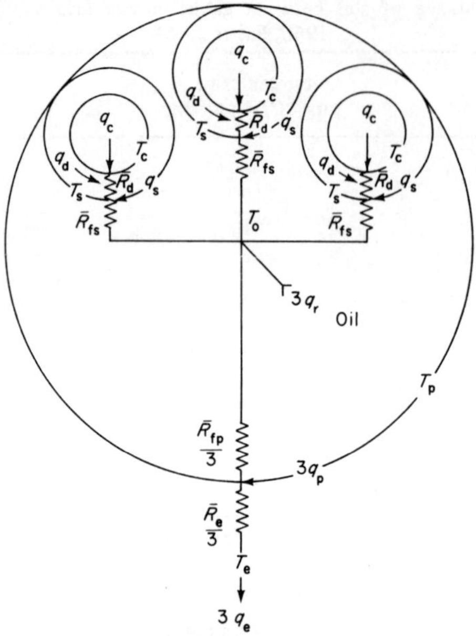

Figure 3.29 Equivalent thermal circuit of pipe-type cable HPOF system with forced cooling by passage of oil. q_c = conductor loss; R_i = insulation thermal resistance; T_c = conductor temperature; T_s = sheath temperature; T_o = oil temperature; T_e = earth temperature; T_p = temperature of pipe; R_{fs} = thermal resistance sheath to oil; R_{fp} = heat-transfer resistance oil to pipe walls; R_f = thermal resistance of pipe; R_e = effective thermal resistance of soil.[50] (Reproduced by permission of IEEE)

$$T_{o(x)} - T_e = (T_o - T_p) + (T_p - T_e) = R_{fp}q_e + R_e q_e \qquad (3.28)$$

$$CW\frac{dT_o}{dx} = 3[q_c + q_d + q_s] + q_p - q_e \qquad (3.29)$$

where T is temperature (°C), q is loss, and

c = conductor	s = sheat or screen
d = dielectric	p = pipe
e = soil	fs = sheath to oil
f = oil film	fp = oil to pipe.
o = oil	

From these equations,

$$T_{o(x)} = T_{o(in)} \exp\left(-\frac{x}{\lambda'}\right) + \frac{b}{a}\left[\exp\left(-\frac{x}{\lambda'}\right) - 1\right] \qquad (3.30)$$

where

$$b = q_c + q_d + q_s - \frac{1}{a}(R_e q_p + T_e)$$

$$a = -\left(\frac{1}{R_e + R_{fp}}\right)$$

$$\lambda' = WC(R_e + R_{fp})/3.$$

As the screen and conductor losses depend on (current)^2, the current rating may be found. With low oil velocities the flow may be laminar and the heat-transfer resistance (R_{fs}) between the core surface and oil may be difficult to determine because the heat transfer due to forced convection may be comparable to that due to free (without forced) flow. With the geometry of the pipe-type system this mixed heat transfer regime must be quantified by experiment.

Relationships[52] have been obtained both for the trefoil and cradle arrangements. The most consistent result is obtained by expressing the Nusselt Number (hD/k) in terms of ($N_{Re}N_{pr}^{0.4}$) where the Reynolds Number (N_{Re}) is based on the hydraulic diameter (D_H) and the Nusselt Number on the cable diameter. The relationship for a number of oils is shown in Figure 3.30. The measured Nusselt Number increases along the pipe from the entrance for a value of ($N_{gr}N_{Pr}$) above 50 000 and the results in Figure 3.30 are for an (x/D_H value of 473, where x is the distance from the pipe entrance. Heat transfer is seen to increase dramatically as the oil flow changes from laminar ($N_{Re} \simeq 1000$) to turbulent ($N_{Re} \simeq 10^4$), the transition occurring at $N_{Re} \simeq 1400$.

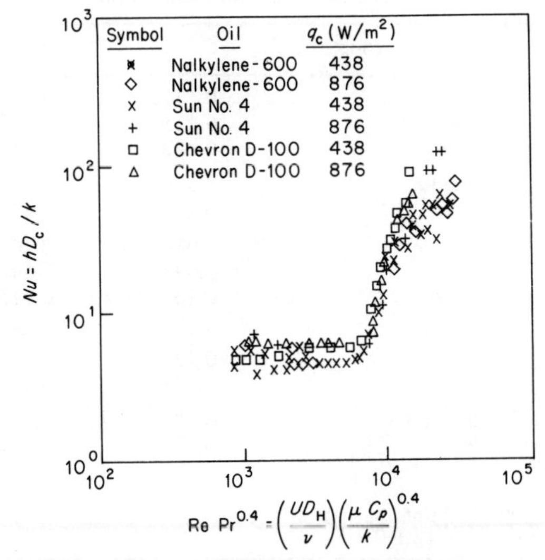

Figure 3.30 Forced oil cooling of pipe-type cables:[52] Nu–(RePr$^{0.4}$). (Reproduced by permission of IEEE)

3.10 CABLES IN TUNNELS AND SHAFTS[10,54,55,56,57,58,59]

Cables are sometimes installed in existing or purpose-built tunnels. Several river crossings have been accomplished by the construction of a tunnel of, say, 3–4 m diameter under the river bed. On a smaller scale tunnels have been installed under major roads where overhead circuits are not advisable. Horizontal tunnels need vertical or inclined shafts at the ends and long vertical shafts are used in certain hydro schemes where it is necessary to convey the power by cable up the mountain to the overhead-line circuits.

In these installations the limitation to the current transmittable is often the permissible maximum air temperature rather than the insulation temperature. The former is often restricted to 35 °C, a reasonable maximum temperature for maintenance personnel. Current ratings may be increased by forced flow of air through the tunnel but this is sometimes avoided because of the resulting heightened fire risk.

The heat transfer from cables in various formations has been investigated[10] both for free and forced convection of air. A summary of the suggested empirical relationships is given in Table 3.4. The computation of the maximum conductor and air temperatures follows the methods already discussed but is complicated by the considerable effect of thermal radiation.

Assuming an effective emissivity of 0.9 and natural convection, the heat dissipated from a cable/pipe in a tunnel or shaft is given by[38] the following empirical formula:

$$q = 3.23 \times 10^{-11}\, m \frac{D}{2} ({}_{a}T_{s}^{4} - {}_{a}T_{w}^{4}) + 1.21\, C \frac{D}{2} \Delta\theta^{5/4} \text{ (W/m)} \qquad (3.31)$$

Table 3.4 Parameters for convection formulae

Grouping	Cable of group	C	F	P
○		0.130	0.725	0.24
○ ○ ○	middle	0.115	0.785	0.21
$S/D > 2$	outer	0.123	0.815	0.21
○○○ Touching	middle	0.086	0.55	0.20
	outer	0.105	0.69	0.20
8 (vertical, three) Touching	middle	0.086	0.40	0.215
	upper	0.090	0.54	0.215
	lower	0.105	0.715	0.215
○ ↕ S	middle	0.115	0.60	0.225
○	upper	0.123	0.70	0.225
○ $S/D > 2$	lower	0.123	0.78	0.225
ꝏ (trefoil) Touching	upper	0.070	0.98	0.20
	lower	0.078	0.68	0.20

Cable diameter = D
Forced convection: $N_{Nu} = C(N_{Re})^{0.65}$
Natural convection: $N_{Nu} = F(N_{Gr,D}N_{Pr})^{P}$

From Weedy, B. M. and Zayyat, H. M., IEEE Conference Paper C72.[10] (Reproduced by permission of IEEE)

where

D = outside diameter of cable or pipe (m)

m = proportion of surface available for radiation (affected by obstructions and proximity of other cables)

${}_aT_s$ and ${}_aT_w$ = absolute temperatures of cable (pipe) surfaces and tunnel walls respectively (K)

$\Delta\theta$ = temperature difference between cable surface and air (K)

C = 1.3 for horizontal cables in tunnels

C = 1.0 for vertical cables in shafts.

The above temperatures depend on the inlet temperature and the heat evolved and hence the rating process is iterative in nature. The large thermal inertia associated with tunnels makes the response of the tunnel walls to ambient surface changes very slow. In Reference 59 a time constant of 6 months is quoted for tunnels of modest depth, implying that the mass of earth around the tunnel will be at summer temperatures whilst the cooling air is at winter levels and vice versa. With very deep tunnels the thermal situation never reaches steady state, with a slow increase in wall temperature occurring over the years.

In practice it is likely that natural ventilation will provide a draught of air through a tunnel. However the prediction of the velocity of this flow is difficult and the tendency at present is to rate the cables on natural convection or on the calculated velocity when fans are provided. More detailed calculations of temperatures in cables in tunnels may be carried out by means of the equivalent thermal networks, typical of which is shown in Figure 3.31 for steady-state conditions. The variation of temperature with time in a deep tunnel with seasonal changes in air inlet temperature is often regarded as sinusoidal in form

Although cooling by movement of the tunnel air would seem to be the obvious method, sometimes other forms are required. With self-contained cables a surface-cooled integral pipe scheme has been used,[46] as also have water troughs in which

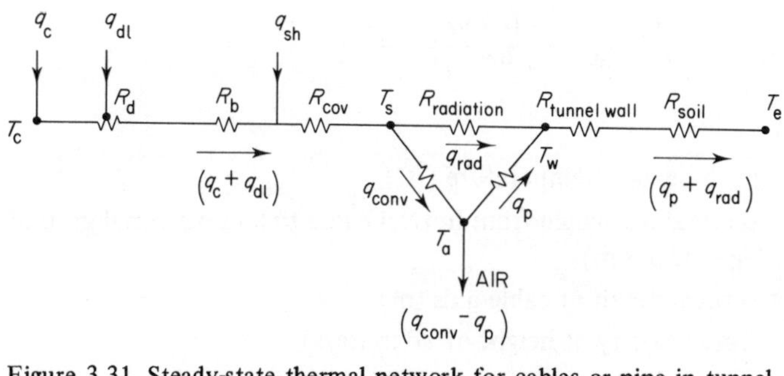

Figure 3.31 Steady-state thermal network for cables or pipe in tunnel with forced air cooling. Losses: q_p, heat transferred air to tunnel wall; q_c, conductor; q_{dl}, dielectric; q_{sh}, sheath; q_{rad}, heat transfer by radiation; q_{conv}, heat transfer by convection. T_c, T_s, T_a, T_e are temperatures of conductor, sheath, air, and ambient.

the water is in an open trough and is allowed to flow past the cables installed in the trough. The water flows by means of the tunnel gradient and is either discharged into the river or passed through a heat exchanger.

Calculation of Air Flows

One of the most difficult problems in tunnel schemes is the calculation of the air flows. This involves a knowledge of the pneumatic conditions such as wall roughness, bends, etc. Pressure drops occur with the friction of the walls, at entry to the system, at expansions and contractions, and at discharge. The pressure drop due to the tunnel walls is proportional to (Length) $(\text{Velocity})^2$ $(\text{Diameter})^{-1}$ and an empirical friction factor. Because of the additional losses due to the cable supports and joints the total calculated pressure drop is often doubled. The capacity of the required fan is proportional to $(\text{air velocity})^3$ and hence pneumatic losses are a severe design constraint.

Surface Troughs

A typical installation is shown in Figure 3.32. This type of installation has been used in Britain using LPOF cables for routes along railways, canals, in substations, and on certain cross-country routes. The cables are placed with small separation between phases and at a depth of 0.3 m in a concrete trough with a reinforced concrete lid for mechanical protection. The important thermal resistances external to the cable are those associated with the trough filling and between the trough surface and the air. The value of the former can be reduced by use of a special backfill material; the contact between the trough fill and the lid is also important since up to one half of the total heat dissipated may flow through the top of the trough. For depths of less than 0.6 m cable temperatures may show significant daily or even hourly variations due to changes in solar radiation, wind velocity, and air temperature. The maximum cable surface temperature in a surface trough system derived from an analogue investigation[38] can be written as:

$$T_s = q_c R_s + \frac{0.33 q_c}{U^{0.74} h^{0.2}} + \frac{0.29 q_{sr}}{U^{0.89} h^{0.07}} + T_a \qquad (3.32)$$

where

T_s = cable surface temperature (°C)
R_s = thermal resistance from hottest cable to an isothermal ground surface (deg C/W per m)
h = vertical depth of cable axis (m)
U = wind velocity at height of 10 m (m/s)
q_{sr} = solar radiation to ground surface (W/m^2)
q_c = cable heat dissipation (W/m)
T_a = ambient temperature (°C).

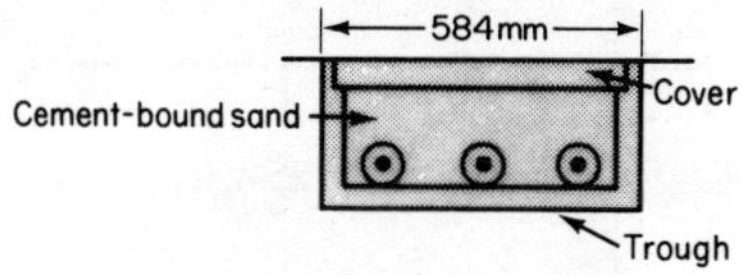

Figure 3.32 Surface trough for 132 kV, 1000 mm², self-contained cables.[8] (Reproduced by permission of IEE)

Example 15.4

Calculate the conductor temperature rise resulting from a stoppage in coolant flow for a system of self-contained cables each contained in a separate cooling pipe through which water is pumped.

The cable specification is as follows: 400 kV, 2000 mm², LPOF,
Radius over conductor = 2.88 cm
Radius over insulation = 5.34 cm
Mean diameter of sheath = 11.5 cm, thickness 0.82 cm
Thermal capacity of water in pipe (per metre length) = 84 kJ/deg C
Heat dissipation (3200 A with surface water cooling) = 140 W/m.

Solution

Representation of cable thermal capacities:

$$\text{conductor } C_c = \frac{2000}{100} \times 1 \times 8.93 \times 0.39 = 67.4 \text{ J/deg C per cm}$$

$$\text{dielectric } C_i = \pi(5.34^2 - 2.88^2) \times 1 \times 1.6 \times 1.05$$
$$= 106.7 \text{ J/deg C per cm}$$

$$\text{sheath } C_{sh} = \pi 11.5 \times 0.82 \times 0.13 \times 11.37$$
$$= 44.0 \text{ J/deg C per cm.}$$

Thermal resistance of dielectric

$$R = \frac{5.0}{2\pi} \ln\left(\frac{R}{r}\right) = 0.57 \text{ deg C-m/W.}$$

Using the equivalent network shown in Figure 3.33,

$$p = 0.4$$
$$C_0 = C_c + pC_i = 11 \text{ kJ/deg C per m}$$
$$C_1 = (0.6 \times 10.6) + 4.4 + 84 = 95 \text{ kJ/deg C per m.}$$

When the coolant flows, T_1 = temperature of cable outer surface above the coolant. Before coolant stoppage, $T_1 = 0$, T_0 (conductor temperature) = qR, $q_c = q_0$ = total loss (assumed at conductor). With water stoppage, superpose the effect of a step

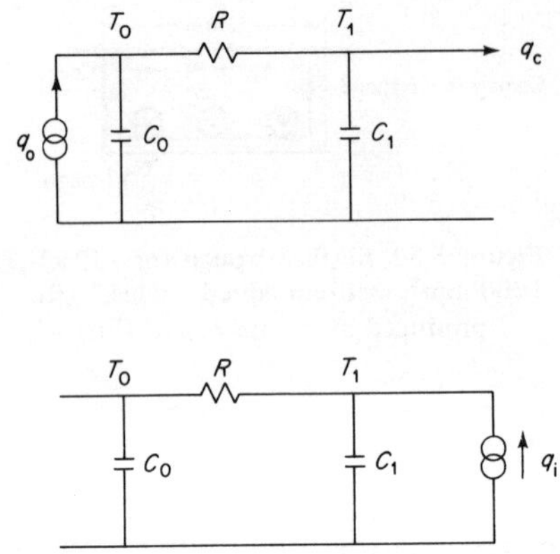

Figure 3.33 Thermal network for Example 15.4: (a) with water flow in pipe; (b) without water flow.

function of heat input from the coolant, $q_i = q_0$. Hence,

$$\frac{q_i}{s} = sC_1 T_1 + \frac{(T_1 - T_0)}{R}$$

and

$$0 = \frac{T_1 - T_0}{R} - sC_0 T_0.$$

Using the Laplace Transformation,

$$T_0 = \frac{q_i}{\beta(C_1 + C_0)} \left\{ e^{-\beta t} + \beta t - 1 \right\}$$

and

$$T_1 = q_i \left\{ \frac{C_0}{\beta C_1 (C_1 + C_0)} (1 - e^{-\beta t}) + \frac{k}{C_1 + C_0} \right\}$$

where

$$\beta = \frac{C_1 + C_0}{RC_1 C_0} = \frac{(95 + 11) \times 10^3}{0.57 \times 11 \times 95 \times 10^6} = 0.18 \times 10^{-3}.$$

The temperature of the conductor 1 hour after water stoppage is

$$\frac{140}{0.18 \times 106 \times 10^{-3} \times 10^3} \{\exp(-0.00018 \times 3600) + (0.00018 \times 3600) - 1\} = 1.2\ ^\circ\text{C}$$

T_0 after 10 hours is 40 °C.

This general approach may be applied to a pipe-type HPOF cable.

Example 15.5

A horizontal tunnel contains 6 trefoil groups of 138 kV, 290 mm^2 cables. Each cable has external diameter 6.25 cm. Calculate the heat transfer coefficient and equivalent thermal resistance for the whole of the cable: (*a*) assuming natural convection; (*b*) with forced air flow of 3 m/s.

If the heat developed in each cable is 0.5 W/cm calculate the temperature rise of the hottest cable-surface above the air temperature.

Solution

(*a*) Heat transfer from the surface of the hottest cable of the group is given by,[10]

$$N_{Nu,D} = \frac{hD}{k} = 0.48(N_{Gr}N_{Pr})^{0.2}$$

$$= 0.48(9 \times 10^5 \times 0.71)^{0.2} = 0.48 \times 14.5$$

from which,

$$h = \frac{0.48 \times 14.5 \times 2.4 \times 10^{-4}}{6.25} = 2.78 \times 10^{-4} \text{ W/cm}^2 \text{ deg C.}$$

Equivalent thermal resistance for 18 cables is

$$\frac{1}{h\text{ (surface area)} \times 18} = \frac{10^4}{2.78 \times \pi \times 6.25 \times 18} = 10.2 \text{ deg C/W cm.}$$

From Figure 3.34,[10] the ratio

$$\frac{\text{convective heat transfer}}{\text{total heat transfer}} = 0.4.$$

Heat to the air = 0.4 x 0.5 = 0.2 W/cm. Temperature rise of hottest cable above the air = 18 x 0.2 x 10.2 = 36 deg C.

(b) Forced air convection

For hottest cable,[10]

$$N_{Nu} = \frac{hD}{k} = 0.07(N_{Re})^{0.65}$$

$$= 0.07(12400)^{0.65}$$

$$= 32$$

and

$$h = 12.8 \times 10^{-4} \text{ W/cm}^2 \text{ deg C.}$$

Overall heat transfer resistance for 18 cables = 2.2 K/W-cm. Heat absorbed by

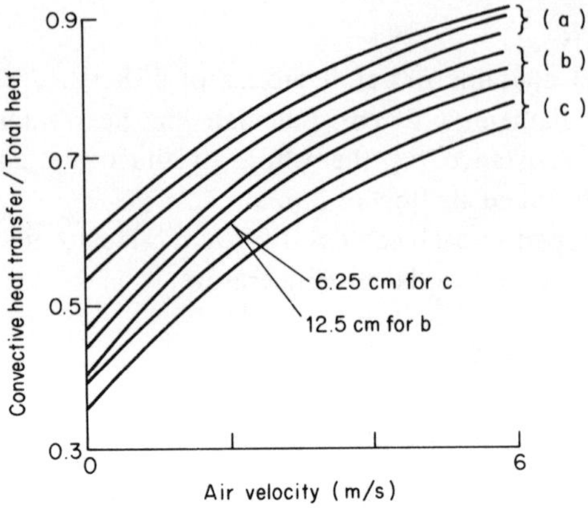

Figure 3.34 Relationships between (Convective heat transfer/total heat transfer) and air velocity in tunnel schemes: (a) single cable and spaced horizontal and vertical cables; (b) touching horizontal and vertical; (c) trefoil touching. Upper curve 6.25 cm D, middle 9 cm D, lower 12.5 cm D, for each group.[10] (Reproduced by permission of IEEE.)

air = 0.68 x 0.5 x 18 W per cm (factor 0.68 from Figure 3.34[10]). Temperature rise of hottest cable-surface above air is

$$0.68 \times 0.5 \times 18 \times 2.2 \text{ deg C} = 13.5 \text{ deg C}.$$

Heat transferred by radiation

$$(1 - 0.68) \times 0.5 \times 18 = 2.9 \text{ W per cm}.$$

3.11 ACCESSORIES

The current rating of an underground circuit will be determined by the lowest of the limiting currents either of the cable itself or of the various accessories such as joints and terminations. With natural cooling little problem has been experienced but with intense artificial cooling the higher currents possible in the cable may cause unacceptable temperatures in the accessories. The latter have different thermal characteristics to the cable and the analysis of temperatures is complicated by the presence of longitudinal as well as radial heat flows. With joints the basic problem is caused by the extra layers of insulation required compared with those needed for the cable, which represents higher radial thermal resistance and hence higher temperatures. Finite difference or finite-element computer methods may be used to determine the joint temperatures.

In a HPOF 138 kV system with forced oil flow the cable to joint centre differential is 15 °C. The differential could be minimized by creating turbulent flow at the joint surface, the flow becoming laminar towards the cable. Normally for

HPOF joints the radial stress in joint is restricted to, say, two-thirds of that in the cable, giving approximately 1.5 times the cable thermal resistance. Special through-joints have been developed using low-loss polypropylene–paper laminate for the joint tapes, which gives smaller joint diameters compared with flat or crepe paper. The reduction in dielectric loss is also helpful. Other developments (LPOF) use 'Voltalit' as a part of the joint insulation. This epoxy has a thermal resistivity of approximately 0.7 deg C-m/W which reduces the radial thermal resistance.

Terminations (Pot-heads or Sealing Ends)

Relatively little has been published regarding the rating of terminations. In Reference 19 a lumped constant network has been used to establish ratings for 345 kV capacitor pot-heads with aluminium and copper conductors of various cross-sectional areas. Being above ground there are heat injections into the network due to solar radiation and there is heat transfer by forced convection (wind), natural convection, and radiation.

Joints in LPOF Systems[18,38,61]

With naturally-cooled systems the greater spacing in the joint bays leads to smaller mutual heating than for the remote cables, and the joint–cable temperature differential is not a major problem. With intense water cooling, however, particularly the integral or surface type, the problem becomes acute. However, it may be alleviated since stop joints incorporate substantial proportions of epoxy-resin castings and also the presence of an embedded metal stress-control electrode at the centre of the joint reduces the thermal resistance from the ferrule (see Figure 3.14).[38]

Alleviation of Hot-spots by Oil Oscillation

The temperature differentials reported above and hot spots in cables caused by other external factors, such as proximity of steam pipes and regions of high soil thermal resistivity, may be reduced in self-contained systems without recourse to the pumping of oil by the slow oscillation of the insulating oil[62,63] in the conductor duct. This oscillation is achieved by the use of time-pressure relays at the oil reservoirs which produce a slow (laminar) flow in one direction and then after a prescribed time (e.g. several minutes) the flow is reversed.

The result is a smoothing of temperature hot spots by the conveyance of heat from them to the adjacent cooler regions. A detailed study[63] for a 400 kV, LPOF system with a conductor area of 2000 mm^2 (intensely cooled by surface or integral cooling and hence with a large joint–conductor temperature differential) has shown that the period of oil oscillation, flow rate, and reduction in joint temperature are related by the following equation (joints 350 m apart):

$$\text{Reduction in joint temperature} = \frac{35.5W\lambda}{(W + 1.35 \times 10^{-6})(\lambda + 2.0)} \text{ deg C}$$

where W = oil flow rate (m^3/s) and λ = time for one cycle of oil oscillation (minutes).

For the 400 kV, LPOF system the optimum value for λ is 20 minutes and for W, 20×10^{-6} m^3/s. The majority of this temperature reduction is achieved in 2 to 4 hours after the start of the oil oscillation. The 400 kV scheme reported in Reference 46 uses oil oscillation to reduce joint temperatures. To achieve the required reduction special low-viscosity oil is used, 25 mm conductor duct and laminar flow conditions in the cable with turbulence created artifically in the joint. A flow rate of 0.035 litre/s is used. In the pipe-type system flow reversal of the oil in the pipe may be arranged by suitable control of the oil pumps. Results of an application in the USA to a 138 kV system have been reported.[64]

3.12 USE OF COMPUTERS

The use of computers and methods involved have already been mentioned. With high capacity circuits the large sums of money involved for the cables and installation are such as to make worthwhile as detailed an analysis as possible.

Two main approaches are used:

(*a*) Finite difference methods.[35]
(*b*) Finite element methods.[34]

In both, the region under examination is subdivided into small elemental volumes each represented by a node, thus forming a mesh of nodes. In (*a*) the mesh is square or rectanglular in form; in (*b*) the mesh often consists of interconnected triangles. The relevant equations for the nodes are formulated as the mesh is scanned by the computer. Direct or iterative methods for the solution of the large number of simultaneous equations may be used. It is not proposed to go into any detail here but to merely state the methods available.

For the steady-state analysis the finite-difference methods likely to be useful are:

(i) Successive over-relaxation (SOR), a very well established and much used method.
(ii) SOR with block relaxation.
(iii) Gaussian elimination.
(iv) Alternating direction implicit (ADI).

(i), (ii), and (iv) are iterative methods; (iii) involves the inversion of a matrix and is termed a direct method. The ADI method basically yields a transient solution and for steady-state solutions as long a time step as possible is used, commensurate with satisfactory small truncation errors.

For transient temperature solutions popular finite difference methods are:

(i) Step-by-step explicit – simple to apply but can have severe limitations because of numerical instability.

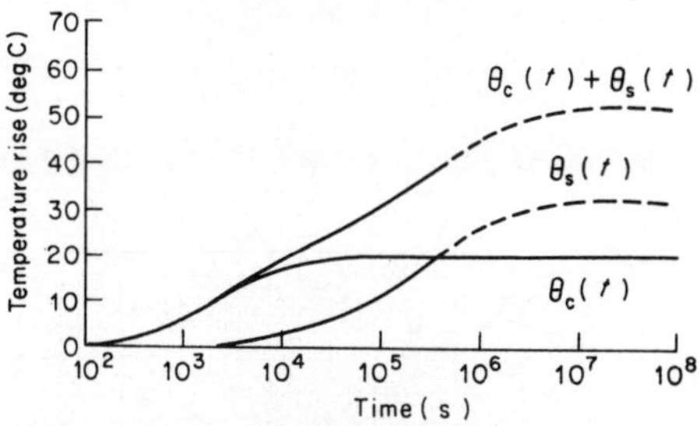

Figure 3.35 Transient temperatures in a direct buried self-contained cable system. 400 kV, 630 mm² copper conductor, 775 A, 250 mm cable spacing. $\theta_c(t)$, temperature rise across cable; $\theta_s(t)$, temperature rise across soil.

(ii) Implicit.
(iii) Crank–Nicolson.[32]
(iv) ADI

If both steady-state and transient solutions are required the ADI method has much to recommend it, although its performance depends on the nature of the problem and it is difficult to generalize. The methods are available is most modern texts on numerical methods.

A typical purely radial nodal division, for a normally buried LPOF cable, allows the oil, sheath, and servings to be represented by single nodes and the dielectric to be divided into five nodes with equal thermal time-constants. The soil region is divided into 28 nodes (annular rings). The transient curves produced by this method agree with the exponential-integral formula which is correct for long times. Figure 3.35 shows the transient temperature response of a 400 kV, 630 mm^2 cable group, determined as described above, to a step-change in load of 775 A.[45]

Water-cooled Cables

In order to obtain the load-change transient temperature response of a cable group externally cooled with separate cooling water pipes, a program capable of modelling this more complicated geometry is required. A heat conduction program[65] based on the finite element method has been found suitable for doing this. A section of the cables and water pipes buried in the soil is represented by a two-dimensional mesh of finite elements as shown in Figure 3.36. The finite elements used are triangles and quadrilaterals. The nodes are positioned at the corner points of the elements and on the perpendicular bisectors of each pair of corner nodes. As the preparation of the mesh can be very time consuming when undertaken by hand, a special mesh generation program is available to accomplish this automatically. The input information requires the Cartesian coordinates of the centres of all cables and water pipes, which may be arbitrarily positioned within the soil medium, together with the radii of the internal boundaries.

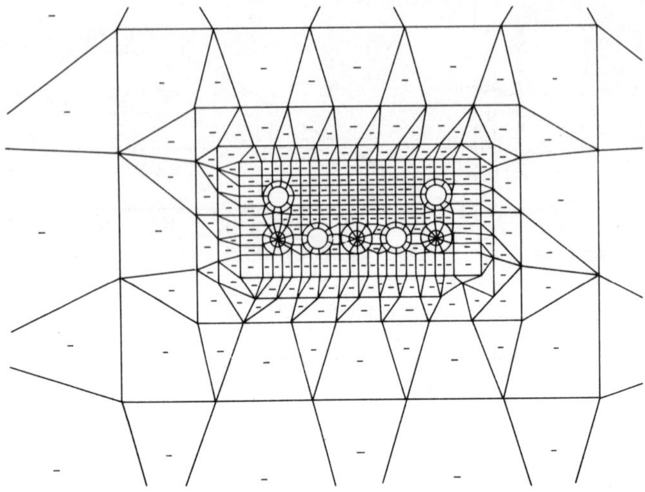

Figure 3.36 Part of finite element mesh representing thermal conditions in an externally cooled cable group.

Boundary conditions which may be time varying are applied to the ground surface and the inside walls of the water pipes, representing heat-transfer to heat sinks at specified temperatures. Dielectric, conductor, and sheath losses are applied as internal heat injections and these may be time dependent. Non-homogeneous soil may be represented if required.

Application to the Operation of Cable Systems

The computer methods described provide an acceptably accurate solution of the temperature and heat-flow distribution for most cable systems provided the boundary conditions can be defined. For design purposes, most controlling parameters can usually be defined. During operation it is possible that certain parameters will have changed and it is necessary to re-assess the cable rating. If experience shows that the loading is cyclic a new rating could be ascribed. There is a need therefore for rating evaluation methods using computer programs that have simple input format to enable operations staff to assess cable ratings. Further simplification of existing programs for transient performance is desirable as there is always a need to obtain information quickly for a particular situation. Since a transient rating is dependent on the previous loading history, the determination of ratings by computer requires updating, either on a regular basis or each time there is significant change in the load pattern. Future developments are likely to produce simplification of the existing methods, especially with transient-rating studies where the data preparation, checking, and computing costs are at present excessive for normal design purposes. The analysis of water-cooled systems with many heat sinks requires careful development.

REFERENCES

1. I.E.C. Report 287, 'The Calculation of the Continuous current rating of cables (100% load factor)' (1969).
2. Arman, A. N., Cherry, D. M., Gosland, L., and Hollingsworth, P. M., 'Influence of soil-moisture migration on power rating of cables in h.v. transmission systems,' *Proc. I.E.E.*, **111**, (5), 1000–1016 (1964).
3. Milne, A. G. and Mochlinski, K., 'Characteristics of soil affecting cable ratings', Proc. I.E.E., **111**, (5), 1017–1039.
4. Fink, F. H., 'Control of thermal environment of buried cable systems', *Trans. Amer. Inst. Elect. Engrs.*, **73**, Pt. III, 406–410 (1954).
5. King, S. Y. and Halfter, N. A., *Underground Cables – Some Aspects of Their Thermal Environment*, Hong Kong University Press, 1977.
6. Orchard, R. S. *et al.*, CIGRE Paper No. 214, 'Soil Thermal Resistivity: A Practical Approach to its Assessment and its Influence on the Current Rating of Buried Cables' (Paris 1960).
7. A.I.E.E. Committee Report, *Trans. A.I.E.E.*, **79**, III, 792–856 (1960).
8. Cox. H. N. *et al.*, 'Developments in U.K. cable-installation techniques to take account of environmental thermal resistivities', *Proc. I.E.E.*, **122**, 1253–1259 (1975).
9. Schmill, J. V., 'Variable Soil Thermal Resistivity–Steady State Analysis', *I.E.E.E. Trans. PAS-86*, pp. 215–223 (1967).
10. Weedy, B. M. and Zayyat, H. M., 'Heat Transfer from Cables in Tunnels and Shafts', *I.E.E.E., Conference Paper C72506–4 (1972).*
11. Carter, G. W., *The Electromagnetic Field in its Engineering Aspects*, Longmans, London, 1958.
12. Brockbank, J. *et al.*, 'Sheath and Armour Losses in Single Core Cables for Single and 3-Phase Transmission', *Journal I.E.E.*, **67**, 354–360 (1928).
13. McAdams, W. H., *Heat Transmission*, 3rd Edition, McGraw Hill, New York, 1954.
14. Neher, J. H. and McGrath, M. H. 'The Calculation of the Temperature Rise and Load Capability of Cable Systems', *Trans. A.I.E.E.*, **76**, III, 752–764 (1957).
15. Neher, J. H., 'The Temperature Rise of Buried Cables and Pipes', *Trans. A.I.E.E.*, **68**, 9–21 (1949).
16. Bullar, F. H. and Neher, J. H. 'The Thermal Resistance Between Cables and a Surrounding Pipe or Duct Wall', *Trans A.I.E.E.*, **69**, I, 342–349 (1950).
17. Shanklin, G. B., Buller, F. H., Neher, J. H., and Wiseman, R. S. A committee report and papers from Symposium on Temperature Rise of Cables: 'The Effect of Loss Factor on the Temperature Rise of Pipe Cable and Buried Cables', *Trans. A.I.E.E.*, **72**, III, 530–562 (1953).
18. Weedy, B. M. and Perkins, J. B., 'Steady State Thermal Analysis of a 400 kV cable through joint', *Proc. I.E.E.*, **114**, 109–115 (1967).
19. Nradkarni, J. and Nicholas, J. H., 'Amplicity of High Voltage Potheads', *Proc. I.E.E.E. Conference Underground Transmission and Distribution, 1974*, pp. 507–510 (1974).
20. Weedy, B. M., 'Thermal aspects of changes in the environment of underground cables', Proc. I.E.E. **114**, 519–524 (1967).
21. Baldwin, C. I. *et al.*, 'Computational Technique for Evaluating the Thermal Performance of Underground Transmission Systems.', *CIGRE*, Report 21–04 (1972).
22. Buller, F. H., 'Thermal Transients on Buried Cables', *Trans A.I.E.E.*, **70**, 45–55 (1951).
23. Neher, J. H., 'The Transient Temperature Rise of Buried Cable Systems', *I.E.E.E., Trans. PAS-83*, pp. 102–112 (1964).

24. Weedy, B. M. 'Thermal Transients in a High-Voltage Cable System with Natural and Artificial Cooling', *Proc. I.E.E.*, **109A** 461–470 (1962).
25. Weedy, B. M., 'Effects of environment on transient thermal performance of underground cables', *Proc. I.E.E.*, **119**, 225–230 (1972).
26. Goldenberg, H., 'Emergency loading of buried cable with temperature-dependent conductor resistance', *Proc. I.E.E.*, **118**, 1807–1810 (1971).
27. Mildner, R. C., 'The short circuit rating of thin metal tape cable shields', *I.E.E.E. Trans. PAS-87*, pp. 749–759 (1968).
28. Wormer van, F. C., 'An improved approximate technique for calculating temperature transients', *Trans. A.I.E.E.*, **74**, 277–280 (1955).
29. Goldenberg, H., 'Transient Heating of Buried Power Cables', *Proc. I.E.E.*, **114**, 809–814 (1967).
30. Morello, A., 'Transient Temperature Variations in Power Cables', *Elettrotecnica*, **45**, 213–218 (1958).
31. Hitchcock, I. A. and Thelwell, M. J., 'The Cooling of Underground EMV Transmission Cables', *I.E.E.E. Trans. PAS-87*, pp. 128–134 (1968).
32. Crank, J. and Nicholson, P., 1947, 'A Practical Method for Numerical Evaluation of Solutions of Partial Different Equations of Heat Conduction Type', *Proc. Camb. Phil. Soc.*, **43**, No. 50, 50–67 (1947).
33. Dufort, E. C. and Frankel, S. P., 1953, 'Stability Conditions in the Numerical Treatment of Parabolic Differential Equations', *Maths, Tables Aids Comp.*, **7**, 135 (1953).
34. Zienkiewicz, O. C., *The Finite Element Method*, McGraw Hill, New York, 1967.
35. Dusinberre, G. M., *Heat Transfer Calculations by Finite Differences*, International Text Book Co., 1961.
36. Ralston, P. and West, G. H., 'The Artificial Cooling by Water of Underground Cables', *CIGRE*, Paper 215 (1960).
37. Flamand, C. A. and Terramorisi, G., 'Etude du refroidissement des Cables d'Energie par Circulation Forcée', *Revue Générale de l'Electricité*, pp. 415–429 (April 1968).
38. Endacott, J. D. *et al.*, 'Thermal Design Parameters used for High Capacity EHV Cable Circuits.' *CIGRE*, Report 21–03 (1970).
39. Buller, F. H., 'Artificial Cooling of Power Cable', *A.I.E.E.E. Trans.* **71**, 634–640 (1952).
40. Heinemann, H. J., 'Increasing the transmitting capacity of power cables by artificial cooling', *V.D.E.Fachber*, **22**, 60–65 (1962).
41. Burnbreier, H. *et al.*, 'High-Power Cables with Internal Water Cooling', *CIGRE* , Report 21–09 (1974).
42. Flamand, C. A., 'Forced Cooling of High-Voltage Feeders', *I.E.E.E. Trans. PAS-85*, pp. 980–986 (1966).
43. Oudin, J. M. *et al.*, 'French Research and Development in the field of Extra High Voltage cables', *CIGRE*, 230 (1964).
44. Arnaud, U. C. *et al.*, 'Development and Trials of the 400 kV integral pipe-cooled system', *Proc. I.E.E.E.*, **124**, 286–294 (1977).
45. Luoni, C. *et al.*, 'Study of Mechanical Behaviour of Cables in Integral-pipe water-cooled Systems', *Proc. I.E.E.E.*, **124**, 295–302 (1977).
46. Arkel, C. A. *et al.*, 'Design and Construction of the 400 kV System for the Severn Tunnel', *Proc. I.E.E.*, **124**, 303–316 (1977).
47. Arkel C. A. *et al.*, 'Development of Internally Oil-Cooled Systems', *Proc. I.E.E.*, **124**, 317–325 (1977).
48. Botherton, W. *et al.*, 'Field Trials of 400 kV internally oil-cooled cables', *Proc. I.E.E.*, **124**, 326–333 (1977).
49. Rogers, E. C., 'Internally Force-Cooled Power Cables: optimum working temperature in the range 70–358 k', *Proc. I.E.E.*, **122**, 908–914 (1975).

50. Notaro, I. and Webster, D. J., 'Thermal Analysis of Forced Cooled Cables', *I.E.E.E. Trans. PAS-90* (1971).
51. Burrell, R. W., 'Application of Oil Cooling in High Pressure Oil Filled Pipe-Cable Circuits', *I.E.E.E. Trans. PAS-84*, pp. 795–802 (1965).
52. Abdulhadi, R. S. and Chato, J. C., 'Combined Natural and Forced Convective Cooling of Underground Electric Cables', *I.E.E.E. Paper F76*, 312–9 (1976).
53. Moulding, J. D., 'Application of Heat Exchangers to Existing High Pressure Oil Filled, Pipe-Type Cable Lines', *I.E.E.E. Paper No. 31*, pp. 67–75 (1967).
54. Giaro, J. A., 'Temperature Rise of Power Cables in a Gallery with Forced Ventilation', *CIGRE*, Paper N1 213 (1960).
55. Kitagawa, K., 'Forced Cooling of Power Cables in Japan', *CIGRE* Paper No. 213 (1964).
56 Burrell, R. W., Falcone, A. J., and Roberts, W. J., 'Forced Air Cooling for Station Cables', *Trans. A.I.E.E.*, III, **70**, 1798–1803 (1951).
57. Germany, N., 'Calculation of the Temperature Rise of Cables in Gallery with Forced Ventilation'. *Rev. Electr.*, suppl. to *Bull. Soc. Roy. Belge Electr. Brussels*, **4**, No. 1. 3–13 (1963).
58. Salinka, P. 'External Thermal Resistance of Air-Installed Power Cables'., *Proc. I.E.E.*, **116**, 1547–1552 (1969).
59. Weedy, B. M. and El Zayyat, H. M., 'Current Rating of Cables in Tunnels and Shafts', *I.E.E.E. Trans. PAS-92*. pp. 298–307 (1973).
60. Balget, A. F., 'Appraisal of Cooling-System Design for Water-Cooled Underground High Voltage Cables', *Ontario Hydro Research Quarterly*, pp. 8–16 (June 1964).
61. Weedy, B. M. and Thelwell, M. J., 'Transient Thermal Performance of e.h.v. Cable Joints', *Proc. I.E.E.*, **119**, 472–478 (1972).
62. Matsuura, D., 'Practical Means of Suppressing Temperature Increase at Joints in Direct Water Cooled Oil-Filled System', *Surritomo Elec. Tech. Rev.*, pp. 9–16 (January 1970).
63. Horrocks, D. J. and Weedy, B. M., 'Reduction of Cable-Joint Temperatures by Oscillation of Oil in the Conductor Duct', *Proc. I.E.E.*, **120**, 593–594 (1973).
64. Brookes, A. S. and Starrs, T. E., 'Thermal and Mechanical Problems on 138 kV Pipe Cable in New Jersey', *Trans. Am. Inst. Elec. Eng., PAS-76*, p. 773 (1957).
65. Thelwell, M. J. and Soulsby, D., 'Computer Methods for Studying the Cooling of Underground Cables.' *Universities Power Engineering Conference, Portsmouth, U.K.* (1976).

PROBLEMS

Problem 3.1

An installation consists of equally loaded cables of 100 mm outer diameter buried at a depth of 1 m. Calculate the percentage increase over the value for a single cable in the effective thermal resistance as seen by the hottest cable of the group when:

(*a*) Three cables are laid horizontally with an axial spacing of 10 mm.
(*b*) Six cables are laid horizontally with an axial spacing of 10 mm.
(*c*) Twelve cables are laid horizontally with an axial spacing of 10 mm.

(Answers: 88%, 160%, 250%.)

Problem 3.2

Three 132 kV single core cables rated at 650 A are laid at a depth of 1.06 m in horizontal formation 50 mm apart. The thermal resistivity of the soil is 90 deg C/W. The cable details are as follows:

Copper conductor	Cross-section 290 mm^2
Insulation	10.6 mm radial thickness
	Outer radius 23.3 mm, $g = 500$ deg C-cm/W
Lead sheath	Thickness 2.3 mm, $g = 2.9$ deg C-cm/W
Covering	Thickness 5.6 mm, $g = 500$ deg C-cm/W

Rated-load copper loss at 85 °C is 0.323 W/cm
Dielectric loss per cm is 0.0166 W/cm
Cables cross-bonded–sheath loss negligible.

Calculate the temperature of the conductor. (Answer: 87.5 °C if ambient is 15 °C.)

Problem 3.3

Two parallel circuits comprising 69 kV, 760 mm^2 (1500 kcM) self-contained oil-filled cables are installed in a concrete duct of the type shown in Figure 3.9a. The dimensions and cable data are given below. Determine the current rating (cables equally loaded) for an ambient temperature of 25 °C assuming a loss factor of 80%. Maximum conductor temperature, 75 °C. Data:

$g_{\text{soil}} = 1.2$ deg C-m/W $g_{\text{concrete}} = 0.85$ deg C-m/W
$g_{\text{dielectric}} = 5.5$ deg C-m/W $g_{\text{fibre duct}} = 4.8$ deg C-m/W
Fibre duct wall thickness 3.3 mm
R_{dc} of conductor 28.2 $\mu\Omega$/m, R_{dc} of sheath 426.23 $\mu\Omega$/m at 50 °C
Assume sheaths open circuited
Conductor: inside diameter 15.24 mm, outer diameter 39.2 mm
Insulation – outside diameter 53.67 mm
Sheath – outside diameter 69.42
ϵ_r (insulation) 3.5, tan δ (insulation) 0.003
Configuration dimensions (as in Figure 3.9a)
Spacing between centre lines of ducts,
vertical and horizontal = 0.229 m = $d_{ab} = d_{ac}$ etc.
y and x = 0.686 m and 0.457 m respectively
Depth to centre of middle duct (h) = 1.105 m
Inside radius of each duct = 0.127 m.

(Answer: 700 A – for solution see Reference 14.)

Problem 3.4

Determine the equation relating the conductor temperature rise with time for a 400 kV, 1935 mm^2, self-contained, LPOF cable, following the applicaction of a step-function of conductor heating of 1 W per cm length. The cable is contained in a water pipe and the cooling is such as to keep the cable sheath at constant temperature. The cable parameters are as follows:

Diameter over conductor	58 mm
Diameter over insulation	107 mm
Diameter over lead sheath	115 mm
Diameter over PVC covering	125 mm.

Material parameters:

Conductor $g = 0.0026$ deg C-m/W, $c_p = 390$ J/deg C kg, $\rho = 8930$ kg/m^3
Insulation $g = 5$ deg C-m/W, $c_p = 1600$ J/deg C kg, $\rho = 1050$ kg/m^3
Sheath g 0.029 deg C-m/W, $c_p = 130$ J/deg C kg, $\rho = 11370$ kg/m^3
Insulation $\epsilon_r = 3.5$, $\tan\delta = 0.002$
Conductor R_{ac} 12.4 $\mu\Omega$/m at 85 °C.

(Answer: $\theta = 0.23(1 - e^{-0.00166t}) + 53.64(1 - e^{-0.00016t})$ where t is in seconds.)

Problem 3.5

Determine the summer rating (normal and emergency) of a 230 kV, HPOF pipe type cable. The relevant data is as follows:

Pipe depth 0.91 m, pipe diameter 254 mm
$T_c = 80$ °C (normal) 95 °C (300 hr emergency)
$T_{amb.}$ (summer) = 25 °C load factor 100%
1000 mm^2 (1970) kc mil) Copper conductor (outer diameter 91.5 mm)
Pipe coating – extruded polyethylene,
Trench backfill, thermal resistivity 0.4 deg C-m/W,
Insulation thickness 19.3 mm.

(Answer: 406 MVA.)

Problem 3.6

Determine the MVA capacity of a proposed 500 kV HPOF pipe-type cable buried at a depth of 0.92 m in soil of thermal resistivity 0.9 deg C-m/W:

(*a*) when naturally cooled
(*b*) when forced cooled with an oil inlet temperature of 10 °C, a flow rate of 0.3 m/s and a cooled distance of 1 mile.

The conductors are insulated with polypropylene–paper laminate tape. Relevant data:

Conductor, 1015 mm^2 (2000 kc mil), diameter 41.45 mm
Conductor a.c. resistance (in pipe) 0.07 $\mu\Omega$/ph/mile
Pipe diameter 254 mm (10 in)

Insulation thickness 21.6 mm
Insulation tan $\delta = 0.001$
Thermal resistivity of insulation 5.5 deg C-m/W
Maximum conductor temperature (naturally cooled) 75 °C
Maximum conductor temperature (forced cooled) 85 °C
Ambient 25 °C, Load factor 100%
Oil parameters: c_p 2 kJ/kg, $k = 0.1316$ W/mK, $\rho = 848$ kg/m^3,
kinetic viscosity $\gamma = 0.6 \times 10^{-5}$ m^2/s.

(Answers: (*a*) 740 MVA, (*b*) 1600 MVA.)

Problem 3.7

Determine the MVA rating for a proposed 765 kV HPOF pipe-type cable buried at a depth of 0.92 m in soil of thermal resistivity 0.9 K-m/W

(*a*) when naturally cooled
(*b*) when forced cooled with an oil inlet temperature of 10 °C and a flow rate of 0.30 m/s over a cooled length of 1 mile (1.6 km).

Relevant data:

Conductor 1230 mm^2 (2500 kc mil)
Conductor diameter 46.2 mm Pipe diameter (internal) 355 mm
Insulation thickness 39.6 mm
Insulation tan δ 0.002 deg C-m/W
Thermal resistivity of insulation 5.5 deg C-m/W
Maximum conductor temperature 85 °C
Ambient (soil) temperature 25 °C
Load factor 100%
Oil parameters: $c_p = 2$ kJ/kg-K, $\rho = 848$ kg/m^3,
kinetic viscosity $\gamma = 0.6 \times 10^{-5}$ m^2/s.

(Answers: (*a*) 645 MVA, (*b*) 2140 MVA.)

Problem 3.8

A step function of load is applied to a cable. Determine the pipe temperature rise after one hour and 10 000 hours, above ambient per watt of loss per m length. The cable is 138 kV, 1000 mm^2, single-circuit, high-pressure, oil-filled pipe-type. The insulation thickness is 12.8 mm, the pipe diameter 129 mm. The soil thermal resistivity is 1.2 deg C-m/W. (Answers: 0.39 deg C at 1 hour, 2.065 deg C at 10 000 hours.)

Problem 3.9

Three self-contained 138 kV cables are buried in soil of g 0.9 deg C-m/W in horizontal formation at a depth of 1 m. The horizontal axial spacing is 0.3 m. The thermal difusivity of the soil is 0.0047 cm^2/s. The diameter over the conductor is 25 mm, over the insulation 46.7 mm, and over the sheath 55 mm. Neglect sheath and dielectric losses. Determine the conductor temperature rise above ambient for a step-function of heat input per cable 1 W/cm at 0.5, 1.0, and 5.0 hours after the application of load. Assume the cable to be initially unloaded. (Answers: 0.5 hour 53.5 deg C; 1 hour 70 deg C; 5 hours 105 deg C.)

Problem 3.10

A cable has a solid conductor of 40 mm diameter surrounded by insulation of thickness 10 mm and of thermal resistivity 500 °C-cm/W. The specific heat of the insulation is 1.6 J/°C gm and the density 1.05 gm^3. The insulation is enclosed by a lead sheath of thickness 5 mm. The thermal parameters for the conductor and sheath are: specific heat (J/°C gm) 0.39 and 0.13 density (gm/cm^3). Determine a thermal R–C ladder network to simulate the cable, the dielectric being subdivided into three rings of equal radial thickness. The full-load conductor heat loss per cm is 0.3 watts, and the dielectric loss is 0.1 watts/cm. Discuss briefly how the soil surrounding the cable (if it is buried) could be represented.

Problem 3.11

A self-contained cable under a fault condition carries a fault current of 50 kA in its sheath and bronze tape reinforcement. The sheath outer radius is 115 mm and the inner 107.5 mm and the resistivity 20 $\mu\Omega$-cm. The bronze tapes are 0.3 mm (12 mil) in thickness and have a resistivity of 6 $\mu\Omega$-cm. Determine the division of current and the total heat evolved. (Answers: Current in sheath 38.8 kA; current in reinforcement 11.2 kA; total loss 3085 W/cm.)

Problem 3.12

In a cable system with artificial cooling by water pipes, calculate the total resistance to heat flow presented by each pipe, i.e. heat transfer resistance and pipe wall resistance. The pipe details are as follows:

Internal diameter	19 mm
Material polyethylene	$g = 570$ °C cm/W
Wall thickness	31.9 mm

Coolant details (water) at 20 °C

$\eta = 0.994 \times 10^{-2}$ g/cm-sec

$k = 0.00598$ (W/cm^2) (°C/cm)

$\rho = 1$ g/cm^3

$c = 4.2$ J/g per °C.

Velocity of flow 2.38 m/s.

(Answer: 25.9 deg C/W per cm.)

Problem 3.13

An underground cable system is cooled by water pipes laid along the route. On normal operation with the water cooling in action the cable conductors attain a steady temperature rise of 20 °C; this value increases to 120 °C without water cooling. The temperature–time relation without water cooling is exponential with a time constant of 7 hours while with water cooling the time constant is 0.05 hours. If the water cooling is out of action for 7 hours calculate the time that will elapse after the cooling is restored for the conductor temperature rise to be 80 °C. (Answer: 0.75 hour).

Problem 3.14

In a 275 kV self-contained LPOF cable system to attain a current rating of 1600 A, four water-pipes of 75 mm bore are laid up along the cables as shown in Figure 15.22 forming a Go and Return system. The water inlet temperature is 15 °C and the flow rate 166 litres per minute. If the total heat entering the water is 1.45 W/cm estimate the turn-round and outlet temperatures for a route length of 1.6 km; also the heat-exchanger rating (kW). (Answers: 30 °C, 42 °C, 235 kW.)

Problem 3.15

A self-contained 230 kV, 60 Hz cable system comprises three cables in horizontal formation. For emergency purposes it is to be artificially cooled by two plastic water pipes laid between the inner cable and outer cables. Assuming an available water pressure of 3.8 x 10^5 N/m^2 (55 psi), examine the feasibility of obtaining an emergency summer rating of 780 A for a route length of 825 m. Assume a flat route profile and that the hydraulic pressure drop is given by $2f\rho V^2 L/gd$ where f, the friction factor, is 0.008.

Cable data:

R_{ac} conductor	35.9 μΩ/m	10.94 μΩ/ft
R_{sheath}	41.0 μΩ/m	12.5 μΩ/ft

Ratio (total loss/conductor loss) = 1.25
Conductor area 635 mm^2 (1250 kc mil)
Normal operating temperature 80 °C: Ambient 23 °C
Emergency operating temperature 100 °C
Thermal resistances

Dielectric	0.615 deg C/m
Jacket	0.45 deg C/m
Soil (centre cable)	1.42 deg C/m
Conductor to ambient	2.08 deg C/m
Sheath to ambient	1.47 deg C/m

Dielectric loss = 5.35 W/m per cable
Sheath loss = 0.33 W/m

$g_{soil} = 1.0$ deg C-m/W
Spacing between cables (axial) = 22.9 cm
Depth = 127 cm
Cable outside diameter = 8.74 cm (3.44 in)
Use pipes of 38 mm (1.5 in) internal diameter (d_{pi})

(Answer: For a town main system with Go flows in both pipes with a water velocity of 3.35 ft/s (maximum flow for pressure) the rating is about 1 kA. Hence the emergency rating is attainable. The normal rating is attainable with no water cooling. For a circulatory system using the same head as above with one pipe as Go and the other as Return the calculation contains many assumptions and iterative procedures. The maximum water velocity is 2.37 ft/s. The emergency load is now 960 A. Considerable assistance in this problem will be found in Reference 60.)

Problem 3.16

For the self-contained cable system described in Problem 3.15 calculate the current rating obtainable in winter for

(*a*) normal conductor temperature of 80 °C: Ambient 8 °C
(*b*) emergency conductor temperature of 100 °C.

Assume two plastic pipes of 38 m internal diameter and an open-ended flow system fed from town mains at 3.8×10^3 N/m^2 55 psi. Assume a route with no gradients and of length 1.6 km. Calculate also the naturally-cooled ratings. (Answers: 1180 A and 1320 A; 905 A and 1050 A.)

Problem 3.17

A 400 kV, 2000 mm^2 LPOF self-contained cable circuit is externally cooled by placing the cables in pipes and pumping water through the pipes. It is necessary to reduce the high maximum temperature in the through joints (which are 350 m apart) and the slow oscillation of the oil in the conductor duct is used to smooth out these hotspots. Determine the oil flow rate required for an oscillation time of 20 minutes in order to reduce the joint temperature by 10 deg C. (Answer: 0.6 cm^3/s.)

Problems 3.18

Determine the MVA capacity of an internally cooled (i.e. flow of oil in conductor duct) LPOF self-contained cable system rated at 400 kV and with

(*a*) an axial separation between cores of 152 mm
(*b*) an axial separation between cores of 406 mm.

Assume constant heat dissipation along the cables. Data:

Cooled length 915 m, oil flow rate 2831.7 cm^3/s
Sheath temperature 85 °C
Thermal resistivity of insulation 5.5 deg C-m/W
Thermal resistivity of soil 0.9 deg C-m/W
Load factor 100%
Trefoil formation of cores
Conductor duct diameter 53.7 mm
conductor diameter 86.3
conductor area 2000 mm^2
R_{ac} 12.5 $\mu\Omega$/m at 85 °C

Insulation

Diameter over, 122 mm, ϵ_r 3.5, tan δ 0.0025

Sheath

Diameter 132 mm
Mean diameter 127 mm
Thickness 5.1 mm
Effective resistance (electrical) 0.14 $\mu\Omega$/cm (152 mm) and 0.0085 $\mu\Omega$/cm (406 mm)

Oil

Density ρ 0.9 gm/cm^3
Thermal conductivity k 0.001316 W/cm K
Specific heat c_p 2.0 J/gm-K
Viscosity (η) 0.055 gm/cm-s

(Answers: (*a*) 1120, conductor temperature 69.5 °C; (*b*) 1440, conductor temperature 79.5 °C.)

Problem 3.19

A single pipe-type cable of diameter 0.35 m is installed above ground and is to be rated on the following two conditions:

(*a*) natural (free convection only) cooling
(*b*) a cross wind of 0.45 m/s.

If the heat developed by the losses is 1 W/cm calculate the pipe-surface temperature rise above ambient, neglecting the effects of radiation for both conditions. Parameters for air over the temperature range:

ρ	0.706 kg/m^3
N_{Pr}	0.68
k	2.07×10^{-5} kW/m-K
β	3.32×10^{-3}/K
γ	15.7×10^{-6} m^2/s

(Answers: (*a*) 17 deg C, (*b*) 3 deg C.)

Problem 3.20

A cable circuit comprising 3 self-contained cables is installed in a surface trough of the type shown in Figure 3.32. The thermal resistance (R_s) from the hottest cable to the ground surface is 40 deg C/W per cm length and the cables' axes are 30 cm below the ground surface (h). Estimate the maximum cable surface temperature for a dissipation per cable q_c of 0.5 W/cm. Assume an ambient temperature T_a of 18 °C, a solar energy incidence q_s of 9 m W/cm² and a wind velocity V of 1.34 m/s.

The following formula has been given for this type of installation

$$\text{Maximum cable temperature} = q_c R_s + \frac{152 q_c}{V^{0.74} h^{0.2}} + \frac{8.36 q_s}{V^{0.89} h^{0.07}} + T_a \ (°\text{C}).$$

(Answer: 60 °C.)

Problem 3.21

An underground tunnel has a cross section of 5.4 m² and is 250 m in length. A group of cables laid along the length of the tunnel dissipates a total of 10 W/cm. The effective thermal resistance from the air to ambient (air to tunnel wall heat-transfer and transfer through wall and soil) is 6 °C cm/W. If the maximum steady-state temperature rise of the air blown through the tunnel is to be not greater than 32 °C determine the air velocity required. (Answer: 1.3 m/s).

Problem 3.22

Two d.c. cable circuits are installed in a very deep tunnel of equivalent diameter 6 m and of length 18 km. At full load the total heat dissipated from the cables is 90 W/m. The cables are each of 9 cm outer diameter. The thermal parameters of the rock surrounding the tunnel are: ρ 2500 kg/m², c_p 0.8 kJ/kg, g 0.37 deg C-m/W.

(*a*) If the heat received by the tunnel wall is 27 W/m determine the temperature rise of the wall at the following times after the energization of the cables: 1 year, 5 years.

(*b*) Assuming a maximum air temperature in the tunnel of 35 °C calculate for forced air cooling the air velocities required (i) for a summer inlet air temperature of 13 °C, (ii) a winter air inlet temperature of –10 °C. State clearly any assumptions made.

(*c*) Assuming that without forced cooling the mode of cooling is purely natural convection, calculate the temperature rise of the cable outer surfaces above the mean tunnel air temperature.

(Answers: (*a*) 2 deg C, 3.5 deg C; (*b*) 3 m/s (air flow 66 m³/s); 1.35 m/s (30 m³/s); (*c*) 9.25 deg C (use Reference 59).)

4

Conventional A.C. Cable Systems

As mentioned in the chapter on insulation, the solid-type oil impregnated-paper cable is viable up to a.c. line voltages of the order of 33 kV. The systems currently in use and well established for higher voltages are designed to eliminate the effects of internal partial (corona) discharges. The ways of achieving this have already been outlined. Extruded, crosslinked polyethylene insulation, in which the treeing and internal discharge problem has been limiting, is of great interest because of its potentialities. At the time of writing these cables are commercially available for voltages up to 138 kV. Installations with low density polyethylene insulation at 225 KV exist in France. For high capacity cables, copper is still largely used because of the lower conductor loss.

When designing an installation many factors have to be considered. Firstly, the rated and maximum operating voltages and the grounding conditions in the transmission circuit.. Next the magnitude of overvoltages, both atmospheric and system generated. These will be influenced by the physical location of the cable, e.g. between two overhead lines or between a line and a terminal. Other essential system parameters are short-circuit magnitudes and clearing times, load to be expected, and whether it is cyclic or largely continuous. The nature of the soil and climatic conditions are important, and if ducts are used the extent to which they are already occupied by existing cables. This summary is by no means exhaustive, but it is hoped that it will give some idea of the many factors involved when designing an underground transmission link. Methods of installation and their effect on ratings have already been discussed.

The insulation design is normally governed by the allowable impulse stress at the conductor. The Basic Insulation Levels for the various system voltages are as follows:

USA: 138 kV–650 kV, 230 kV–1050 kV, 345 kV–1175 kV
500 kV–1475 kV, 765 kV–2100 kV, 1200 kV–2400 kV
British: 132 kV–640 kV, 275 kV–1050 kV, 400 kV–1425 kV.

British 400 kV LPOF cables have a maximum a.c. stress at the conductor of 15 kV/mm with impulse stresses at 92.5 kV/mm (conductor) and 56.0 kV/mm (dielectric screen). Present breakdown strengths for existing HPOF cables are 60 kV/mm on a.c. and 130 kV on impulse. 550 kV, 1270 mm^2 HPOF pipe-type cables will have an a.c. stress at the conductor of 16 kV/mm, and a proposed 750 kV HPOF cable a maximum a.c. stress of 19 kV/mm. The maximum a.c. stress in mass-impregnated (solid type) cables is of the order of 4 kV/mm.

4.1 HIGH-PRESSURE OIL-FILLED (PIPE-TYPE) SYSTEMS

As mentioned in the introductory chapter the pipe-type cable has today largely superseded the self-contained cable in the USA. The latter when used in the USA tends to be drawn into duct systems in which manholes are a major feature. These although provided for splicing, must also be designed to accommodate cable movements due to thermal changes with load variations.

A cross-section of the pipe-type system is shown in Figure 1.5. Corona discharges are suppressed by the high oil pressure (1.57×10^6 N/m^2). The screened insulated conductors are helically wound with skid wires to protect the insulation on installation. Because of the absence of a protective metal sheath, long lengths (up to 1000 m) may be pulled in, resulting in fewer splices. Pumping plant (duplicated) maintains the oil pressure on the whole length of cable and eliminates the need for stop joints and reservoirs. This also overcomes the problem (for self-contained oil-filled cables) of route profile. 'Pulling-in' is commenced with the pipe (which has been tested for dryness) filled with nitrogen gas at 2.07×10^6 N/m^2. A diaphragm known as a 'mouse' is propelled down the pipe by nitrogen pressure with a cable pulling-wire attached. This is then attached to the insulated core which is pulled through, assisted by lubrication from insulating oil. Cables may be in either cradle (Figure 4.1) or trefoil (Figure 1.5) formation. If the ratio (pipe diameter D/core diameter d) is less than 2.5, because of the jamming of the cores, 'pulling-in' is difficult.[1]

The costs of the pipe-type system may be roughly apportioned as follows:

Conductor: material 30%, fabrication 4%
Insulation and screens: material 10%, taping 12%
Drying and impregnation 18%, wires 5%, sealing 5%, shipping 10%.

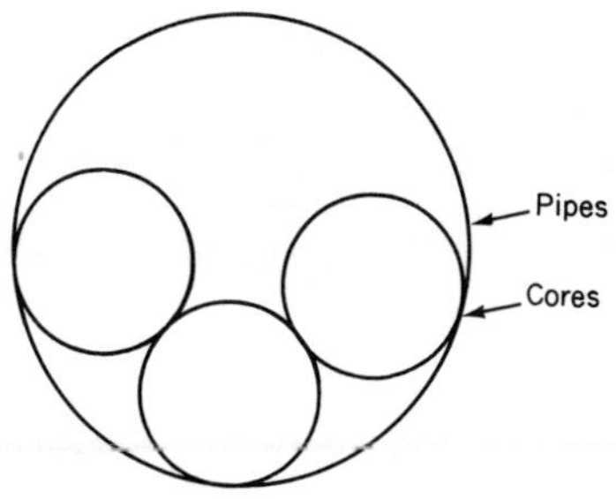

Figure 4.1. Pipe-type cable – cables in cradle formation.

Calculation of Oil Flow Rate for Forced Cooling Cables

A procedure not dependent on traditional heat transfer data has been reported in Chapter 3 for the thermal calculations in pipe-type cables. The heat transfer increases dramatically as the oil-flow changes from laminar (N_{Re} of 1000) to turbulent (N_{Re} above 10^4), the actual transition occurring at a Reynolds number of about 1400. This change in flow condition also greatly increases the pressure drop, and these two factors must be carefully balanced. In the laminar range the pressure drop is proportional to the flow-velocity and in the turbulent range it is proportional to $(\text{velocity})^2$. The skid-wire roughness ratio is defined as (D_s/D_h) where D_s is the diameter of the skid wire and D_h the overall hydraulic diameter. This has no influence in the laminar region but has a strong influence on the transition Reynolds number. In the laminar region the mode of heat transfer is a mix of forced and free connection and the N_{Nu} is independent of N_{Re}. The upper limit of the laminar friction factor $(D_e \Delta P/2LUe^2)$ can be represented by $(fN_{Re}) = 30$. Asymptotic values for the turbulent friction factors are 0.013 and 0.021 corresponding to skid wire roughness ratios of 0.0216 and 0.0293 respectively. It is wasteful to increase N_{Re} beyond the minimum value for turbulence because above this the increases in pressure drop (αN_{Re}^2) outweigh increases in N_{Nu}. The relationship between the friction factor and Reynolds number is shown in Figure 4.2.

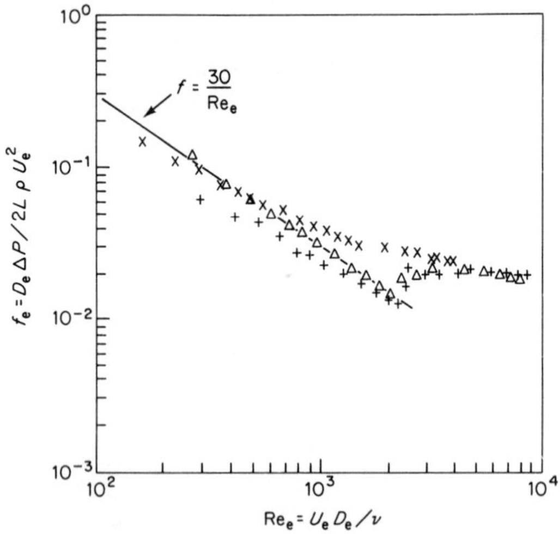

Figure 4.2 Friction factor in pipe-type cable with oil flow. Based on hydraulic diameter. Conductor heat 438 W/m². Distance from pipe entrance/hydraulic diameter D_H = 400.[2] (Reproduced by permission of IEEE.)

Example 4.1

A 550 kV HPOF, pipe-type cable has a 1000 mm² (2000 kc mil) conductor and is forced cooled by the pumping of oil in the pipe. Determine the oil flow rate and heat transfer coefficient. The cable parameters are as follows: pipe size (D_p) 305 mm (12 inches); insulation thickness 30.7 mm (1.25 inches); dielectric loss 13.1 W/conductor/m; resistive losses 82.68 W/conductor/m; pipe temperature 32.4 °C; oil temperature 40 °C; conductor temperature 85 °C; cable diameter (D_c) 100.5 mm; length 1.6 km; oil kinematic viscosity $(\gamma) = 0.6 \times 10^{-5}$ m²/s; oil density $(\rho) =$ 848 kg/m³; oil specific heat (c_p) = 2.021 kJ/kg-deg C; oil thermal conductivity $(k) = 0.15$ W/m-deg C.

Solution

Cross-sectional area of flow is

$$\frac{\pi}{4}(D_p^2 - 3D_c^2) = \frac{\pi}{4}(0.305^2 - 0.1005^2) = 0.06513 \text{ m}^2.$$

Wetted perimeter,

$$P_w = \pi(D_p + 3D_c) = 1.905 \text{ m}.$$

Hydraulic diameter,

$$D_H = 4 \times \text{Area of flow}/P_w = 0.13675 \text{ m}.$$

Heat per cable per m²,

$$q_c = \frac{95.78}{0.1005} = 303.36 \text{ w/m}^2.$$

Select N_{Re} to be 3000 as a reasonable compromise between high heat transfer and high pressure drop. The velocity of the oil is

$$\frac{\gamma N_{Re}}{D_H} = 3000 \times 0.6 \times 10^{-5}/0.13675 = 0.1316 \text{ m/s}.$$

Volumetric flow rate is 0.00857 m³/s = 8.57 litres/s. Pressure drop per m is $2f\rho$ $(\text{Vel})^2/D_H$.

From Figure 4.2, $f = 0.02$ for $N_{Re} = 3000$, hence,

$$\Delta p = 2 \times 0.02 \times \frac{1}{0.13675}(0.1316)^2 848$$

$$= 4.296 \text{ N/m}^2 \text{ per metre length}$$

$$N_{Pr} = \frac{\mu c_p}{k} = \frac{\rho\gamma c_p}{k} = \frac{(848)(0.6 \times 10^{-5})(2.021)10^3}{0.15} = 68.55$$

$$(N_{Re} N_{Pr}^{0.4}) = 3000(68.55)^{0.4} = 16.27 \times 10^3.$$

From Figure 3.22, $N_{Nu} = 50$, hence

$$h = \frac{N_{Nu} k}{D_c} = \frac{50 \times 0.15}{0.1005} = 74.627 \text{ W/m}^2\text{-deg C}.$$

Cable surface temperature

$$T_s = \frac{q_c}{h} + T_{oil} = \frac{303.36}{74.627} + 40 = 44\ ^\circ\mathrm{C}.$$

4.2 SELF-CONTAINED LOW-PRESSURE OIL-FILLED SYSTEMS

There is a tendency to standardize on segmental construction for the conductor and for areas above 1000 mm^2; six-segment Milliken construction is used to reduce the skin effect. At the present time conductor areas up to 2600 mm^2 are manufactured.

Conductor a.c. resistance decreases with increase in axial spacing by as much as 20% for large conductors. For example, in a 400 kV cable 1935 mm^2 conductor (R_{dc}, 0.00925 Ω/km at 20 °C) has an R_{ac} of 0.0165 Ω/km with 50 mm axial horizontal spacing which decreases to 0.0128 Ω/km at 610 mm. Similarly for a 1300 mm^2 conductor (R_{dc} = 0.014 Ω/km at 20 °C), R_{ac} falls from 0.0203 Ω/km at 50 mm to 0.0184 Ω/km at 610 mm spacing. For 645 mm^2 the values are R_{dc} 0.026 Ω/km, R_{ac} 0.037 Ω/km at 50 mm to 0.343 Ω/km at 610 mm spacing. All values of R_{ac} related to 85 °C.

Conductor and core screens are of carbon-black paper and metallized insulating paper tapes. In particular, today, two-ply tapes comprising carbon paper and metallized carbon paper are used. The reinforced lead sheaths originally used have been largely replaced by corrugated aluminium, and extruded thermoplastic over-sheaths, usually of PVC or polyethylene, have replaced lapped protective coverings. The insulated conductor which has been dried under vacuum is sheathed under light oil pressure.

The majority of single-conductor cables operate with fully insulated sheaths using crossbonding or single-point bonding to minimize sheath losses (see Figure 4.3). Connections between the sheath sections are made by means of PVC insulated bonding leads to link boxes which also house nonlinear resistors which limit to a safe value (for the sheath insulation) steep-fronted overvoltages. Oil pressure tanks are mounted above ground where possible. Feed pipes are of armoured lead or corrugated aluminium with PVC protection for burial. A survey

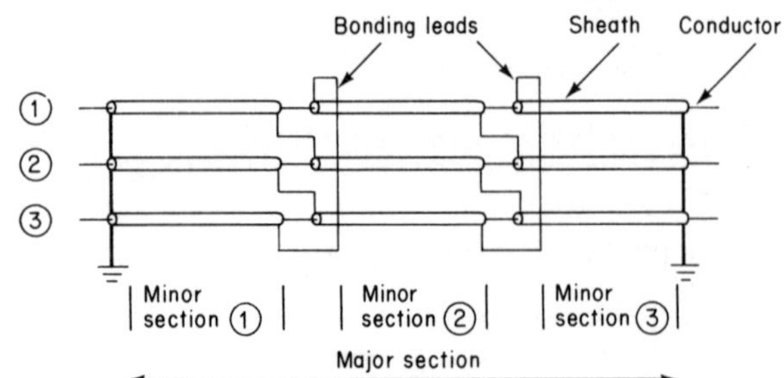

Figure 4.3. Crossbonded sheaths in self-contained cable system.

of progess in the design of the LPOF cable is given in References 4 and 5, and of installation and testing in reference 6.

Hydraulic Design

The oil is maintained under pressure by reservoirs of the form shown in Figure 4.4. In the tank are a number of circular containers comprised of two circular diaphragms and a short cylinder. Before insertion these are filled with gas at 1 atm (10^5 N/m^2) for standard tanks and $1\frac{1}{2}$–2 atm for pre-pressurized tanks. The space between the gas containers and tank is filled with cable oil under vacuum. When the oil pressure in the cable increases, the diaphragms flex towards each other and reduce the volume thus compressing the contained gas. Oil from the cable is then able to enter the tank.

If V_0 is the capacity of a tank in litres of gas at atmospheric pressure and 15 °C, then,

$$V_0 = \frac{CT}{P} = (273 + 15)\frac{C}{1}, \tag{4.1}$$

where P is in atmospheres. If the upper and lower tank pressures are P_1 and P_2 and the corresponding temperatures are T_1 and T_2, then the change in volume (ΔV) is given by,

$$\frac{\Delta V}{V_0} = \frac{T_1}{288P_1} - \frac{T_2}{288P_2}. \tag{4.2}$$

Tanks are installed at high points along the cable route to avoid partial compression and loss of efficiency due to the static head of oil in the cable. If this cannot be arranged, pre-pressurized tanks are used. The length of cable to be fed from a tank depends upon:

(*a*) the maximum head of oil at the lowest point of cable route for this section
(*b*) the maximum transient pressure at the lowest point – this is caused by the oil flow and duct friction and is additional to (*a*)
(*c*) the minimum static and transient pressures (on cooling).

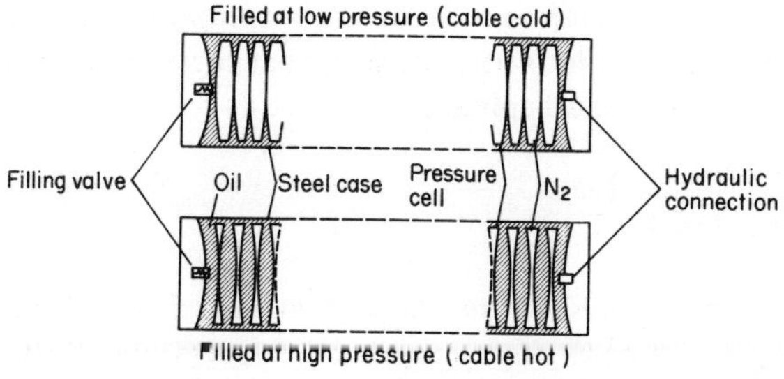

Figure 4.4. Arrangement of cells in oil reservoirs for LPOF cables.

When the cable cools, oil returns to the cable from the reservoir and it is necessary to know the maximum and minimum pressures created by the oil-flow in order to determine the length of cable allowable per tank. Let

x = distance (m) of the point from the reservoir
α = quantity of oil absorbed by cable at a given instant per unit length per second
β = oil flow in cable duct (m^3/s).

If total length of cable is L and the flow at the remote end is zero, then $\beta = \alpha(L - x)$.

Let f be the coefficient of friction in the duct and let the pressure drop be dp over dx. For flat route-profiles,

$$\frac{dp}{dx} = -f\beta = -\alpha f(L - x)$$

and

$$P_x = -\alpha f\left[Lx - \frac{x^2}{2}\right]_0^x + P_0$$

where P_0 = pressure at cable infeed. Let

h_x = the difference in levels between point x and the feed point, and
ρ_0 = the density of oil.

Then

$$P = h_x\rho_0 + P_0 - \alpha f\left[Lx - \frac{x^2}{2}\right]$$

and

$$h = h_x + H_0 - \frac{\alpha f}{\rho_0}\left[Lx - \frac{x^2}{2}\right]. \tag{4.3}$$

It is seen that under transient condition (change of load) above a certain length from the feed point the pressure falls below atmospheric (causing a vacuum in the core and dielectric). It is therefore necessary to use *stop joints* containing hydraulic barriers to isolate the cable into sections such that the vacuum is not created, i.e. $h > 0$. Poiseuilles Law for flow applies, and

$$f = \frac{8 \times \text{Viscosity}}{\pi(\text{duct internal radius})^4}. \tag{4.4}$$

The calculation of α depends on the thermal characteristics of the cable and the volumetric changes accompanying changes in temperature. During heating the transient pressures may be calculated in a similar way.[3] The variation of pressures along a cable route in terms of head for a given route profile is shown in Figure 4.5.

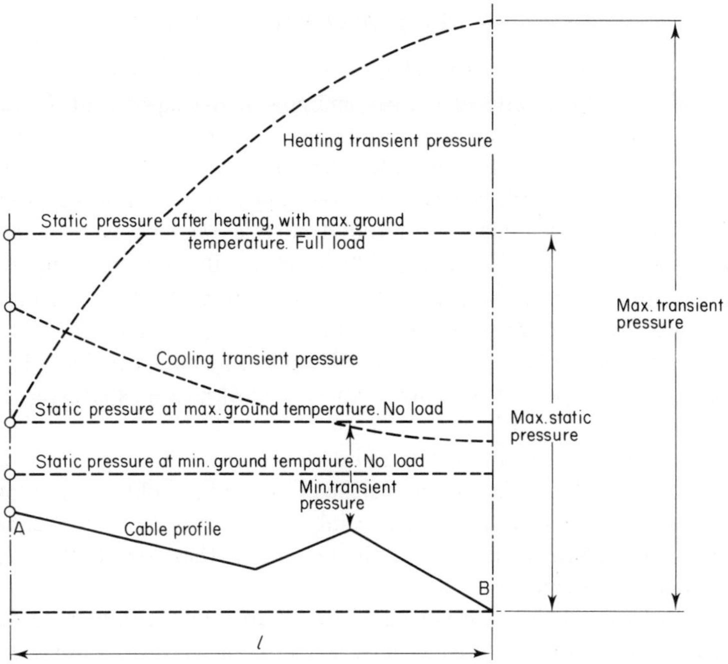

Figure 4.5. Variation of oil pressure along LPOF cable fed from reservoir at one end.

In practice the limiting oil pressures are: minimum static 20.7×10^3 N/m^2 (3 psi), maximum static 520×10^3 N/m^2 (75 psi), and maximum transient 790×10^3 N/m^2 (115 psi).

A stop joint is shown in Figure 4.6; this divides the cable into independent lengths from an oil-flow aspect and hence the static pressures due to route profile are kept within acceptable limits. The less elevated the profile the greater the distance between stop joints. In certain cases the sheath must be reinforced to withstand much higher oil pressures, e.g. in vertical shafts as found in hydro-stations where the effective head of oil may be in the order of 300 m without a stop joint.

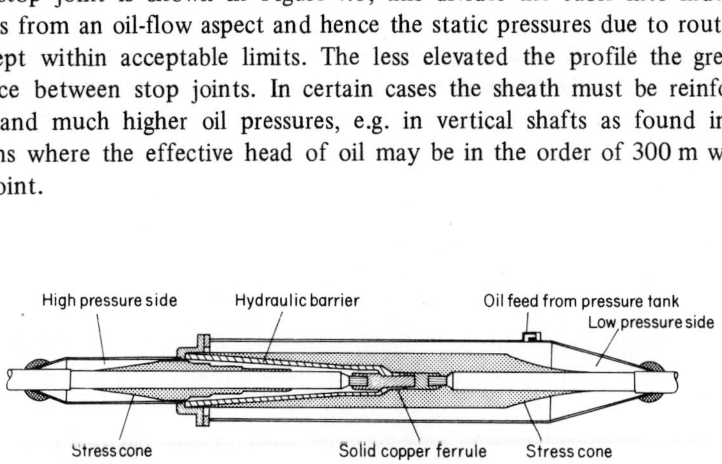

Figure 4.6. Stop-joint for LPOF self-contained single core cable. (Reproduced by permission of IEE.)

4.3 EXTRUDED DIELECTRIC CABLES[7,8]

The extruded dielectric cable is mechanically and electrically the simplest form of cable, free from the necessity of oil, gas, and pressurizing equipment. It consists of a standard conductor (aluminium or copper) covered by an extruded semi-conducting screen over which is extruded the dielectric. Finally, a further screen is extruded and covered with helically wound copper tapes or wires. The screen material is carbon-loaded plastic. Polyethylene is a thermoplastic material that becomes plastic at 75 °C and melts at 100 °C. Hence the current rating is severely limited. Crosslinked polyethylene (XLPE) can operate at 90 °C with short term (8 hours per year) overload temperatures of 130 °C and a fault-temperature of 250 °C. Ethylene-propylene rubber (EPR) insulated cable is manufactured at 132 kV and has given successful operational service. However, most interest appears to be in the XLPE cable.

Extruded cables are widely used up to 69 kV but above this, e.g. at 138 kV, only a modest circuit mileage has been installed. Electrical problems in crosslinked polyethylene cables have been caused mainly by protrusions on the screens and voids and contaminants in the bulk of the insulation. These lead to the creation of trees and breakdown paths.

The traditional process of producing chemical crosslinking in polyethylene uses steam heating and additives such as small amounts of peroxide. In new extrusion processes steam is not used, heat being applied via nitrogen or oil. This has brought about an improvement in dielectric quality, as has the triple extrusion of the conductor shield, insulation, and insulation shield. Further attempts at improvement in the dielectric performance involve the addition of semiconducting organic fluids, voltage stabilizers, and gas impregnation. Experience to date indicates that many additives tend to diffuse out of the insulation and the beneficial effects are short lived. Voids ranging from less than 1 μm up to 30 μm are observed in conventional steam-cured crosslinked polyethylene with about 10^5–10^6 voids per m^3.

The extrusion plant used (see Figure 4.7) is expensive and complex, and often the extrusion rates low. Crosslinked polyethylene cables are passed through long catenary vulcanizing pipes and a temperature of 150–200 °C is required for the

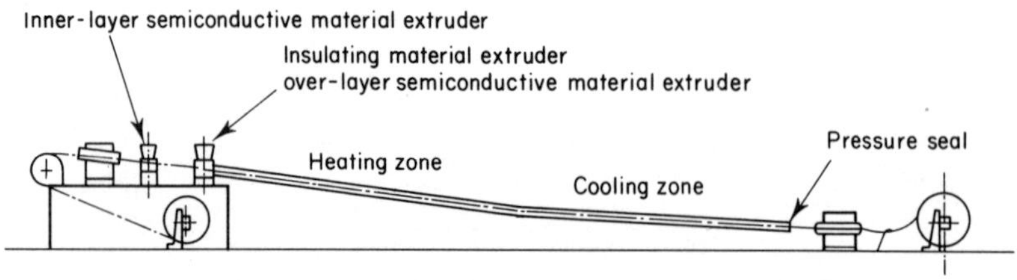

Figure 4.7 Manufacture of conventional XLPE cable – catenary continuous volcanization process.[11] (Reproduced by permission of IEE.)

reaction to proceed. The process is known as catenary continuous vulcanization (CCV).

During the crosslinking reaction, decomposition products in gaseous form are released and these will create voids in the insulation unless high steam or gas pressure is used. Because of the conductor weight for the larger, high-voltage cables, vertical extrusion pipes may be necessary, e.g. the French 225 kV cable extrusion tower is 55 m in height. The extrusion of thick-walled insulation in intimate contact with the two extruded semiconducting screens is a complex process and the machinery used comparable in scope with the lapping and impregnation plant used in paper/oil cables.

Reports on the performance of extruded polyethylene cables installed in wet environments indicate that their electrical performance is worse than in dry conditions. This involves decrease in breakdown strength and insulation resistance and increase in tan δ.[9] Transparent tree-like growths have been found, i.e. water trees. In view of the fact that the mechanism of ageing in insulation with ingress of moisture is not well understood it is advisable to use results from short-length model tests as the design basis for real cables with caution and service experience should be utilized as far as possible. Inspection of cables removed from service indicates that treeing occurs under service conditions and could be associated with unsatisfactory electrical performance. The treeing is related to the quality of the cable manufacture and materials used.[10] The provision of a plastic jacket around the insulation does not provide major protection against water.

General criteria[9] for the life of cables are as follows. (Note that life period is proportional to $(\text{stress})^{-n}$.)

(a) Life-exponent (n) has a value 8 for XLPE and 11 for EPR for a wet environment.
(b) When the cable is protected by a metallic sheath the ageing is similar to that for cables in dry situations; n may be taken as 13 for XLPE and 15 for EPR.

Generally, present-day designs[11] for extruded dielectrics use an *average* stress of 2.35 kV/mm (60 V/mil) up to 69 kV and 3.75 kV/mm (95 V/mil) at 138 kV. These values compare unfavourably with values for cellulose/oil taped insulation and even more unfavourably with the intrinsic strength of polyethylene of 0.79 MV/mm (20 kV/mil). The present solution to this poor dielectric performance is to overdesign. However this becomes more difficult at higher voltages because of the increased cost of manufacture, splicing problems, and increased thermal resistance.

Present development work is aimed at the use of purer material, high-pressure dry curing to prevent microvoid formation, uniform dispersion of crosslinking agents, and the alleviation of shield asperities. Further in the future is irradiation crosslinking, where the lower temperature avoids shrinkage-voids and other byproducts due to chemical crosslinking. Overall it would seem that if current development work is successful, *average* design stresses of 7.8 kV/mm (200 V/mil) and perhaps above are possible leading to much thinner walls, improved current capability, and lower cost. Possibly then cable circuits rated up to 500 kV will become available.

Sodium Conductor Cables[1 2]

Sodium as a conductor material possesses many advantages. In addition it exists abundantly in the Earth's crust and is cheaply extracted from sodium chloride by electrolysis; for a given current rating it is less than half the cost of aluminium. Other properties of interest are: good malleability and ductility giving flexibility when encased by polyethylene insulation, low melting point (98 °C) and increased resistivity when molten, and, very important, a violent chemical reaction when exposed to water. At the present time sodium conductor cables have been manufactured for working voltages up to 34.5 kV, the insulant being polyethylene. They are produced by the extrusion of a polyethylene tube which is simultaneously filled with sodium. Virtually all of the mechanical strength of the cable is in the insulation and this is normally adequate owing to the light weight and flexibility. Although the change in phase resulting from a short circuit results in a storage of heat and a temporary delay of further temperature rise, the high electrical resistance of the molten metal may cause thermal instability. A total of some 50 circuits have been installed in the USA at voltages up to 34.5 kV but manufacture has stopped, presumably because the economic incentives were not sufficient to encourage the installation of such a novel cable on a large scale.

4.4 COMPARISON BETWEEN PIPE-TYPE AND SELF-CONTAINED CABLE SYSTEMS

It is difficult to generalize the cost difference between these systems, and each installation should be examined on its own merits. Pipe cable is not so sensitive to trenching costs as self-contained, the capacity of the latter being dependent on the cable spacing. Hence in an area of low trenching costs the self-contained system may show to advantage. Similarly the pipe cable is not sensitive to the profile of the route provided the height difference of terminals is less than roughly 300 m. With self-contained oil-filled systems change in elevation is limited by stop joints and the ability of the mechanical strength of the sheath to withstand oil pressure.

Material costs for the cables themselves are also difficult to generalize. The oil requirement for the pipe system is greater. Although the volume of paper tends to be greater in the self-contained cable because of the presence of the conductor duct, this is offset by the reduction in the maximum stress at the larger conductor radius. The terminal equipment for the oil-filled self-contained system consists of low-pressure sealing ends which tend to be less expensive than the pressure-resistant pot-heads and trifurcation devices.

Direct-buried, self-contained cables dissipate heat more readily, although this advantage is offset when installation is in ducts. An impediment to their use in many urban areas is the need to keep open trenches of length at least that of the unjointed cable during installation. They are also more vulnerable to damage by construction and building activities than the pipe-enclosed cables.

4.5 ELECTRICAL PARAMETERS OF SELF-CONTAINED CABLES

A knowledge of electrical parameters is essential for system design and performance studies. Current rating, being dependent on losses, is also influenced by electromagnetic induction in sheaths and armouring. In Chapter 1 the losses resulting from the conductor a.c. resistance were formulated; this will now be extended to include the sheath.

Inductance

For two conductors of radius r (m) with an axial separation of s (m) the inductance of each cable is

$$2.0\left(1+\ln\left(\frac{s}{r}\right)\right)\times 10^{-7}\ \text{H/m} = 2\times 10^{-7}\ln\left(\frac{1}{0.779}\right) + 2\times 10^{-7}\ln\left(\frac{s}{r}\right)$$

$$= 2\times 10^{-7}\ln\left(\frac{s}{0.779r}\right)\ \text{H/m.} \qquad (4.5)$$

where $0.779r$ is the *geometric mean radius* (r_e) for a solid circular conductor of radius r. This allows for the internal flux in the conductor.

For a trefoil arrangement the inductance of the conductors with respect to the hypothetical neutral (i.e. per phase) are equal and of magnitude $2\ln(s/r_e)\times 10^{-7}$ H/m. For three conductors in non-equilateral formation the mean inductance to neutral of the conductors is $2\times 10^{-7}\ln(s_{eq}/r_e)$ H/m; where s_{eq} is the mean geometric spacing (i.e. of the equivalent trefoil) and is given by $(s_{ab}\,s_{bc}\,s_{ca})^{1/3}$. For three cables in horizontal formation and of spacing s, $s_{eq} = (2s^3)^{1/3}$.

The inductances of the individual conductors a, b, and c in horizontal formation are as follows:

$$\left.\begin{aligned} L_a &= 2\left(\ln\left(\frac{\sqrt{2}s}{r_e}\right) - j\sqrt{3}\ln\sqrt{2}\right)\times 10^{-7}\ \text{H/m} \\ L_b &= 2\ln\left(\frac{s}{r_e}\right)\times 10^{-7}\ \text{H/m} \\ L_c &= 2\left(\ln\left(\frac{\sqrt{2}s}{r_e}\right) + j\sqrt{3}\ln\sqrt{2}\right)\times 10^{-7}\ \text{H/m} \end{aligned}\right\}. \qquad (4.6)$$

These formulae apply for cables without metal sheaths or for cables with sheaths grounded at one point only.

Capacitance

The capacitance between conductor and screen or sheath is given by

$$\frac{2\pi\epsilon\epsilon_r}{\ln(R/r)}\ \text{F/m}$$

where r and R are the radii over the conductor and dielectric screen respectively. This formula also obtains for the screened cores in the pipe-type cable.

Currents and Voltages in Sheaths

Two cables will be considered, as shown in Figure 4.8a. The sheaths are isolated, i.e. not grounded at the ends, and the fluxes set up by one cable cut the sheath of the other and induce eddy currents. Although the loss in this condition is very small except when the cables are close or touching, voltages are set up along the sheath which may become excessive. To avoid this the sheaths are bonded to ground at the ends forming the end-connections indicated in Figure 4.8b.

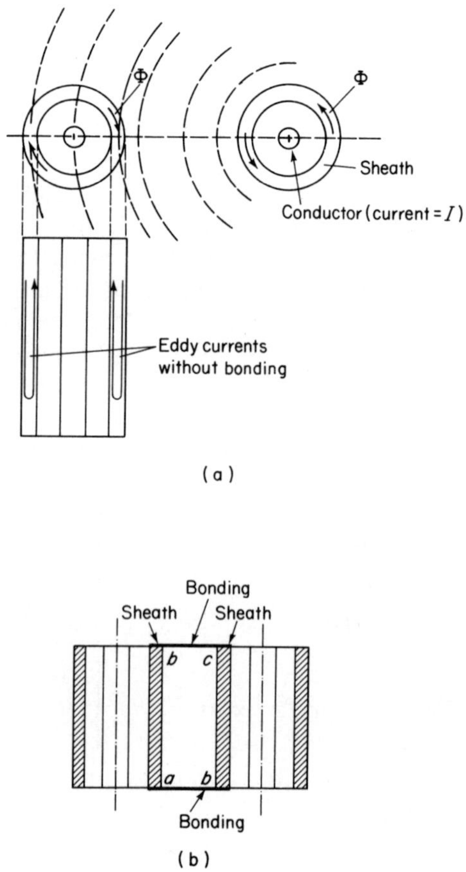

Figure 4.8(a) Currents and fluxes in sheaths of self-contained cables without bonding (short circuiting) of sheaths to ground. (b) Two cables with sheaths bonded. Spacing between centres of conductors s(m), r and r_s are radii of conductor (solid) and sheath (mean value) in metres.

The flux through the loop *abcd* forms a mutual inductance M between sheath and conductor equal to

$$M = 4 \times 10^{-7} \ln\left(\frac{s}{r_s}\right) \text{ H/m}$$

where s is the spacing between the sheath axial centres (m) and r_s is the mean radius of the sheath (m).

For a three-phase system with equilateral spacing the effective value of M is

$$M = 2 \times 10^{-7} \ln\left(\frac{s}{r_s}\right) \text{ H/m.} \tag{4.7}$$

The self-inductance of the sheath L_s is approximately equal to the mutual inductance M and, letting the resistance of the current path, i.e. the sheath, be R_s, the sheath current,

$$I_s = \frac{I\omega M}{\sqrt{(R_s^2 + (\omega M)^2)}} \tag{4.8}$$

and the sheath loss per phase is

$$I_s^2 R_s = R_s\left(\frac{I^2\omega^2 M^2}{R_s^2 + \omega^2 M^2}\right). \tag{4.9}$$

The total effective resistance,

$$R_{ac}^* = R_{dc}(1 + y_s + y_p) + \left(\frac{\omega^2 M^2}{R_s^2 + \omega^2 M^2}\right) R_s$$

or

$$R_{ac}^* = R_{ac} + \left(\frac{\omega^2 M^2}{R_s^2 + \omega^2 M^2}\right) R_s. \tag{4.10}$$

The equivalent circuit is shown in Figure 4.9; the conductor and sheath form a coupled circuit.

Consider a metre length of the circuit. The voltage drop per metre is given by $\mathbf{\Delta V}$ where

$$\mathbf{\Delta V} = \mathbf{I}(R_c + j\omega L_c) + \mathbf{I}_s j\omega M$$

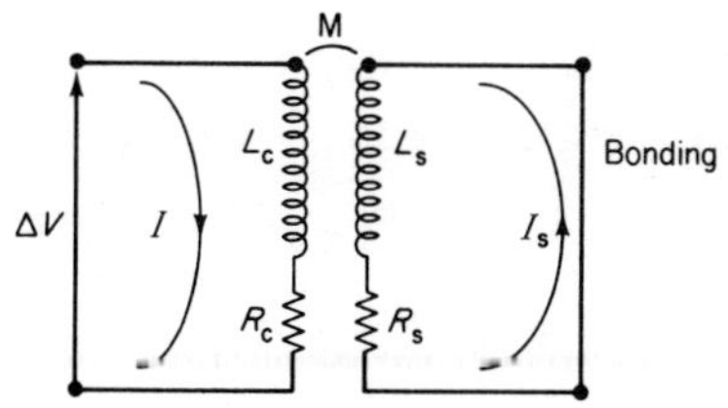

Figure 4.9. Equivalent circuit formed by conductor and bonded sheaths.

and in the sheath,

$$0 = \mathbf{I}_s(R_s + j\omega L_s) + \mathbf{I}\, j\omega M.$$

From these two equations

$$\mathbf{Z} = \frac{\Delta \mathbf{V}}{\mathbf{I}} = \left[R_c + R_s\left(\frac{\omega^2 M^2}{R_s^2 + \omega^2 M^2}\right)\right] + j\left(\omega L_c - \frac{\omega^3 M^3}{R_s^2 + \omega^2 M^2}\right) \tag{4.11}$$

$$= R_{\text{effective}} + jX_{\text{effective}}$$

where $L_s = M$.

Although the equivalent circuit shown is for a single-phase circuit, equation (4.11) applies also for a 3-phase system, where

$$\omega L_c = X_c = \omega\left(2 \ln \frac{s}{r} + \frac{1}{2}\right) \times 10^7\ \Omega/\text{m}$$

$$X_s = \omega M = \omega \left(2 \ln \frac{s}{r_s}\right) \times 10^{-7}\ \Omega/\text{m} \tag{4.12}$$

where s is the spacing between the conductor and sheath centres (m).

Crossbonding

The high losses occurring with sheaths short-circuited have resulted in the widespread use of a system of sheath transposition known as *crossbonding* which is illustrated in Figure 4.3. At each joint along the route the sheath is interrupted and transposed, and after three lengths have been traversed the sheaths are bonded to ground as shown. The effect is to add the sheath voltages in the adjacent sections and for balanced loads the sum will be zero. High sheath currents are avoided and induced voltages kept in reasonable proportions. High power circuits will be crossbonded or transposed, and if the sheaths are bonded at every third transposition for single-core cables in horizontal formation, the ratio of losses is given by:[13]

$$\left(\frac{\text{Circulating-current loss in sheath}}{\text{Loss in conductor}}\right) = \frac{R_s}{R_{ac}} \frac{1}{1 + (R_s/X_s)^2}. \tag{4.13}$$

This is a loss caused by circulating currents in the sheath, and eddy-currents are neglected. When the eddy-current loss cannot be ignored, i.e. for cables bonded at a single point or crossbonded, the ratio:[13]

$$\left(\frac{\text{Eddy-loss in sheath}}{\text{Loss in conductor}}\right) = A_1 \frac{R_s}{R_{ac}} \frac{(d/2s)^2}{(R_s \times 10^9/\omega) + \frac{1}{5}(2s/d)} \left[1 + A_2\left(\frac{d}{2s}\right)^2\right]. \tag{4.14}$$

Values of A_1 and A_2 are as follows.[13] Trefoil formation: A_1 3.0, A_2 0.417. Horizontal formation; $A_1 6$, A_2 0.083 (centre cable) and $A_1 1.5$, A_2 0.27 (outer cable). For large conductors constructed of insulated segmental subconductors the above eddy-loss formula is modified by multiplication by a factor F given by:[13]

$$F = \frac{4M^2N^2 + (M+N)^2}{4(M^2+1)(N^2+1)}$$

where

$$M = N = \frac{R_s}{X} \text{ for trefoil formation}$$

and

$$M = \frac{R_s}{X + X_m} \qquad N = \frac{R_s}{X - X_m/3} \text{ for flat formation cables.}$$

X is the sheath reactance with trefoil formation and X_m is the mutual reactance between the sheath of one cable and the conductors of the others when in horizontal formation.

Sequence Impedances

The zero-sequence impedance Z_0 depends upon the cable environment and is often difficult to determine precisely. Several situations are possible for the return current in an unbalanced current-flow situation:

(*a*) Return through the sheath alone.
(*b*) Return through the ground alone.
(*c*) Return through the ground and sheath in parallel.
(*d*) Return through the ground and sheath of adjacent cables.

The presence of railway lines, pipe lines, and other buried equipment will also influence the situation.

The self impedance (Ω per metre) of a cylindrical conductor near the ground surface with a return *through ground* is given by

$$R_{ac} + 1.00 \times 10^{-6} f + j\,2.89 \times 10^{-6} f \log_{10}\left(658.37 \frac{\sqrt{(\rho_s/f)}}{r_c}\right) \tag{4.15}$$

where

ρ_s = electrical resistivity of the soil (Ω-m)
f = frequency (Hz)
r_c = geometric mean radius of the conductor (m).

The quantity, $658.37\sqrt{(\rho_s/f)}$ (m) is considered as the effective depth D_e of the current in the return path in the ground. An expression for the self impedance in a commonly used mixture of units is:

$$R_{ac} + 0.00159f + 0.004657f \log_{10}\left(2160 \frac{\sqrt{(\rho_s/f)}}{r_c}\right) \ \Omega/\text{mile}$$

where R_{ac} is in Ω/mile, ρ_s is in Ω-m, r_c is in feet, and

$$D_e = 2160 \frac{\rho_s}{f} \text{ feet.}$$

The mutual impedance between conductor and sheath with ground return is given by:

$$1.0 \times 10^{-6} f + \mathrm{j}\, 2.89 \times 10^{-6} f \log_{10} \left(\frac{D_e}{r_s} \right) \ \Omega/\text{m} \tag{4.16}$$

where r_s is the geometric mean radius of the sheath in m. Alternatively, it is given by

$$0.0015 f + 0.004657 f \log_{10} \left(\frac{D_e}{r_s} \right) \Omega/\text{mile}$$

with D_e and r_s in feet.

For a cable current return through the sheath only, the impedance between conductor and sheath,

$$\mathbf{Z}_{cs} = R_{ac} + R_s + \mathrm{j}\, 5.78 \times 10^{-6} f \log_{10} \left(\frac{r_s}{\sqrt{(r_s r_c)}} \right) \ \Omega/\text{m} \tag{4.17}$$

$$= R_{ac} + R_s + \mathrm{j}\, 0.009314 f \log_{10} \left(\frac{r_s}{\sqrt{(r_s r_c)}} \right) \ \Omega/\text{mile}$$

where r_s, r_c are in feet and R_{ac}, R_s in Ω/mile. Note that $\mathbf{Z}_0$, the zero-sequence impedance, is $3 \times \mathbf{Z}_{cs}$.

The equivalent circuit for a cable with return current through its own sheath and the sheath of an adjacent cable and ground is shown in Figure 4.10.

The self impedance for an aluminium sheath at 50 °C,

$$\mathbf{Z}_{ss} = \frac{410}{(r_o + r_i)(r_o - r_i)} + \mathrm{j}\, 2.89 \times 10^{-6} f \log_{10} \left(\frac{D_e}{D_{cs}} \right) \ \Omega/\text{m}$$

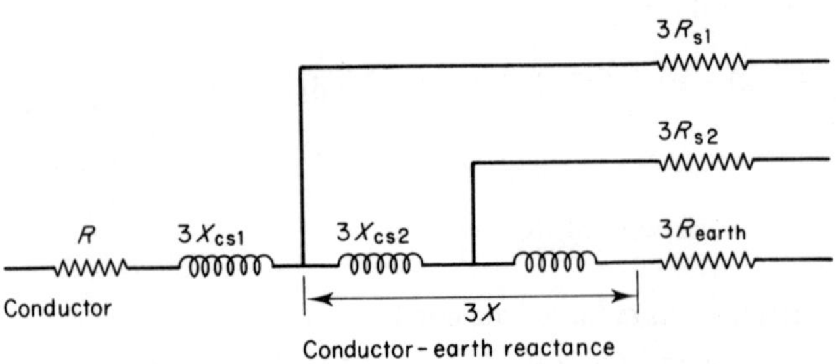

Figure 4.10. Zero sequence network of self-contained system with return current in two sheaths in parallel with the ground. $\mathbf{X}_{cs1}$ = Reactance between conductor (1) to sheath of cable (1). $\mathbf{X}_{cs2}$ = Reactance between conductor (1) to sheath of cable (2). $\mathbf{R}_{s1}$ = resistance of sheath of cable (1).

where

D_{cs} = geometric mean distance between conductor and sheath (m)
r_o = outer radius of sheath (m)
r_i = inner radius of sheath (m)

Sequence Parameters for Crossbonded Systems

The mutual impedances existing in the 6-conductor, 3-phase system, formed by the conductors and sheaths are designated $\mathbf{Z}_{c1s1}$, $\mathbf{Z}_{c1s2}$, $\mathbf{Z}_{c1s3}$, $\mathbf{Z}_{c2s1}$, etc, where

$$\mathbf{Z}_{c1s1} = 1.0 \times 10^{-6} f + j\,2.89 \times 10^{-6} f \log_{10} \frac{(D_e)}{(D_{c1s1})} \ \Omega/\text{m}$$

$$\mathbf{Z}_{c1s3} = 1.0 \times 10^{-6} f + j\,2.89 \times 10^{-6} f \log_{10} \frac{(D_e)}{(D_{c1s3})} \ \Omega/\text{m}$$

$$\mathbf{Z}_{c1s2} = 1.0 \times 10^{-6} f + j\,2.89 \times 10^{-6} f \log_{10} \frac{(D_e)}{(D_{c1s2})} \ \Omega/\text{m}$$

D_{c1s1} = Geometric mean distance (GMD) between conductor 1 and sheath 1 (m)
D_{c1s3} = GMD between conductor 1 and sheath 3 (m) (see Figure 4.11)
D_{c1s2} = GMD between conductor 1 and sheath 2 (m).

The matrix equation $\mathbf{V}_n = \mathbf{Z}_{nm}\mathbf{I}_m$ can be expanded generally in terms of conductor (suffix c) and sheath (suffix s) quantities as follows:

$$\begin{bmatrix} \mathbf{V}_c \\ \mathbf{V}_s \end{bmatrix} = \begin{bmatrix} \mathbf{Z}_{cs} & \mathbf{Z}_{cc} \\ \mathbf{Z}_{ss} & \mathbf{Z}_{sc} \end{bmatrix} \begin{bmatrix} \mathbf{I}_s \\ \mathbf{I}_c \end{bmatrix}.$$

Assuming three identical cables laid in flat formation (Figure 4.11) and at equal spacings, then from certain equalities due to symmetry the impedance matrix can be expanded for 3-phase cable systems as follows.[14] Let,

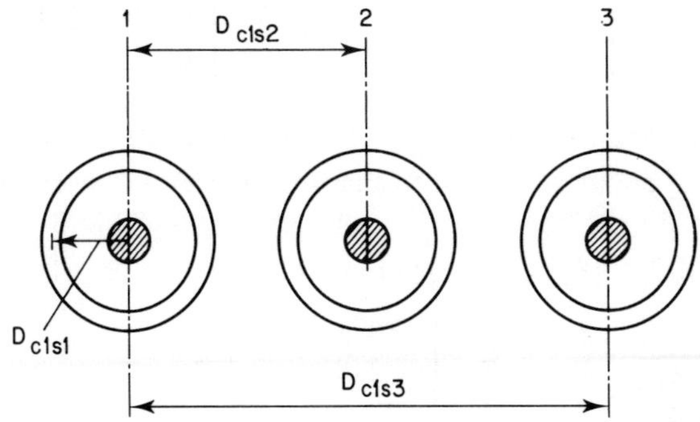

Figure 4.11. Self-contained cables in horizontal formation.

$$\mathbf{Z}_{c1s1} = \mathbf{Z}_{c2s2} = \mathbf{Z}_{c3s3} = \mathbf{a}$$

$$\mathbf{Z}_{c1c1} = \mathbf{Z}_{c2c2} = \mathbf{Z}_{c3c3} = \mathbf{d}$$

$$\mathbf{Z}_{s1s1} = \mathbf{Z}_{s2s2} = \mathbf{Z}_{s3s3} = \mathbf{e}$$

$$\mathbf{Z}_{c1s3} = \mathbf{Z}_{c3s1} = \mathbf{Z}_{c1c3} = \mathbf{b}$$

$$\mathbf{Z}_{c1s2} = \mathbf{Z}_{c3s2} = \mathbf{Z}_{c1c2} = \mathbf{c}$$

$$[\mathbf{Z}] = \left[\begin{array}{ccc|ccc} a & c & b & d & c & b \\ c & a & c & c & d & c \\ b & c & a & b & c & d \\ \hline e & c & b & a & c & b \\ c & e & c & c & a & c \\ b & c & e & b & c & a \end{array}\right] \tag{4.19}$$

Figure 4.3 illustrates the connections of a major section of a transposed, sheath crossbonded 3-phase system and equation (4.19) is modified by combining three minor sections to give the impedance matrix of a major section as follows:

$$[\mathbf{Z}] = \frac{1}{3}\left[\begin{array}{ccc|ccc} a+b+c & 2c+a & a+b+c & 3d & 2c+b & 2c+b \\ a+b+c & 2c+a & a+b+c & 2c+b & 3d & 2c+b \\ a+b+c & 2c+a & a+b+c & 2c+b & 2c+b & 3d \\ \hline 3e & 3c & 3b & a+b+c & a+b+c & a+b+c \\ 3c & 3e & 3c & 2c+a & 2c+a & 2c+a \\ 3b & 3c & 3e & a+b+c & a+b+c & a+b+c \end{array}\right]$$

The above matrix can be partitioned, giving:

$$\begin{bmatrix} \mathbf{V}_1 \\ \mathbf{V}_2 \end{bmatrix} = \left[\begin{array}{c|c} \mathbf{Z}_1 & \mathbf{Z}_2 \\ \hline \mathbf{Z}_3 & \mathbf{Z}_4 \end{array}\right] \begin{bmatrix} \mathbf{I}_1 \\ \mathbf{I}_2 \end{bmatrix}$$

where

$$[\mathbf{V}_1] = \begin{bmatrix} \mathbf{V}_{c1} \\ \mathbf{V}_{c2} \\ \mathbf{V}_{c3} \end{bmatrix} \quad [\mathbf{V}_2] = \begin{bmatrix} \mathbf{V}_{s1} \\ \mathbf{V}_{s2} \\ \mathbf{V}_{s3} \end{bmatrix} \quad [\mathbf{I}_1] = \begin{bmatrix} \mathbf{I}_{s1} \\ \mathbf{I}_{s2} \\ \mathbf{I}_{s3} \end{bmatrix} \quad [\mathbf{I}_2] = \begin{bmatrix} \mathbf{I}_{c1} \\ \mathbf{I}_{c2} \\ \mathbf{I}_{c3} \end{bmatrix}. \tag{4.20}$$

The phase impedance matrix $[\mathbf{Z}_p]$ can be derived for a major section. Noting that the bonding points are grounded and therefore $\mathbf{V}_2 = 0$, it can be shown that[15]

$$[\mathbf{Z}_p] = [\mathbf{Z}_1] - [\mathbf{Z}_2][\mathbf{Z}_4]^{-1}[\mathbf{Z}_3] \tag{4.21}$$

$$[\mathbf{Z}_p] = \begin{bmatrix} \mathbf{Z}_{aa} & \mathbf{Z}_{ab} & \mathbf{Z}_{ac} \\ \mathbf{Z}_{ba} & \mathbf{Z}_{bb} & \mathbf{Z}_{bc} \\ \mathbf{Z}_{ca} & \mathbf{Z}_{cb} & \mathbf{Z}_{cc} \end{bmatrix}$$

where

$\mathbf{Z}_{aa}, \mathbf{Z}_{bb}$ = phase self impedances
$\mathbf{Z}_{ab}, \mathbf{Z}_{ac}$ = phase mutual impedances.

Figure 4.12 gives the equivalent circuit for the three phases of a major section. $\mathbf{Z}_s$ represents the self impedance of the phases and $\mathbf{Z}_m$ the mutual impedance.

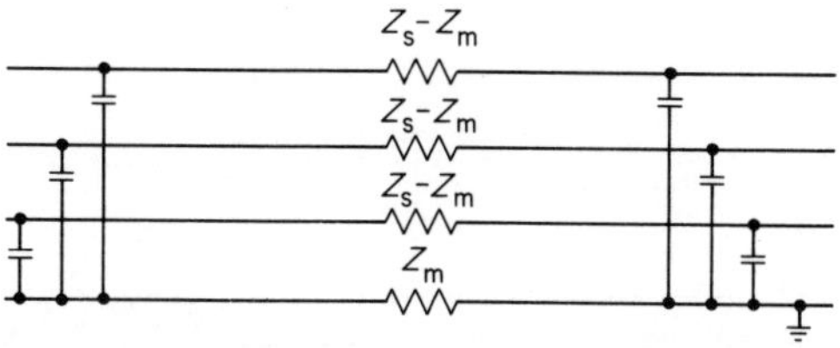

Figure 4.12. Equivalent circuit of section of cross-bonded cable system. $\mathbf{Z}_s - \mathbf{Z}_m = \mathbf{Z}_{11} = \mathbf{Z}_{22}$; $\mathbf{Z}_m = (\mathbf{Z}_{00} - \mathbf{Z}_{11})/3$.

In Figures 4.14 and 4.13 positive and zero-sequence parameters of British 400 kV cables[14] are summarized.

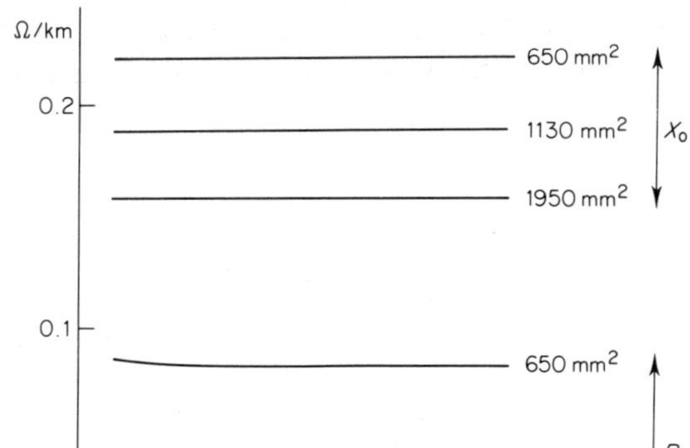

Figure 4.13. Zero sequence impedance of 400 kV LPOF cable system; crossbonded. X_1 positive sequence reactance; R_1 positive sequence resistance.[14] (Reproduced by permission of IEE.)

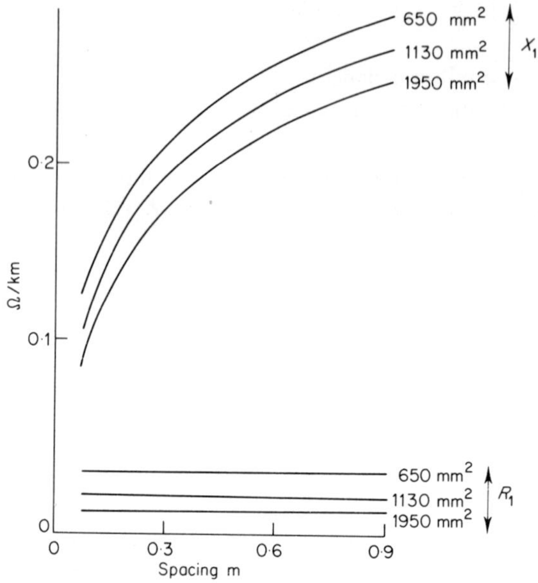

Figure 4.14. Positive sequence impedance – 400 kV LPOF system. Crossbonded or solidly bonded. Seamless aluminium sheaths.[14] (Reproduced by permission of IEE.)

Example 4.2

For a crossbonded system the voltages between bonding positions may be determined for a single-phase fault as follows. Assume that the load currents are negligible compared with the fault current, and that one-third of the fault current flows through each of the sheaths as shown in Figure 4.15. Assume cables to be in trefoil formation.

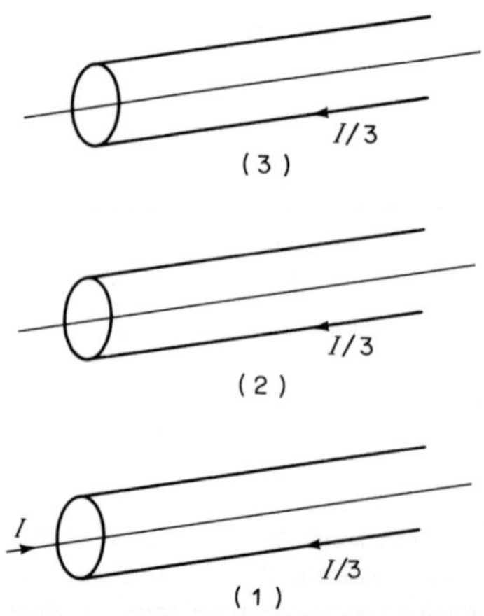

Figure 4.15. Currents in cable sheaths for Example 4.2.

The voltage drop in the sheath of cable (1) is

$$-\frac{I}{3}R_s + j\omega 2\frac{I}{3}\ln\left(\frac{s}{d}\right) + 2\,j\omega\frac{I}{3}\ln\left(\frac{s}{r_s}\right) \text{ V/m.}$$

Similarly in sheaths (2) and (3)

$$\text{Voltage drop} = -\frac{I}{3}R_s - 2\,j\omega\frac{I}{3}\ln\left(\frac{s}{r_s}\right) \text{ V/m}$$

where $I = 13\,000$ A, $R_s = 6.85 \times 10^{-4}$ Ω/m, B = spacing = 6.4 cm, and $r_s = 2.4$ cm.

For a length of 500 m between bonding positions the voltage drop in phase (1) is $-1350 + j240$ V, and the voltage drop in phase (2) is $-1350 - j120$ V. Hence

Voltage across bonds = j360 V.

4.6 ELECTRICAL PARAMETERS OF PIPE-TYPE OIL-FILLED CABLES

The following empirical expression holds for the conductor a.c. resistance:

$$R_{ac} = R_{dc}\,[1 + 1.7\,(y_s + y_p)]. \tag{4.22}$$

For the losses in the dielectric outer screens the self-contained cable relationship is modified as follows:[13]

$$\frac{R_s}{R_{ac}}\left(\frac{1.7}{1 + (R_s/X)^2}\right) \tag{4.23}$$

where X is the sheath reactance. The loss in the steel pipe may be determined analytically[16] but often empirical formulae are used. *It must be stressed that the empirical formulae given*[13] *apply only to 60 Hz and to pipe sizes in use at the time of writing and should not be extrapolated without further experimentation to larger sizes.*

For the triangular touching arrangement of cables:

$$\frac{\text{Loss in pipe}}{\text{Loss in conductor}} = \left(\frac{0.0115s - 0.001485\,d_p}{R_{ac}}\right) \times 10^{-6}. \tag{4.24}$$

For the open or cradled arrangement:

$$\frac{\text{Loss in pipe}}{\text{Loss in conductor}} = \left(\frac{0.00438s + 0.00226\,d_p}{R_{ac}}\right) \times 10^{-6} \tag{4.25}$$

where d_p = internal diameter of pipe (cm), s = axial spacing of adjacent conductors (cm), and R_{ac} = conductor a.c. resistance (Ω/cm).

The change of the effective loss with conductor size is illustrated in Table 4.1.

The inductance of pipe-type cables is determined as for self-contained separate cables in triangular or trefoil formation. However, the reactance is increased by 15%[13] to account for the presence of the steel pipe.

Table 4.1 Loss ratios for copper segmental conductors in steel pipe–cradle formation at 70 °C

Conductor size				$\left(\frac{I^2R_{\text{total}}}{I^2R_{\text{dc}}}\right)$		
cir. mil	mm²	Insulation wall thickness (mm)	Diam. pipe D_p (mm)	At conductor	At shield	At pipe
10^6	510	7.25	128	1.05	1.07	1.19
	510	12.8	178	1.04	1.06	1.23
1.5×10^6	760	7.25	155	1.1	1.14	1.34
	760	12.8	206	1.09	1.12	1.40
2×10^6	1020	7.25	155	1.19	1.23	1.53
	1020	12.8	206	1.16	1.20	1.58
2.5×10^6	1270	7.25	184	1.28	1.33	1.75
	1270	12.8	206	1.25	1.29	1.78

Zero-sequence Impedance

The determination of the zero-sequence impedance of pipe-type cables is more complex than that for self-contained cables because of the electromagnetic effects in the steel pipe. This compounds the normal problems of possible current return in the ground, i.e. whether or not the soil is homogeneous, and the presence of other conducting objects. The out-of-balance current may flow in the individual cable screens and skid wires (if metallic) in the pipe, and in the soil. As the magnetic permeability of the pipe material varies with magnitude of current (and hence with time), even the assumption of zero current return in the ground does not make the calculation precise. In view of this, although theoretical approaches are of value, *experimental verification* on the actual system is essential.

Assuming that the system parameters are known, the following circuit analysis[17] applies for the case of current return through the pipe and screens only. The suffices c, s, and p refer to the conductor, screen, and pipe respectively and cs, cp, and sp to the mutual coupling impedances. From Figure 4.16,

$$\mathbf{I}_c = \mathbf{I}_s + \mathbf{I}_p$$

$$0 = -\mathbf{I}_s\mathbf{Z}_s - \mathbf{I}_p\mathbf{Z}_{sp} - \mathbf{I}_c\mathbf{Z}_{cp} + \mathbf{I}_c\mathbf{Z}_{cs} + \mathbf{I}_p\mathbf{Z}_p + \mathbf{I}_s\mathbf{Z}_{sp}$$

$$= \mathbf{I}_c(\mathbf{Z}_{cs} - \mathbf{Z}_{cp}) - \mathbf{I}_s(\mathbf{Z}_s - \mathbf{Z}_{sp}) + \mathbf{I}_p(\mathbf{Z}_p - \mathbf{Z}_{sp})$$

$$\mathbf{V}_{cp} = \mathbf{I}_c\mathbf{Z}_c - \mathbf{I}_s\mathbf{Z}_{cs} - \mathbf{I}_p\mathbf{Z}_{cp} + \mathbf{I}_p\mathbf{Z}_p + \mathbf{I}_s\mathbf{Z}_{sp} - \mathbf{I}_c\mathbf{Z}_{cp}$$

$$= \mathbf{I}_c(\mathbf{Z}_c - \mathbf{Z}_{cp}) - \mathbf{I}_s(\mathbf{Z}_{cs} - \mathbf{Z}_{sp}) + \mathbf{I}_p(\mathbf{Z}_p - \mathbf{Z}_{cp})$$

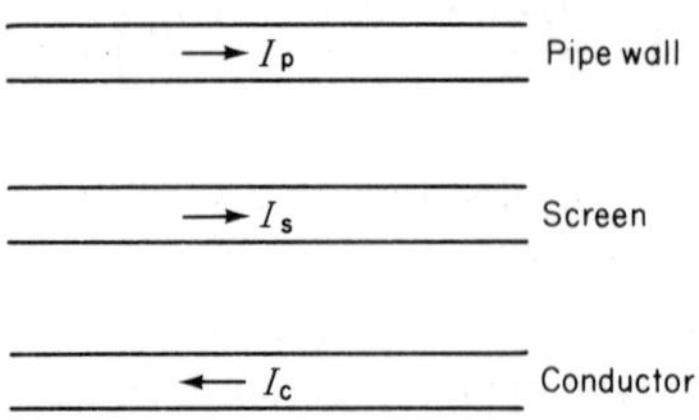

Figure 4.16. Pipe-type cable. Currents flowing.

Hence,

$$\mathbf{I}_c = \mathbf{I}_p \left[\frac{(\mathbf{Z}_s - \mathbf{Z}_{sp}) + (\mathbf{Z}_p - \mathbf{Z}_{sp})}{(\mathbf{Z}_s - \mathbf{Z}_{sp}) - (\mathbf{Z}_{cs} - \mathbf{Z}_{cp})} \right]. \tag{4.26}$$

and

$$\mathbf{V}_{cp} = \mathbf{I}_c \left[(\mathbf{Z}_c - \mathbf{Z}_{cp}) + (\mathbf{Z}_p - \mathbf{Z}_{cp}) - \frac{\{(\mathbf{Z}_{cs} - \mathbf{Z}_{cp}) + (\mathbf{Z}_p - \mathbf{Z}_{cp})\}}{(\mathbf{Z}_s - \mathbf{Z}_{sp}) + (\mathbf{Z}_p - \mathbf{Z}_{sp})} \times \{(\mathbf{Z}_{cs} - \mathbf{Z}_{sp}) + (\mathbf{Z}_p - \mathbf{Z}_{cp})\} \right]. \tag{4.27}$$

Treatment of Pipe Saturation with Unbalanced Fault Current[17]

Consider the wall of a pipe of thickness t_p and of internal diameter D_p. It is assumed that the relative permeability μ_r is constant over the pipe section and that all of the out of balance or return current flows in the pipe. The current density $\mathbf{J}_x$ at a radius x from the centre of the pipe is given by

$$\mathbf{J}_x = \mathbf{J}_0\, e^{-cx}\, e^{-jcx}$$

where $\mathbf{J}_0$ is the current density on the inner surface of the pipe, and

$$c = (\text{skin depth})^{-1} = \frac{1}{(2/\omega\sigma\mu_0\mu_r)^{1/2}} = \sqrt{\left(\frac{\pi f \mu_0 \mu_r}{\rho}\right)}.$$

e^{-cx} represents an attenuation factor whilst cx represents a phase shift in radians.

Consider an annular area of thickness dx. The current in the pipe wall is

$$\mathbf{I}_p = \pi D_p \int_0^{t_p} \mathbf{J}_x\, dx$$

$$= \pi D_p \int_0^{t_p} \mathbf{J}_0\, e^{-cx}\, e^{-jcx}\, dx$$

$$= \pi D_p\, \mathbf{J}_0 \frac{(1 - j)}{2c} [1 - e^{ct_p(1+j)}]. \tag{4.28}$$

The voltage drop along the pipe per unit length

$$= \rho\, \mathbf{J}_0 [1 - e^{-ct_p(1+j)}] \text{ V/m}. \tag{4.29}$$

The impedance of the pipe return path per metre length,

$$\mathbf{Z}_p = \frac{\mathbf{V}_p}{\mathbf{I}_p} = \rho c \frac{(1 + j)}{\pi D_p}\ \Omega \text{ per m}$$

$$= (\rho f \mu_0 \pi)^{1/2} \frac{1 + j}{\pi D_p} \mu_r^{1/2}. \tag{4.30}$$

However, μ_r will vary over the cycle because of the slope of the B–H curve for the pipe steel.

4.7 TRANSIENT RESPONSE TO UNBALANCED FAULTS

Pipe-type HPOF Cables

Simplified model approach

The variation of the zero-sequence impedance $(\mathbf{R}_0 + j\,\mathbf{X}_0)$ with current for a pipe-type cable is shown in Figure 4.17.[18] The models used for transmission lines for both steady-state and transient analysis are usually linear, although frequency-dependent parameters can be included. The nonlinearities indicated in Figure 4.17 would be very difficult to include in a model for analysis and it would be of great utility to use simple linear models with constant parameters corresponding to:

(*a*) low-current unsaturated values of pipe resistance and inductance,
(*b*) high-current (saturated values) of R and L.

The general equations for a system of $(m+1)$ conductors are as follows, the $(m+1)$th conductor being regarded as reference,

$$\frac{\partial \boldsymbol{\psi}(t,x)}{\partial t} = \frac{\partial \mathbf{V}(t,x)}{\partial x} - \mathbf{R}\mathbf{I}(t,x)$$

$$\frac{\partial \mathbf{Q}(t,x)}{\partial t} = \frac{\partial \mathbf{I}(t,x)}{\partial x} - \mathbf{G}\mathbf{V}(t,x)$$

$$\mathbf{I} = \mathbf{L}^{-1}\boldsymbol{\psi} \quad \text{and} \quad \mathbf{V} = \mathbf{C}^{-1}\mathbf{Q}$$

where t is time, x is the distance along the cable, and

$\boldsymbol{\psi}$ is an mth order vector of flux linkages per unit length
$\mathbf{Q}$ is an mth order vector of charges per unit length
$\mathbf{I}$ is an mth order vector of conductor currents
$\mathbf{V}$ is an mth order vector of conductor voltages with respect to the reference conductor.

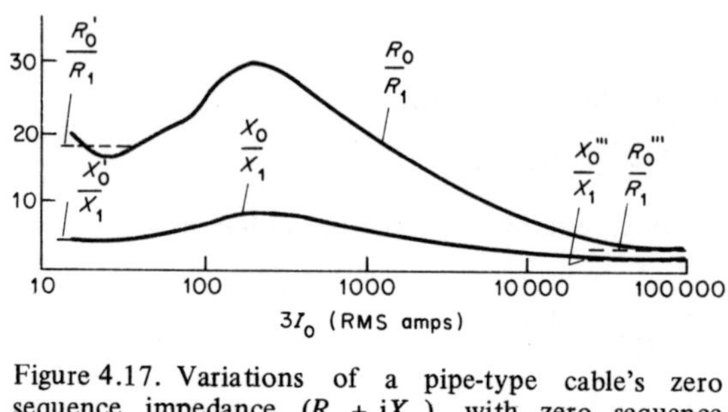

Figure 4.17. Variations of a pipe-type cable's zero sequence impedance $(R_0 + jX_0)$, with zero sequence current, $\mathbf{I}_0$, at 60 Hz. R_1 and X_1 are the cable's positive sequence resistance and reactance and are essentially constant. R_0' and X_0' are the unsaturated values of R_0 and X_0. R_0''' and X_0''' are saturated values of R_0 and X_0.[18] (Reproduced by permission of IEEE.)

R, **L**, **C**, and **G** are $m \times m$ matrices of resistance, inductance, capacitance, and shunt conductance per unit length, which can vary with position and time.

For purposes of computation the above are approximately discretized (cables divided into discrete longitudinal sections) as follows for values of n from 1 to N.

$$\left.\begin{aligned} \frac{\partial \boldsymbol{\psi}_n(t)}{\partial t} &= \frac{\mathbf{V}_n(t) - \mathbf{V}_{n-1}(t)}{\Delta x} - \mathbf{R}_n \mathbf{I}_n(t) \\ \frac{\partial \mathbf{Q}_n(t)}{\partial t} &= \frac{\mathbf{I}_n(t) - \mathbf{I}_{n-1}(t)}{\Delta x} - \mathbf{G}_n \mathbf{V}_n(t) \end{aligned}\right\} \qquad (4.31)$$

$$\mathbf{I}_n = \mathbf{L}_n^{-1} \psi_n \quad \text{and} \quad \mathbf{V}_n = \mathbf{C}_n^{-1} \mathbf{Q}_n$$

where, ψ_n, etc., are the values of ψ, etc., at the point x_n on the cable.

In Reference 18 the results of tests are described using these equations with nonlinear R and L and also with the assumptions (*a*) and (*b*) above. These relate to a 345 kV, 60 Hz, 1270 mm^2 cable with a 250 mm diameter steel pipe. Various fault-current waveforms and the highest unfaulted phase voltages are given in Reference 18. The conclusions from these tests are as follows:

(*a*) The transient voltages are insensitive to pipe saturation.
(*b*) The peak values of the fault currents, but not the current waveshapes, are sensitive to pipe saturation.
(*c*) It is suggested that discrete values of zero-sequence impedance be used corresponding to low currents (unsaturated), intermediate currents, and high currents (saturated). This will ease computation compared with the use of the nonlinear characteristics.

Crossbonded Systems – Self-contained Cables

The equivalent circuit of one section of the system with common ground reference is shown in Figure 4.18. The ground return path is assumed to be concentrated in a

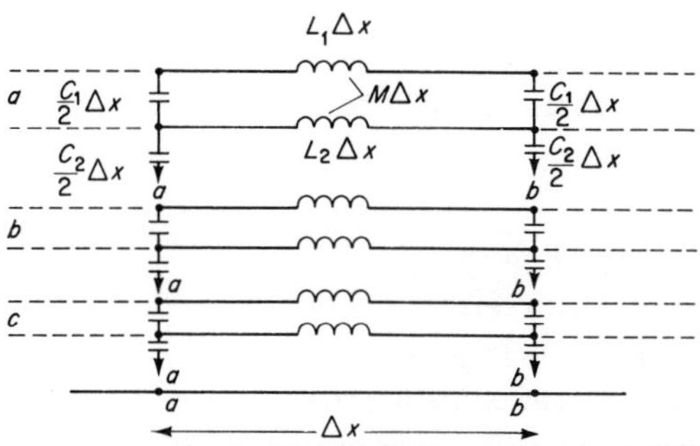

Figure 4.18. Equivalent circuit of symmetrical 3-phase self-contained cable system. Common ground reference.

thin shell at the outer surface of the cable. The following equations apply per phase:

$$-\frac{\partial e_1}{\partial x} = L_1 \frac{\partial i_1}{\partial t} + M \frac{\partial i_2}{\partial t}$$

$$-\frac{\partial e_2}{\partial x} = M \frac{\partial i_1}{\partial t} + L_2 \frac{\partial i_2}{\partial t}$$

and

$$-\frac{\partial i_1}{\partial x} = C_1 \frac{\partial e_1}{\partial t} - C_1 \frac{\partial e_2}{\partial t}$$

$$-\frac{\partial e_2}{\partial t} = C_1 \frac{\partial e_1}{\partial t} + (C_1 + C_2) \frac{\partial e_2}{\partial t}$$

where

M = Mutual inductance between sheath and conductor (H/m)
= L_2 for single-core cables (i.e. self inductance of sheath with respect to ground).

Suffix (1) refers to the conductor; suffix (2) refers to the sheath.

Solving the above equations and using the surge impedance concept[19]

$$[e] = \begin{bmatrix} \left(\frac{L_1 - L_2}{C_1}\right)^{1/2} \left(\frac{L_2}{C_2}\right)^{1/2} & \left(\frac{L_2}{C_2}\right)^{1/2} \\ \left(\frac{L_2}{C_2}\right)^{1/2} & \left(\frac{L_2}{C_2}\right)^{1/2} \end{bmatrix} [i] \tag{4.32}$$

where

$\left(\frac{L_1 - L_2}{C_1}\right)^{1/2}$ = characteristic or surge impedance between conductor and sheath

and

$\left(\frac{L_2}{C_2}\right)^{1/2}$ = characteristic impedance between sheath and ground.

Let

r = radius of conductor (m)
r_{2i} = inner radius of sheath (m)
r_{2o} = outer radius of sheath (m)
r_3 = outer radius of sheath insulating jacket (m).

Then

$$(L_1 - L_2) = 2 \times 10^{-7} \ln\left(\frac{r_{2i}}{r}\right) \text{ H/m}$$

and

$$L_2 = 2 \times 10^{-7} \ln\left(\frac{r_3}{r_{2o}}\right) \text{ H/m}$$

$$C_1 \text{(conductor/sheath capacitance)} = \frac{2\epsilon_0 \epsilon_1}{\ln(r_{2i}/r)} \text{ F/m}$$

$$C_2 \text{(sheath/ground capacitance)} = \frac{2\epsilon_0 \epsilon_1}{\ln(r_3/r_{2o})} \text{ F/m.}$$

A wave travelling along a conductor of surge impedance $\mathbf{Z}_0$ is reflected and refracted when it reaches a discontinuity such as a connection to a network of impedance $\mathbf{Z}$. It can be shown that,[15]

$$2\mathbf{e} - \mathbf{E} = \mathbf{Z}_0\mathbf{I} \quad \text{and} \quad \mathbf{E} = \mathbf{ZI}$$

where $\mathbf{e}$ is the amplitude of the incident wave and $\mathbf{E}$ and $\mathbf{I}$ the voltage and current of the refracted wave. A cable with ground return is a three-conductor system which can be represented by the equivalent circuit[20] shown in Figure 4.19. This assumes that the sheath acts as a perfect screen between conductor and ground. In Figure 4.19 the right-hand impedances refer to the connected or load part of the system. All the voltages are referred to conductor (0), i.e. the sheath. Travelling incident waves exist both between the conductor and sheath ($\mathbf{e}_1$), and between sheath and ground ($\mathbf{e}_2$). Reflected waves $\mathbf{e}_1' = \mathbf{E}_1 - \mathbf{e}_1$ and $\mathbf{e}_2' = \mathbf{E}_2 - \mathbf{e}_2$, where $\mathbf{E}_1$ and $\mathbf{E}_2$ are the refracted waves into the connected network.

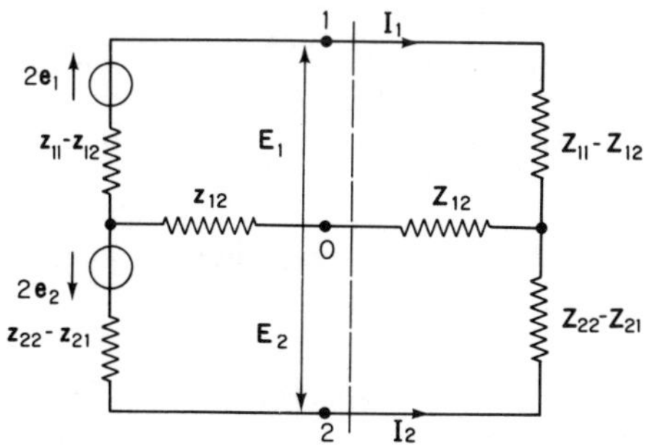

Figure 4.19. Equivalent circuit of connection between self-contained cable and connected network (right-hand side of diagram).[20] (Reproduced by permission of IEEE.)

Examples of sheath discontinuities in non-crossbonded cables are shown in Figure 4.20, illustrating combinations of sheath bonding. In this, $\mathbf{Z}_1$ is the characteristic impedance from conductor to sheath, and $\mathbf{Z}_2$ the characteristic impedance from ground to sheath.

With crossbonded cables the other phase must be included,[20] as indicated in

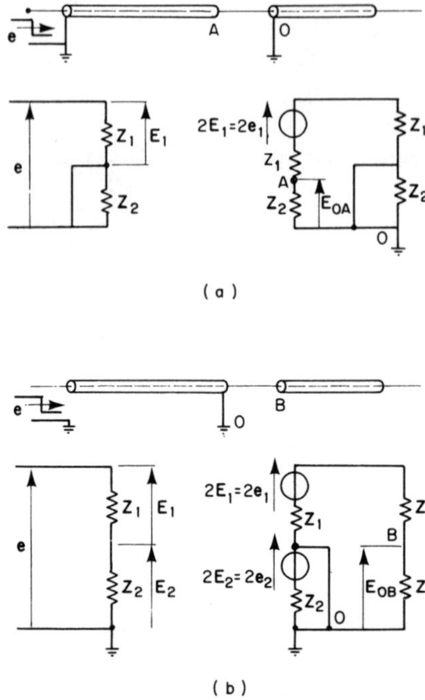

Figure 20. Equivalent circuit of sheath discontinuities.[20]

(a) $$E_{OA} = \frac{2Z_2}{2Z_1 + Z_2}\, e$$

(b) $$E_{OB} = \frac{2Z_1 Z_2}{(Z_1 + Z_2)(2Z_1 + Z_2)}\, e$$

(Reproduced by permission of IEEE).

Figure 4.21. The voltage $\mathbf{E}_{HG}$, $(\mathbf{E}_H - \mathbf{E}_G)$ across a joint sleeve insulation for an incident wave $(\mathbf{e}_A)$ (conductor–sheath) is:

$$\mathbf{E}_{HG} = \frac{2\mathbf{Z}_2}{2\mathbf{Z}_1 + 3\mathbf{Z}_2/2}\, \mathbf{e}_A \text{ (first reflection only).}$$

This ignores the impedance of the bonding leads, which will have an influence for steep-fronted waves. The voltage is of course for the first reflection only and for subsequent reflections the calculations become complex and either a digital computer[20,26] or transient network analyser[26] needs to be used.

Results of a computer study[20] are shown in Figure 4.22 for a 275 kV LPOF circuit about 6000 m in length. The voltage $\mathbf{E}_a$ occurs across the insulated portion of the sheath for the first sheath break on phase (*a*) (i.e. first crossbonded position) for a step surge entering the conductor of phase (*a*) from an overhead line. The ratio of surge impedance, conductor to sheath $(\mathbf{Z}_1/\mathbf{Z}_2)$, is two. $\mathbf{Z}_2$ will have a

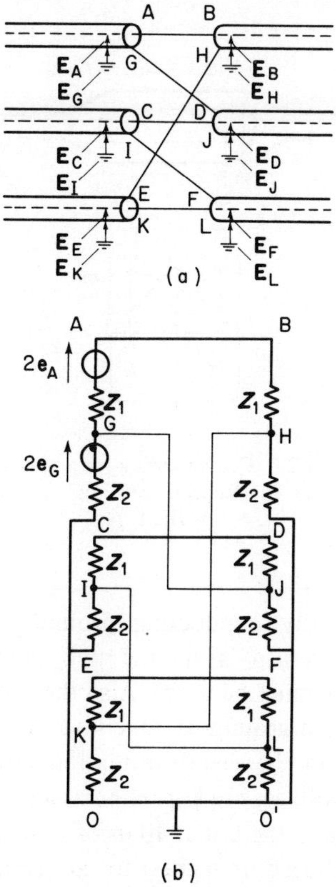

Figure 4.21. Single crossbonding connection: (a) physical arrangement; (b) equivalent circuit.

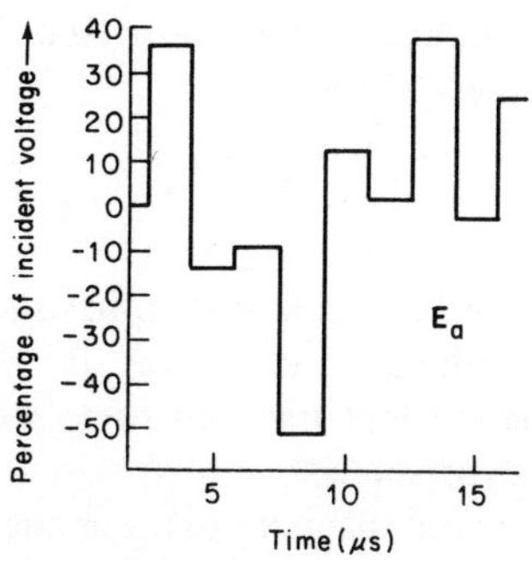

Figure 4.22. Voltage across joint-sleeve insulation for crossbonded section resulting from surge from overhead line on conductor (a).[20] (Reproduced by permission of IEEE.)

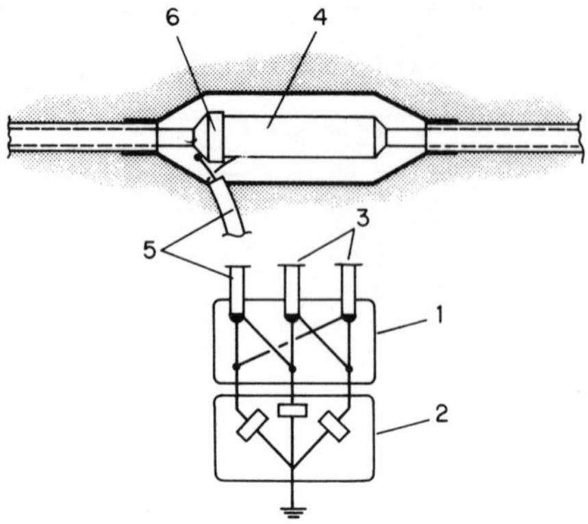

Figure 4.23. Crossbonding arrangements in joint bay.[20] (Reproduced by permission of IEEE.)

minimum value for perfectly conducting ground and a maximum for nonconducting ground (when return is via the two other sheaths and the mutual impedance between sheaths must be used). Also the ratio of propagation velocities, conductor/sheath and sheath/ground is taken to be unity. These velocities are proportional to $\epsilon_r^{-1/2}$; ϵ_r for the sheath insulation varies from 2 to 6 and for the ground from 4 to 80. The voltages are expressed as a percentage of the amplitude of the entering incident wave and the initial 40 μs of time is shown.

Generally the sheath voltages are highest for nonconducting ground and increase with the velocity ratio. The physical layout of a joint bay is shown in Figure 4.23, the bonding leads being in the order of 10 m in length. To reduce transient sheath voltages set up by faults, surge diverters are connected as shown. These comprise discs of nonlinear resistance material connected in 'Y' formation without series spark gaps. The maximum voltage across the sheath insulation is then determined by the disc, typically 7 kV for a surge current of 10 kA (8/20 μs wave) for a disc of diameter 150 mm and thickness 25 mm.

Example 4.3

A 525 kV self-contained oil-filled cable system has sheath discontinuities and the sheaths are single-point grounded as shown in Figure 4.20. Determine the proportion of the incoming incident transient surge set up across the break in the sheath. The cables are not crossbonded.

Relevant cable parameters are: diameter over conductor, 40 mm; over insulation, 103 mm; over sheath, 114.4 mm; over sheath insulation (PVC), 122 mm. ϵ_r (dielectric) = 3.5, ϵ_r (sheath insulation) = 4. Neglect resistance.

Solution

$$(L_1 - L_2) = 2 \times 10^{-7} \ln\left(\frac{r_{2i}}{r}\right) \text{ H/m}$$

$$= 2 \times 10^{-7} \ln\left(\frac{51.5}{20}\right) = 1.89 \times 10^{-7} \text{ H/m}$$

$$L_2 = 2 \times 10^{-7} \ln\left(\frac{r_3}{r_{2o}}\right) = 2 \times 10^{-7} \ln\left(\frac{61}{57.2}\right) = 0.129 \times 10^{-7} \text{ H/m}$$

$$C_1 = \frac{2\pi\epsilon_1 \times 10^{-12} \times 8.85}{\ln(r_{2i}/r)} = \frac{3.5 \times 10^{-7}}{17.98 \ln(51.5/20)} = 0.206 \times 10^{-9} \text{ F/m}$$

$$C_2 = \frac{2\pi\epsilon_2 \times 10^{-12} \times 8.85}{17.98 \ln(r_3/r_{2o})} = \frac{4 \times 10^{-7}}{17.98 \ln(61/57.2)} = 3.46 \times 10^{-9} \text{ F/m.}$$

The equivalent circuit is shown in Figure 4.20a.

$\mathbf{Z}_{01}$ = Characteristic impedance conductor to sheath

$$= \sqrt{\left(\frac{\mathbf{L}_1 - \mathbf{L}_2}{\mathbf{C}_1}\right)} = \sqrt{\left(\frac{1.89 \times 10^{-7}}{0.206 \times 10^{-9}}\right)} = 30.3$$

$\mathbf{Z}_{02}$ = Characteristic impedance between sheath and ground

$$= \sqrt{\left(\frac{\mathbf{L}_2}{\mathbf{C}_2}\right)} = \sqrt{\left(\frac{0.1286 \times 10^{-7}}{3.46 \times 10^{-9}}\right)} = 1.93.$$

From Figure 4.20a

$$\mathbf{E}_1 = \frac{30.3}{32.23}\mathbf{e} = 0.94\mathbf{e}$$

$\mathbf{E}_{OA}$ = voltage across interrupted sheath

$$= 2 \times 1.93\left(\frac{2 \times 0.94\mathbf{e}}{64.46}\right) = 0.11\text{e}.$$

This assumes ground of zero electrical resistivity and zero lead resistance.

4.8 JOINTS AND TERMINATIONS

The electric design of a coaxial cable is relatively simple. When a discontinuity occurs due to a joint between two sections of cable, or at the end of the cable where the conductor enters the air, the electric field is no longer uniformly radial but has longitudinal and radial components. This makes the design of the insulation more complex. The determination of the electric stress distribution in two or three dimensions may be accomplished by analogue techniques, e.g. electrolytic tanks or carbon-based semiconducting paper, or by digital computer methods involving finite difference or finite element methods. These methods are well documented.[21]

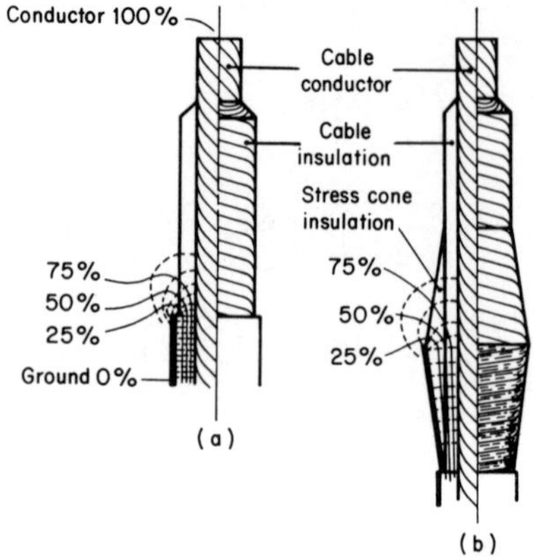

Figure 4.24. Equipotentials between conductor and sheath in termination: (a) no additional insulation; (b) stress cone added.

The nature of the termination or brushing problem is shown in Figure 4.24. Without any measures to modify the electric field the disposition of the flux lines and equipotentials is as shown in Figure 4.24a. High stresses are seen to exist in the longitudinal direction with consequent possiblity of breakdown. By building up the taped insulation to form a stress-relief cone the field is modified to the form shown in Figure 4.24b, with much reduced stress longitudinally.

Terminations for Paper–Oil Cables

The main requirements are as follows:

(1) Uniform voltage distribution.
(2) Coordination of internal and external dielectric strength so that flashover, should it occur, is confined to outside the porcelain.
(3) Minimum diameter of cable assembly to keep the dimensions of the porcelain small in order to limit the hoop and other stresses developed by internal pressure.
(4) Easy assembly and impregnation on site.
(5) Exclusion of air and moisture from cable insulation.
(6) Establishment of a pressure seal to contain the cable fluid.

When the sheath or screen is cut back the stress is non-uniform along the exposed surface of the cable dielectric, the steepest gradient or maximum stress concentration occurring in the vicinity of the end of the screen. The dielectric strength of lapped paper when the electric field is directed along or parallel to the lamination is very low compared with the strength obtained when the field is normal to or through the laminations. In high-voltage oil-filled systems, it is common practice to control the cable-end stress by applying a pre-impregnated paper roll of insulation, adjacent to the end of the sheath. The paper roll is formed to a predetermined cone shape and checked by a template. For system voltages below 66 kV, a paper roll of a straight-line contour is adequate, but above this the profile of the roll is calculated from a logarithmic formula assuming data for the allowable radial and longitudinal stresses, the nominal value of the latter being usually limited to 2.5 kV/mm. In all cases a metal screen, usually consisting of lead wires, is applied from the cable sheath up and over the stress profile and is suitably secured by soldering. A straight portion of paper roll exists beyond the end of the wire. At the top of the stress profile, where the metal screen is terminated, there is still a sharp increase of stress in the vicinity of the top turn of the lead wire. For system voltages above 138 kV, therefore, the stress at this region must be further controlled in order to ensure that no internal flashover occurs during the impulse and a.c. voltage tests.

Oil-impregnated bushings are made by the lapping of paper tape with impregnation after vacuum drying. In them internal discharges occur at higher inception levels than in bushings comprised of synthetic resin-bonded paper and the latter are not used at the higher voltages. Three basic methods exist for stress control, and, although capacitance grading is predominantly used, control[22] can be accomplished by resistance grading and by the use of suitable configurations of electrodes, which are used to guide and modify the electric field.

Capacitance grading

There are two methods: conducting foils or equipotential layers are inserted in appropriate radial positions during the paper lapping process, or grading capacitors completely separate from the cable build up are used. Figure 4.25 shows diagrammatically the physical arrangement of the foil electrodes and the various capacitances involved in a capacitor bushing termination. Neglecting the capacitance of the foils to earth, the arrangement can be divided into series and shunt capacitances. For uniform voltage distribution the capacitance between any two successive foils are so graded that equal voltage differences result. Uniform voltage distribution may be obtained in three alternative ways in the capacitor bushing:

(*a*) By varying the insulation thickness between two successive foils while maintaining the projected steps of equal length.
(*b*) By adjusting the length of projected steps while keeping the insulation thickness between two successive foils constant.
(*c*) By arranging for equal projected steps and equal insulation thickness between two successive foils.

Design (*c*) is the simplest, although it may introduce somewhat higher stresses at the two ends compared with those at the intermediate steps. For a given cable size, however, by suitable choice of the foil dimensions, the potential distribution can be made sufficiently uniform for all practical purposes.

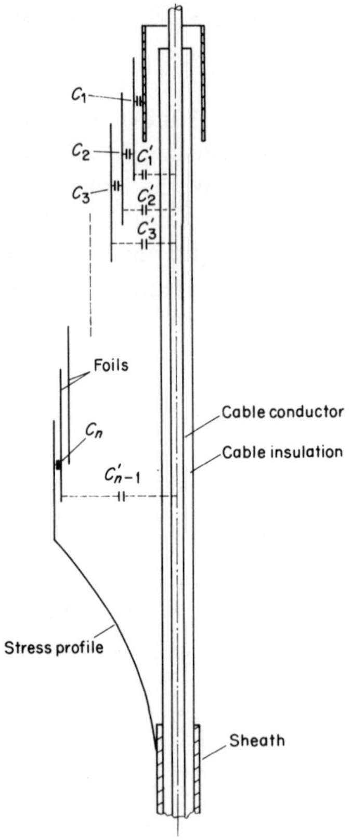

Figure 4.25. Capacitor type cable termination using foils and stepped insulation – schematic diagram.

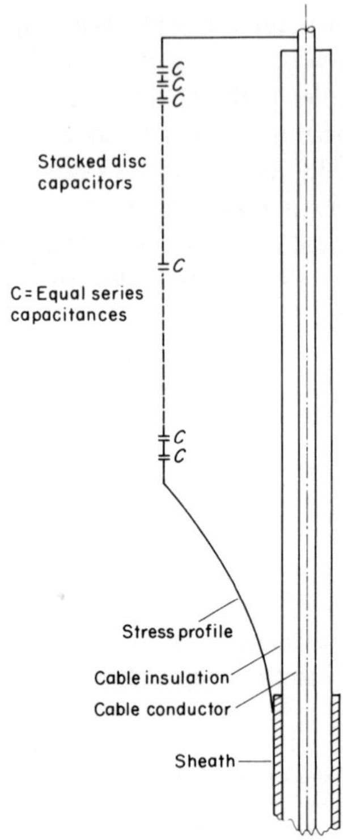

Figure 4.26. Capacitor type termination with grading by disc capacitors – schematic diagram.

All three types of design can either be prefabricated or formed by hand on site. A sealing end is shown in Figure 1.20 for a LPOF cable. The cone is fitted as tightly as possible over the top straight portion of the stress profile with the outermost foil connected to the lead wire on the profile. The advantage of using prefabricated capacitor cones is that it is possible to manufacture, inspect, and test them in the factory under controlled conditions. The cable termination can, moreover, be completed in a short time.

The system using separate capacitors is shown schematically in Figure 4.26. Disc capacitors are stacked on top of each other, producing smooth, longitudinal stress-grading. The electrical design is straightforward since the shunt capacitances between the capacitors and the conductor are negligible. All the series capacitance steps can be made equal, and as long as this is achieved the actual value of the capacitance chosen is not critical. The lowest capacitor electrode is connected to the lead wire of the stress profile, and the top capacitor is connected to the conductor. The series capacitors are stacked around a central tube assembled and

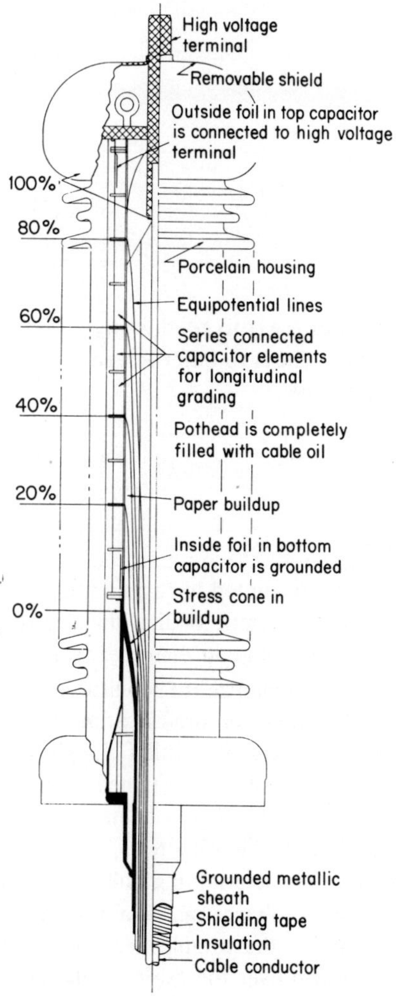

Figure 4.27. Capacitance graded pot-head (termination) for HPOF pipe-type cable core. Schematic diagram – equipotentials as percentage of conductor voltage shown. (From Ohio Brass Catalog 69; reproduced by permission of the Ohio Brass Company.)

impregnated in the factory and immersed in oil for shipment to the installation site. The external bushing surface can be fairly uniformly graded for power frequency and impulse voltages, the stresses being of the order of 2.8 kV/mm and 5.5 kV/mm respectively. A diagram of a pot-head used for pipe-type cables employing separate disc capacitors is shown in Figure 4.27, the voltage levels being indicated.

The maximum stress in capacitor-type (foil-graded) bushings is given by the relationship:

$$\frac{V}{2Er_c} = \frac{\alpha \ln \alpha}{\alpha + 1}$$

where

$$\alpha = \frac{l_1}{l_n}$$

and

l_1 = length of first capacitor (next to conductor)
l_n = length of last capacitor (next to outer screen)
V = conductor voltage ground
E = maximum stress
r_c = radius of conductor.

Joints for Taped Insulation

When the cable insulation is interrupted to form a splice or termination an increase in insulation radius is necessary because of the low electric strength of the tapes in the longitudinal direction. The change in diameter from cable to joint centre must be achieved in such a way as to minimize the longitudinal stress.[23,24]

The electrical design of a joint or modification to an existing design would normally be achieved with the aid of a full field plot giving the equipotentials assuming radial symmetry. This would be obtained by the use of a finite difference (or element) numerical method and a digital computer. The thermal design would be similarly determined, as described in Chapter 3.

An approximate method for determining the longitudinal stresses of critical positions assumes that the radial voltages can be obtained independently of the axial components of stress. Consider the arrangement in Figure 4.28, in which the hand-applied insulation is shown over the cable insulation which has been stepped

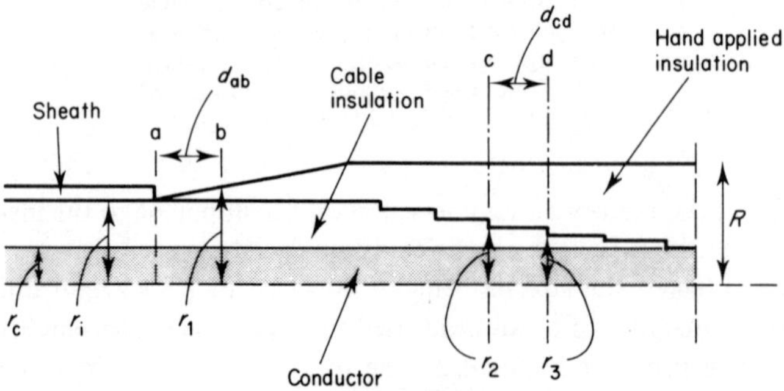

Figure 4.28. Profile of hand-applied insulation in a through (paper/oil) joint or splice.

down to the conductor. The various radii and axial dimensions are indicated in the figure. It is assumed that the relative permittivity of the hand-applied insulation is equal to that of the cable insulation. The longitudinal stress along the boundary of the insulating materials is obtained between, say, the positions a and b (Figure 4.28) by evaluating the voltages at a and b. At b the voltage across the layer of hand-applied insulation (V_1) is given by,

$$V \frac{C}{C_1 + C}$$

where V is the conductor to ground voltage, C is the capacitance per unit length of the cable insulation, where

$$C = \frac{2\pi\epsilon}{\ln(r_i/r_c)}$$

and C_1 is the capacitance of the hand-applied insulation,

$$C_1 = \frac{2\pi\epsilon}{\ln(r_1/r_i)} .$$

Hence,

$$V_1 = V\left[\frac{\ln(r_1/r_i)}{\ln(r_1/r_i) + \ln(r_i/r_c)}\right]. \tag{4.33}$$

As the voltage at a, the cable sheath, is zero the longitudinal stress is approximately $(V_1 - 0)/d_{ab}$. Similarly, the longitudinal stresses at other positions are determined, e.g. between c and d.

At c,

$$V_2 = \frac{V \ln(R/r_2)}{\ln(R/r_2) + \ln(r_2/r_c)}$$

at d,

$$V_3 = \frac{V \ln(R/r_3)}{\ln(R/r_3) + \ln(r_3/r_c)} .$$

Hence

$$E_{cd} = \frac{(V_2 - V_3)}{d_{cd}} .$$

The radial stresses are obtained by the usual formula,

$$E_x = \frac{V}{r_x \ln(R/r_c)} .$$

Although this method is obviously very approximate it gives very readily a rough measure of the electric stresses in the joint.

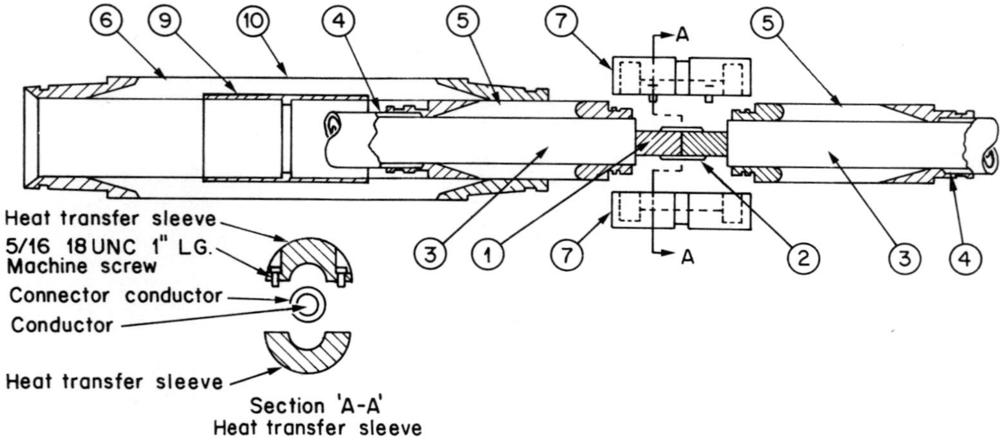

Figure 4.29. Moulded splice assembly for 138 kV extruded solid dielectric cable. 1. Cable conductor, 2. conductor connector, 3. cable insulation, 4. cable insulation shield, 5. cable adaptor, 6. splice sleeve housing, 7. heat transfer sleeve, 9. housing conductive insert, 10. semiconductive shield. (Reproduced by permission of IEEE.)

The cable dielectric may be terminated in steps or by a smooth taper. In pre-computing the profile the change in paper dimensions on being dried must be allowed for, and a shrinkage of 15% is ofen used from humid conditions to vacuum dryness. Cable-end stepping represents a discontinuity which is liable to produce ionization in the butted gaps on the arrival of surges. Under ionization it is likely that the ionization gap will take up a potential which is the mean value of those of the adjacent surfaces. Rough guidelines for design are to take the thickness of the hand-applied tapes as double that of the cable insulation and a maximum radial stress in the joint of two-thirds of that in the cable.

Capacitance grading in joints for oil-filled systems has been reported[25] in which the longitudinal stress is controlled along the joint by the use of printed circuit capacitors. The determination of stress along the joint is indicated in reference 25.

Cables with Extruded Insulation

One attraction of this type of insulation is the relative simplicity of splicing and terminations. Of great utility would be a factory-moulded and pre-tested splice for, say, 138 kV systems: a proposed[24] splice assembly is shown in Figure 4.29. The performance of hand-wrapped and field-mounted joints suffers from varying quality of workmanship and the time required for installation, typically 1 to 3 days per three-phase section.

4.9 FORCES

Short Circuit forces – Conductors in Trefoil

For balanced three-phase currents, the force between two parallel conductors is given by $2I_1I_2 \times 10^{-7}/d$ (N) per metre length where I_1 and I_2 are the currents and d the axial separation (m).

For three-phase currents the forces produced may be resolved as shown in Figure 4.30, from which

$$F_y = (F_{13} + F_{12}) \cos 30°$$
$$F_x = (F_{13} - F_{12}) \sin 30°$$

where

$$F_{13} = (2I^2 \sin \omega t \sin(\omega t + 2\pi/3)) \times 10^{-7}/d$$
$$F_{12} = (2I^2 \sin \omega t \sin(\omega t - 2\pi/3)) \times 10^{-7}/d.$$

Hence,

$$\begin{aligned} F_y &= F(1 - \cos 2\omega t) \\ F_x &= F \sin 2\omega t \end{aligned} \qquad (4.34)$$

where

$$F = \sqrt{3}I^2 \times 10^{-7}/2d.$$

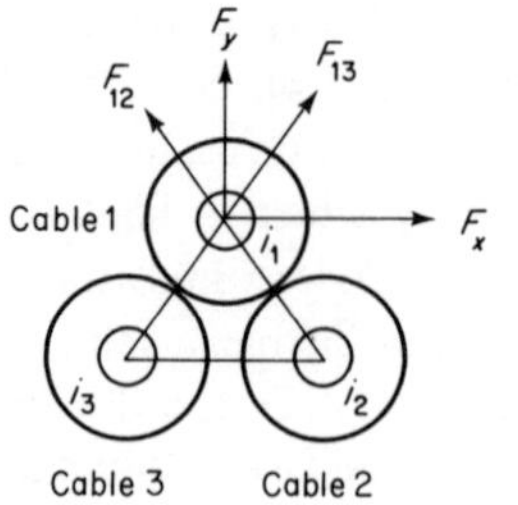

Figure 4.30. Forces on cables in a trefoil group – balanced three-phase currents.

Hence the cable experiences a unidirectional force of twice the power frequency of mean value F and peak value $2F$, plus an alternating force of peak value F at right angles to the unidirectional force. The locus of the resultant force on each conductor is a circle of radius F as shown in Figure 4.31. Generally, under balanced fault conditions, the fault persists long enough for the cable to experience the mean of F_y.

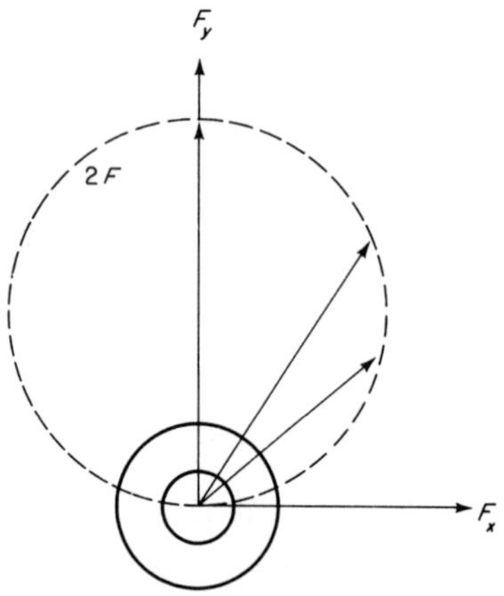

Figure 4.31. Locus of force on one cable of trefoil group – balanced currents.

Unbalanced Faults

Consider a phase to phase fault where $i_1 = i_2$ and $i_3 = 0$. Hence $F_{13} = 0$ and $F_{12} = I^2 \times 10^{-7} (1 - \cos 2\,\omega t)/d$. F_{12} may be resolved into

$$\left.\begin{aligned} F_y &= F(1 - \cos 2\,\omega t) \\ F_x &= F(1 - \cos 2\,\omega t)\sqrt{3}. \end{aligned}\right\} \qquad (4.35)$$

The situation is summarized in Figure 4.32.

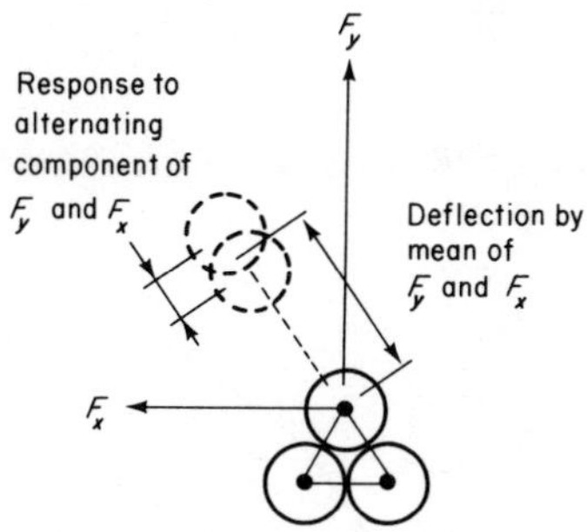

Figure 4.32. Forces on cables in a trefoil group – phase to phase fault current (zero current in third cable).

Asymmetric Waveform

Initially the fault current will contain a d.c. component, and the phase currents are as follows for a balanced fault:

$$i_1 = I(e^{-t/\tau} \sin \alpha + \sin(\omega t - \alpha))$$
$$i_2 = I(e^{-t/\tau} \sin(\alpha - 2\pi/3) + \sin(\omega t - \alpha - 2\pi/3))$$
$$i_3 = I(e^{-t/\tau} \sin(\alpha + 2\pi/3) + \sin(\omega t - \alpha + 2\pi/3))$$

$$\left.\begin{aligned} F_y &= 2F(e^{-t/\tau} \sin \alpha + \sin(\omega t - \alpha))^2 \\ F_x &= 2F(e^{-t/\tau} \sin \alpha + \sin(\omega t - \alpha))(e^{-t/\tau} \cos \alpha + \cos(\omega t - \alpha)) \end{aligned}\right\} \quad (4.36)$$

where τ is the circuit time constant and α the short-circuit phase angle. For a phase to phase fault,

$$F_y = 2F(e^{-t/\tau} \sin \alpha + \sin(\omega t - \alpha))^2$$

and

$$F_x = F_y/\sqrt{3}. \quad (4.37)$$

Thermo-mechanical Forces

A heated conductor attempts to expand, this being resisted by mechanical restraints on the conductor such as the friction forces between the insulating tapes and the conductor and sheath (if present). The middle of a long length of cable is in equilibrium, but towards the ends the frictional forces are unable to resist the conductor expansion and the conductor moves relative to the insulation if it is not prevented from such movement by suitable anchoring. The force developed (without movement) is proportional to the conductor size, expansion coefficient, elasticity modulus of the conductor, and the temperature rise. For a direct-buried self-contained LPOF system[27] where no end movement is permitted the variation

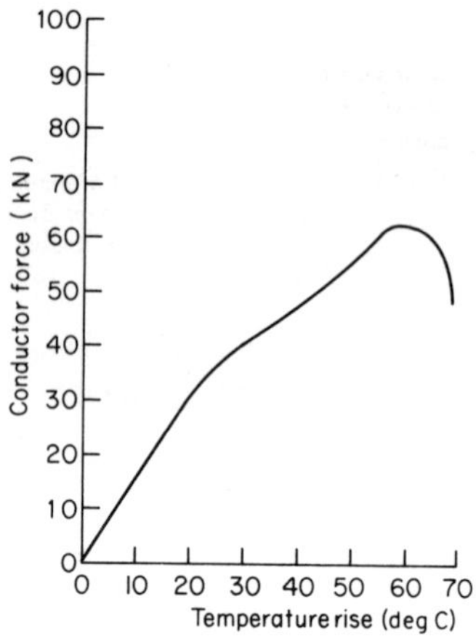

Figure 4.33. Force developed (no movement) in a self-contained cable due to temperature rise on load.[27] (Reproduced by permission of CIGRE from Report 21-05, 1974.)

of conductor thrust with temperature is shown in Figure 4.33. When the conductor movement is restricted, the creep occurring during heating results in tension developing in the conductor on cooling. The effect of this on the conductor thrust is shown in Figure 4.34 for a direct-buried system.[27] If the conductor does move relative to the sheath and insulation, damage to the insulating tapes occurs. In LPOF systems the relative movement which can be repeatedly allowed is 3 mm and with this movement the end thrust can be reduced by 'snaking' the cables when laying.

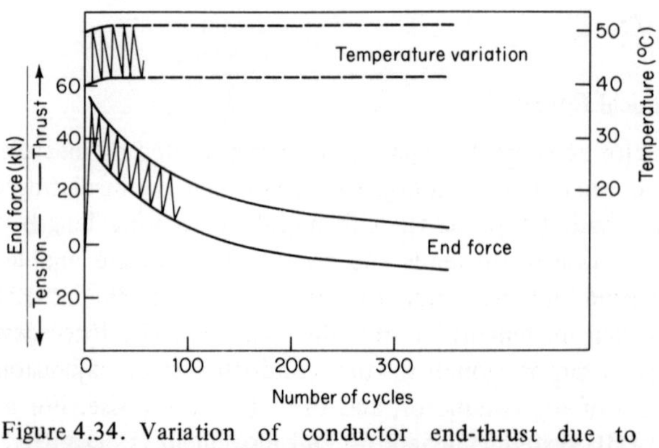

Figure 4.34. Variation of conductor end-thrust due to temperature rise with number of load cycles.[27] (Reproduced by permission of CIGRE from Report 21-05, 1974.)

With self-contained cables in air, two main installation methods are used:

(*a*) cables rigidly supported allowing no lateral movement, resulting in high internal forces
(*b*) cables flexible, where expansion is permitted within limits.

In (*a*) the spacing between the supporting cleats may be calculated using Euler's Buckling Theory.[28] The design of the flexible system has been summarized by Holttum,[28] and this has been since modified to account for aluminium sheaths.

In duct installations movement of the cable into the manhole[29] occurs during heating, followed by a move back during cooling. In self-contained cables this can cause strain in the sheath if the manhole is not large enough. However, the use of aluminium sheaths has eased this problem. In recent years troubles have been experienced with certain pipe-type (HPOF) cables due to conductor movement relative to the insulation papers. A detailed treatment of the mechanical behaviour of integral-pipe water (surface) cooled systems has been reported.[30]

4.10 FUTURE DEVELOPMENTS IN OIL-FILLED CABLES

Introduction

With increase in the amount of transmission at 765 kV (especially in North America) and the possible use of system voltages in the range 1000–1500 kV with power transfers up to 5 GW it is necessary to review the possibility of the use of oil-filled cables for these high ratings. In view of the many years of experience now obtained with the two main types of oil-filled cable their use at these levels would be advantageous.

In Table 4.2, a comparison of the British 400 kV standard overhead line and self-contained cable is given.[31] Although the cable installed length is as yet small, probably more experience exists in Britain on high underground-cable power transfers than elsewhere. From this table the major requirements for cables are seen to be:

(*a*) further reduction in cost
(*b*) improvement in reliability
(*c*) further increase in rating.

Reductions in cost have been achieved in the past through crossbonding (possibly the most important), use of large Milliken conductors, use of aluminium sheaths, increase in maximum stress, and use of intensive cooling. For the rating in Table 4.2 with natural cooling, three cables per phase are required, with separate water-pipe cooling, two, and with integral (surface) cooling, one cable per phase. The corresponding cost ratios are 1.6:1.2:1.0 respectively and the corresponding series–reactance ratios are, 2.2:1.6:1.0.

The basis of rating may be critically examined, especially the precise load

Table 4.2 Basic system parameters of 400 kV circuits–British.

Parameter		Overhead line	Underground cable
Cost		£X/km	~£16X/km
Average non-availability owing to faults		0.2 hrs/yr/km	present ~80 hrs/yr/km future ~8 hrs/yr/km
Average duration of forced outage		1 day	6 weeks
Continuous thermal rating		2.6 → 3.0 GVA	~2.6 GVA
Thermal time constant		10 mins	a few hours
Natural loading		620 MW	4500 MW
Electrical capability 5% regulation, unity p.f. at receiving end	60 km 160 km	2.5 GW 1.1 GW	The limitation is invariably thermal
Positive sequence series reactance (100 MVA base)		0.017%/km	0.01%/km
Positive sequence shunt susceptance (100 MVA base)		0.66%/km	20%/km

conditions. At present a high power cable transmits its rated power only occasionally, namely at peak load conditions, onerous circuit outages, and when system reinforcement is imminent. Also the rather large thermal time constant of the cable may be used to give a rating based on the daily load cycle if the latter is well defined. Artifical cooling with its associated equipment will involve a higher degree of unreliability, although half to two-thirds rating is obtainable with the cooling inoperative. Also, with the use of more than one cable per phase faulty cables may be isolated and the circuit can still operate at reduced power.

At higher voltages it is necessary to reduce the product ($\epsilon_r \times \tan\delta$) of the insulation, as observed in Chapter 2. In spite of many years of development, progress has been disappointing in the development of plastic tapes to replace paper. Polypropylene has a value for ϵ_r of 2.3 which together with paper gives a composite value of about 3 for the polypropylene–paper laminate. As the electric stress in the dielectric varies as $1/\epsilon_r$ it will largely reside in the polypropylene which has a maximum stress of 25.5 kV/mm (650 V/mil) compared with 16.7 kV/mm (425 V/mil) in paper. There would therefore be less thickness of dielectric for a given voltage and less loss resulting from the lower capacitance (ϵ_r) and tan δ. The lower wall thickness also results in less thermal resistance and increase in current rating. However, because of the swelling in the radial direction of the polypropylene (in oil) and the relatively high loss of the laminate (about half that of paper alone), a better approach might be to use cellulose paper near the conductor where the mechanical requirements are severe and a low-loss material for the remainder.

LPOF Self-Contained Cables

By increasing the oil pressure the impulse design strength can be increased from, say, 95 to 110 kV/mm[32] with a.c. stress increasing to 30 kV/mm (this assumes design stress to be approximately half actual breakdown value). In the UHV region the a.c. stress becomes dominant because of the artificially reduced system BIL.[15] There is precedent for such high pressures (up to 30 atmospheres) in submarine links and deep vertical shafts. In a reported[32] design for 765 kV, tape thicknesses (μm) of 80, 110, 140, 170, and 200 are used in that order, from conductor to screen. The oil viscosity is 5 centipoise at 20 °C, the oil having good gas absorption properties. The BIL for 765 kV is 2100 kV and with a conductor of area 1100 mm^2 the outer conductor radius is 43 mm.

Hence,

$$E_{\mathrm{m}} \text{ (taken to be 100 kV/mm)} = \frac{2100}{21.5 \ln(R/21.5)}$$

where R = radius over the dielectric screen. This gives an R of 57 mm over the dielectric and roughly 58 mm over the screen. The corresponding a.c. stress is 21 kV/mm.

The design for maximum capacity is obtained as follows. The temperature rise (conductor to ambient) of a self-contained cable is given approximately by

$$\Delta\theta = I^2 R_{\mathrm{ac}} R_{\mathrm{T}} + \frac{q_{\mathrm{d}}}{2} R_{\mathrm{T}} \tag{4.38}$$

where R_{T} = total thermal resistance from conductor to ambient and q_{d} = dielectric loss = $\omega C \tan\delta\, V_{\mathrm{p}}^2 = (E_{\mathrm{m}} R/e)^2\, \omega C \tan\delta$ per phase where E_{m} = stress at the conductor for minimum diameter. Let $(\omega C \tan\delta)/e^2 = \lambda$, then $q_{\mathrm{d}} = \lambda E_{\mathrm{m}}^2 R^2$. The power transmittable, P, is

$$P = 3V_{\mathrm{p}} I = \frac{3E_{\mathrm{m}} R}{e} \left(\frac{\Delta\theta}{R_{\mathrm{ac}} R_{\mathrm{T}}} - \frac{\lambda E_{\mathrm{m}}^2 R}{2R_{\mathrm{ac}}} \right)^{1/2}. \tag{4.39}$$

Because of the high dielectric loss the use of intensive cooling is essential. For integral cooling $\Delta\theta$ = 70 deg C and if R_{T} = 0.75 deg C-m/W, P_{max} is approximately 5 GVA. This value is approached by an 1100 mm^2, 1100 kV cable. For the 750 kV cable a throughput of 2.2 GVA is possible with integral cooling stations 2–3 km apart. With separate water-pipes the total thermal resistance is 0.95 deg C-m/W of which only 0.15 deg C-m/W is due to the resistance between cable and pipe.

Accessories

Two types of sealing end have been considered; capacitance graded and conventional stress cone. It is considered[32] that both will prove adequate for self-contained cables of 765 kV. For voltages above 765 kV it is probable that only the capacitance graded type will be adequate. Both through and stop joints would be of conventional design, the through-joint having a flush ferrule and large oil gaps at the outer insulation.

Pipe-type Cables

Several possibilities for improved rating and economy are being studied. Artificial cooling via oil circulation enables the average value of soil thermal resistivity (g_s) to be used in rating calculations instead of the maximum value. Further heat dissipation is obtained from the return oil pipe. A total increase in rating compared with natural cooling of the order of 40% is obtainable. As the normal operating pressure is 6.9×10^5 N/m^2 (100 psi), and the design pressure is double this, there is ample allowance for pumping pressure drops. With pipe cables the improvements in current capability resulting from the possible improvements in polypropylene/paper would be masked by the high thermal resistance from the pipe to ground with natural convection. With forced pumping of oil, however, the gains become more prominent. Development is towards the use of pipe-type cables at 500/550 kV and above; much of the testing is being carried out at the Underground Transmission Test Facility at Waltz Mill, Pennsylvania.[33,34,35]

Optimum capacity

The thermal analysis for the pipe-type cable in steady-state operation has been discussed in Chapter 3. An alternative to the self-contained cable approach is as follows.

$$\Delta\theta = K_1 V^2 + K_2 I^2$$

where K_1 and K_2 are constants of proportionality dependent on the cable parameters, and V and I are system voltage (kV) and maximum current (kA).

The rating MVA is $\sqrt{3}VI$ and this becomes a maximum when d(MVA)/dI = 0, for which condition

$$V = \left(\frac{\Delta\theta}{(K_1 K_2 + K_1)}\right)^{1/2}$$

giving

$$(\text{MVA})_{max} = \sqrt{3}\,\Delta\theta \left(\frac{K_1}{K_2}\right)^{1/2} \left(\frac{1}{K_1 K_2 + K_1}\right). \tag{4.40}$$

$(\text{MVA})_{max}$ occurs when the temperature drop due to the dielectric loss equals that due to the I^2R losses, the latter being dependent on the load factor.

A computer study of the relationships between insulation thickness, voltage, and conductor size and the resulting technical limits has been published.[36] The parameters of the paper/oil cable studied follow normal practice, e.g. conductor temperature 85 °C, load factor 75% (loss factor 62.5%), g_{soil} 0.9 deg C-m/W, soil diffusivity 45 cm^2/hour. For forced cooling the constraint is taken to be a fixed shield temperature of 40 °C. The pipe is buried with its centre line 0.92 m (36 inches) below the surface. The variation in system voltage with insulation thickness is shown in Figure 4.35 for standard AEIC wall thicknesses. These standards (Association of Edison Illuminating Companies) recommend *average* stresses in the

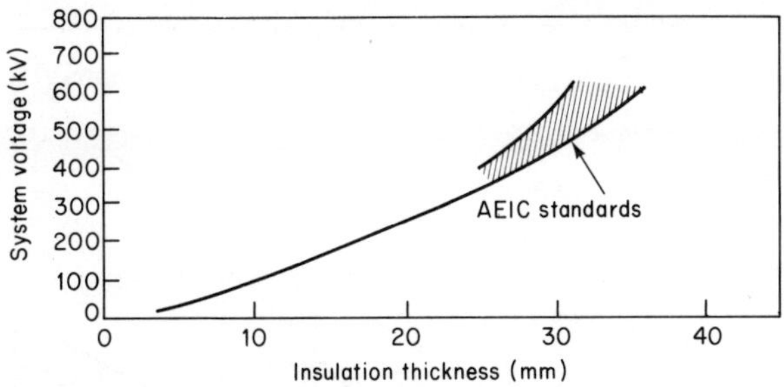

Figure 4.35. Voltage (AEIC standard) as a function of insulation thickness. HPOF pipe-type cables conductor 1000 mm^2, tan δ 0.0018.[36] (Reproduced by permission of IEEE.)

range 5.5–7.9 kV/mm for cables rated from 69–345 kV; a reasonable upper limit would appear to be 11.8 kV/mm. For forced-cooled schemes the use of 11.8 kV/mm would give an insulation thickness of 53.5 mm (2.1) inches) and a voltage of 1085 kV.

A constaint on the insulated core diameter is the minimum bending diameter permissible. Accepting an upper reel diameter of 3 m (10 feet) gives a cable diameter of 150 mm (6 inches). The relation between reel diameter and cable diameter for pipe-type cables according to the AEIC standard is shown in

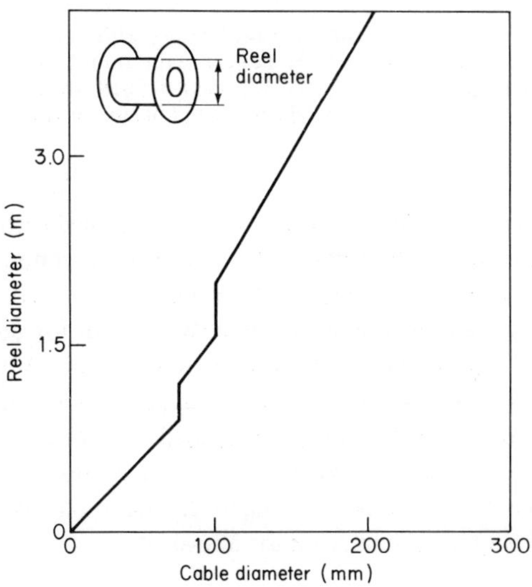

Figure 4.36. Minimum reel diameter as a function of cable diameter for HPOF, pipe-type cables (according to AEIC No 2-67 Section 12.31). Conductor size 1000 mm^2,[36] (Reproduced by permission of IEEE.)

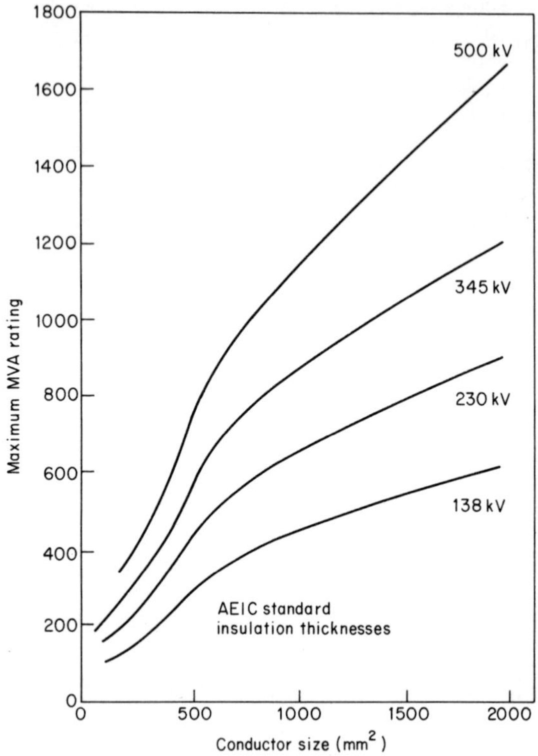

Figure 4.37. Variation of power rating with conductor size for standard AEIC cable designs with forced oil cooling, HPOF pipe-type cables. ϵ_r 3.5, tan δ 0.0018, shield temperature 40 °C. (Reproduced by permission of IEEE.)

Figure 4.36. The maximum MVA ratings for forced-cooled systems based on *AEIC insulation thicknesses* are given in Figure 4.37, for various conductor sizes.

Generally it may be concluded that ratings above 1 GVA are unlikely in self-cooled cables, but about double this is possible with forced cooling. With self cooling there is no object in increasing voltages beyond 500 kV without recourse to plastic or plastic/paper tapes. Forced cooling applied to existing installations will provide ratings up to 1.5 MVA. It must be stressed that the above optimized results are technical and not economic. Also, no account of reactive compensation has been taken. The penalty in terms of rating in the use of pipe-type cables compared with self-contained is very evident from the above.

Tests at Waltz Mill test site

Waltz Mill has a 1100 kV capability, and tests can be performed on underground a.c. transmission systems from 138 kV to 1100 kV (excluding impulse and switching surge tests). The 550 kV design[33] under test have conductors with four segments (alternates insulated), conductor shields of metallized paper or carbon black paper and tinned copper tapes. The insulation thickness is 34 mm and the skid wires are of high density polyethylene. Polybutane or paraffinic oils are used.

REFERENCES

1. *Underground Systems Reference Book*, Edison Electric Institute, N.Y., 1957.
2. ABDULHADI, R. S. and CHATO, J. C., 'Combined natural and forced convective cooling of underground electric cables', *IEEE Trans PAS-96*, pp. 1–8 (1977).
3. BULLER, F. H. and NEHER, J. H., 'Short-Time Transient Temperature Rise of Self Contained Oil-Filled Cable Systems with Particular Reference to Oil Demands', *IEEE Trans PAS-84*, pp. 761–770 (1965).
4. MIRANDA, F. and GAZZANA PRIAROGGIA, P., 'Self Contained Oil-Filled Cables, a Review of Progress, *Proc IEE*, **123**, 229–238 (1976).
5. RAY, J. J. *et al.*, '525 kV Self-Contained Oil-Filled Cable Systems for Grand Coulee Third Power Plant, Design and Development', *IEEE Trans., Paper T73*, 492–6 (1973).
6. ARKELL, C. A., 'Self-Contained Oil-Filled Cable – Installation Design and Techniques', *Proc. IEEE Underground Transmission and Distribution Conference, 1976*, pp. 497–503.
7. NAGASKI, S. *et al.*, 'Recent Developments of High Voltage Cross-linked Polyethylene Power Cables', *IEEE Conference Underground Transmission and Distribution, 1974.*
8. JOCTEUR, R. *et al.*, 'Influence of Surface and Internal Defects on Polyethylene. Electrical Routine Tests on e.h.v. cables', *IEEE Transactions PAS-96*, pp. 513-523 (1977).
9. BAHDER, G. *et al.*, 'Criteria for Determining Performance of Cross-linked Polyethylene Insulated Power Cables', *IEEE Trans. PAS-95*, pp. 1552–1560 (1976).
10. EICHHORN, R. M., 'Treeing in Solid Extruded Electrical Insulation', *IEEE Trans. EI-12*, pp. 2–18 (1976).
11. LAWSON W. G. *et al.*, Developing crosslinked polyethylene cables for high-voltage power transmission, *IEE Electronics and Power*, pp. 116–120 (Feb. 1978).
12. HUMPHREY, L. E. *et al.*, 'Insulated Sodium Conductors', *IEEE Trans. PAS-86*, pp. 876–883 (1967).
13. *Calculation of the Current Rating of Cables (100% load factor)*, I.E.C. Publication 287 (1969).
14. DWEK, M. G. and EGGLETON, M. N., 'Electrical Parameters of 400 kV and 275 Cables in England and Wales', *Proc. IEE Conference on Underground Transmission, London 1968*, pp. 461–479.
15. WEEDY, B. M., *Electric Power Systems*, 3rd Edition, Wiley, Chichester, 1979.
16. EMANUEL, A. and DOEPKIN, H. C., 'Calculation of losses in steel enclosures of three-phase bus or cables', *IEEE Trans PAS-93*, pp. 1758–1767 (1974).
17. NEHER, J. H., 'The Phase Sequence Impedance of Pipe Type Cables', *Trans. IEEE PAS-83*, pp. 795–803 (1964).
18. DUGAN, R. C and TALUKDAR, S. N., 'The effects of saturation on the performance of pipe-type cables', *IEEE Paper A76 340-0.* Presented at PES Summer Meeting, 1976.

19. HEATON, A. G. and ISSA, A. M. H., 'Transient response of cross-bonded cable systems', *Proc. IEE,* **117**, 578–586 (1970).
20. BALL, E. H. *et al.*, 'Sheath Overvoltages in High Voltages Cables Resulting from Special Sheath-Bonding Connections', *IEEE PAS-84*, pp. 964–988 (1965).
21. CHU, S. C., 'E.H.V. Cable Terminations', *Electrical Times (British)*, 8 June 1961.
22. ALSTON, L. L., *High Voltage Engineering*, Oxford University Press, 1968.
23. BRAZIER, L. G., 'Joints, Sealing Ends and Accessories for Pressure Cable', *Proc. IEE,* **93**, II, 415–434 (1946).
24. FULCOMER, R. D. and FOOTE, G. E., 'A simple non-stress cone stress relieving wrap for terminating extruded dielectric shielded power cable', *Proc. IEEE Conference on Underground Transmission and Distribution, 1976*, pp. 367–369.
25. ALLAM, E. M. and McKEAN, A. L., '138 kV Prefab Capacitive-Graded Joint for Oil Filled Cable Systems', *IEEE Trans PAS-96*, pp. 20–26 (1977).
26. GREENWOOD, A., *Electrical Transients in Power Systems*, Wiley, London, 1971.
27. ARKELL, C. A. *et al.*, 'The Thermo-Mechanical Design of High Power Self Contained Cable Systems', *CIGRE Report 21-05*, 1974.
28. HOLTTUM, W., 'The Installation of Metal-Sheathed Cables on Spaced Supports', *Proc. IEE,* **102**, 729–746 (1955).
29. HATA, H. 'On the design of Cable Offsets in Manholes', *Sumitomo Electrical Technical Review*, No. 10, pp. 32–40 (October 1967).
30. LUONI, G. *et al* , 'Study of mechanical behaviour of cables in integral-pipe water-cooled systems', *Proc. IEE,* **124**, 295–302 (1977).
31. COTTRILL, J. E. J., 'System Technical Requirements for Cables in the Developing C.E.G.B. 400 kV System', *Proceedings 11th Universities Power Engineering Conference, UK*, 1976.
32. ARNAUD, U. C. *et al.*, 'Self Contained Oil-Filled Cables for Power Transmission in the 2 to 4 GVA Range', *Proc. IEEE Conference Underground Transmission and Distribution, 1974*, pp. 1–6.
33. BAHDER, G. *et al.*, '550 kV HPOF Pipe Cable Development in the USA', *CIGRE, Report 21.11*, 1976.
34. WILLIAMS, J. L. *et al.*, 'Comprehensive Forced Cooled Tests on Pipe Type Cables at Waltz Mill', *CIGRE Report 21.07*, 1976.
35. BURRELL, R. W. and YOUNG, F. S., 'EEI-Manufacturers 500/550 kV Cable Research Project, Waltz Mill Test Facility', *Trans IEEE PAS-90*, pp. 180–190 (1971).
36. KILAR, L. A. and ENGELHARDT, J. S., 'Ultimate Power Capabilities of HPOF Pipe-Type Cable Systems', *IEEE Trans. PAS-91*, pp. 780–789 (1973).

PROBLEMS

Problem 4.1

Given a BIL of 1425 kV determine the maximum impulse stress in a cable insulation such that the overall diameter is $\not>$140 mm. The diameter over the conductor screen is 50 mm. (Answer: 27.7 kV/mm.)

Problem 4.2

Each core of a self-contained oil-filled cable consists of a hollow conductor of physical area 2500 mm^2 with a central duct. The impulse design strength of the insulation is 110 kV/mm and the outer diameter of the insulation 150 mm. The system voltage is 765 kV and the BIL, 2100 kV. Determine the diameter over the conductor and conductor duct. Calculate the maximum a.c. working stress. (Guess a diameter and then iterate.) (Answers: 51.9 mm, 43.5 mm.)

Problem 4.3

A 3-phase underground circuit consists of three single-core cables each of effective conductor radius of 0.75 cm and spaced 5.1 cm apart (axis to axis) in equilateral formation. The diameter of the lead sheath is 2.3 cm and the sheath thickness 0.15 cm. The specific resistance of lead is 22.0×10^{-6} Ω cm at the working temperature and the conductor resistance per mile is 0.26 Ω at 65 °C. For a cable length of one mile (1.6 km) and a load of 300 A, determine (*a*) the induced sheath voltage without bonding, (*b*) the ratio of sheath loss to conductor loss. (Answers: (*a*) 47 V per sheath, (*b*) 0.029.)

Problems 4.4

Determine the effective electrical parameters of a 3-phase, 50 Hz underground circuit, 65 km in length and comprising three separate single-core 200 mm^2 cables each of conductor radius 0.915 cm. The internal and external radii of the lead sheath are 2.5 and 2.8 cm, respectively. The cables are in touching equilateral formation and the sheaths are bonded to ground at several points. The conductor a.c. resistance per kilometre at 15 °C is 0.0875 Ω/km and the resistivity of lead at the operating temperature may be assumed to be 23.2×10^{-6} Ω cm. (Answers: Conductor 7.2 Ω, effective reactance 7.7 Ω.)

Problem 4.5

In a self-contained cable system, determine the ratio of sheath loss when solidly bonded to sheath loss when bonded through resistance R (Ω) per phase per km. The sheaths are bonded by a star connection of resistors each of resistance RΩ.

Problem 4.6

A phase to phase fault of 40 kA exists between two cables of a self-contained cable system. The cable centres are 15 cm apart. Calculate the force exerted. (Answer: 2.2 kp per cm length.)

Problem 4.7

A 500 kV cable link of length 50 km comprises two identical three-phase cable circuits in parallel of total capacity 2.5 GVA. Both ends of the link are connected to infinite busbars. A three-phase compensatory reactor (reactance j262.5 Ω) is connected to both cable circuits at their midpoints as shown in Figure 4.38. The charging MVA per km for each cable circuit is 19.0 (three-phase value).

Determine the charging current at the cable ends and midpoint, with the reactor (*a*) connected and (*b*) disconnected. Neglect voltage drops along the cables. Investigate the possibility of further reduction of the peak charging current by changing the reactor inductance. (Answers: Current at ends (*a*) 550 A, (*b*) 275 A.)

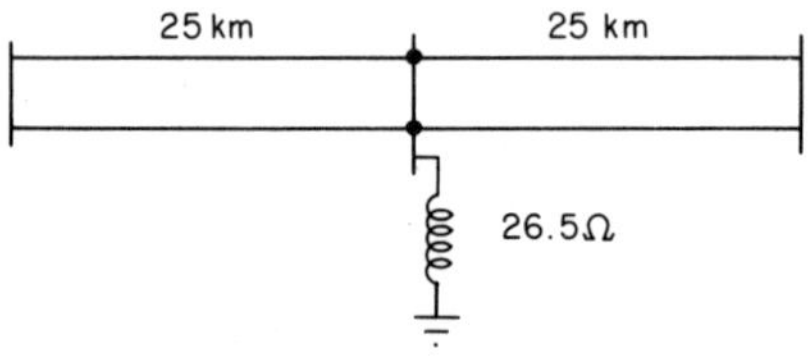

Figure 4.38. System for Problem 4.7.

Problem 4.8

A 750 kV, 60 Hz 3 phase, 20 km long, transmission circuit is comprised of three self-contained oil-filled cables. The cables are rated at 1.5 kA and have the following parameters:

a.c. resistance of conductor	0.0175 Ω/km
diameter over conductor	43 mm
diameter over dielectric	117 mm
relative permittivity	3.5 (paper/oil dielectric)
tan δ	0.002

Determine: (*a*) total loss of the circuit; (*b*) critical length without compensation assuming cable fed from one end only; (*c*) how close the dielectric design is to the optimum. Ignore sheath loss. (Answers: 4 MW, 47 km.)

Problem 4.9

Show that, for a self-contained cable, the MVA capability is given by an expression of the form $(AV_n - BV_n^3)^{1/2}$ where A and B are constants for a given conductor, maximum working stress, conductor temperature rise, and tan δ. V_n is the conductor to neutral voltage. Determine the maximum value of MVA and show that for this condition the temperature rise due to the dielectric is half that due to the conductor loss. Discuss the effects of increase in voltage on MVA transmittable. Assume that thermal resistance outside the cable to be negligible. (*Note*, temperature rise across dielectric due to dielectric loss alone = $g_d\ \epsilon_0\ \epsilon_r\ V_n^2\ \omega \tan \delta/8\pi$, where g_d = thermal resistivity of dielectric.)

Problem 4.10

The sheath of an LPOF self-contained cable is interrupted as shown in Figure 4.20. The surge impedance conductor/sheath is 30 Ω and for sheath/ground 2 Ω. If a surge of peak value 1 MV arrives on the cable conductor calculate the peak voltage across the gap in the sheath. (Answers: (*a*) 60 kV, (*b*) 125 kV.)

Problem 4.11

For a 525 kV three-phase system self-contained oil-filled cables calculate the voltage appearing across the interruption in the sheath (i.e. across joint sleeve insulation) in terms of the incoming voltage transient (peak) for the bonding connections shown in Figure 4.39a,b. Cable parameters are as follows:

Diameter over conductor	40 mm
Diameter over insulation	103 mm
Diameter over sheath	114.4 mm
Diameter over sheath insulation (PVC)	122 mm
ϵ (dielectric) 3.5	ϵ (sheath insulation) 4.

Neglect resistance and state assumptions made.

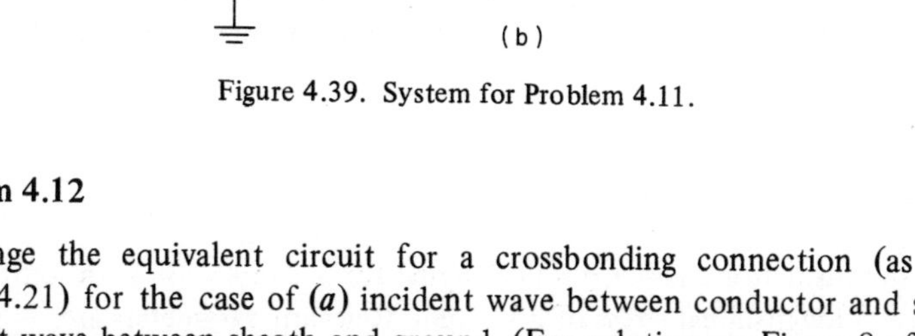

Figure 4.39. System for Problem 4.11.

Problem 4.12

Rearrange the equivalent circuit for a crossbonding connection (as shown in Figure 4.21) for the case of (*a*) incident wave between conductor and sheath, (*b*) incident wave between sheath and ground. (For solution see Figure 8 of Reference 20.)

Problem 4.13

In a 60 Hz pipe-type system the insulated conductors are in a 204 mm diameter pipe. The cables are unshielded and the unbalanced fault current returns through the pipe. Determine the pipe impedance given the variation of μ_r with current shown. $\rho_{steel} = 16.5\ \mu\Omega$-cm.

I_p (A)	250	500	1000	4000	8000
$\sqrt{\mu_r}$	31.4	30.4	26.2	14.5	10.5

(Answers. $17\angle 45$ mΩ at 250 A, $5.7\angle 45$ mΩ at 8 kA per m.)

Problem 4.14

Determine the effective impedance of a self-contained cable system of the following specification, and having current return through the ground.

Diameter of conductor	5.8 cm
Diameter over insulation	10.7 cm
Diameter over sheath	11.5 cm

Spacing between centres 89 cm (horizontal)
Frequency 50 Hz, soil resistivity 150 Ω-m
Conductor a.c. resistance 0.124 $\mu\Omega$/cm at 85 °C.

(Answer: 61.9 + j664/$\mu\Omega$/m.)

Problem 4.15

Determining the effective conductor impedance of a 60 Hz oil-filled self-contained cable system: (*a*) with the sheaths bonded; (*b*) with the sheaths open circuit. Also calculate the zero sequence resistance assuming all return current flows in the sheaths.

The cables are in horizontal formation with a spacing between centres of 33 cm. The mean radius of the sheath is 3.17 cm and the conductor radius is 1.5 cm. The a.c. resistance of the conductor is 0.655 $\mu\Omega$/cm at 65 °C and the sheath resistance is 3.25 $\mu\Omega$/cm. (Answers: (*a*) (204 + j149) $\mu\Omega$/m; (*b*) (65.5 + j269) $\mu\Omega$/m; (*c*) (390.5 $\mu\Omega$.)

Problem 4.16

Determine the insulation profile (radius-length) for a 275 kV through joint assuming a maximum longitudinal stress of 2 kV/mm. Relevant parameters are:

Diameter over conductor	58 mm
Diameter over cable insulation	90 mm
Diameter over joint insulation (maximum)	140 mm
Max. a.c. stress (cable)	150 kV/mm
Max. a.c. stress (joint radial)	100 kV/mm.

Problem 4.17

Design a capacitor bushing having five layers of insulation each of thickness 5 mm separated by aluminium foils. The inner layer is 300 mm in length. The value of maximum stress in each layer is to be approximately constant. The conductor radius is 10 mm and if the maximum a.c. stress (r.m.s.) is to be 5 kV/mm determine the working a.c. voltage. (Answer: lengths of layers (mm) 300, 200, 150, 120, 100, phase voltage 110 kV.)

Problem 4.18

A section of 345 kV, 60 Hz, artificially cooled HPOF cable, 20 km in length, is fed from one end and produces 13.7 MVAr per km. Calculate the volt drop or rise at no load and a full load of 1 GVA at unity p.f. The impedance per phase of the cable is $(15 + j120) \times 10^{-6}$ Ω per km on a 100 MVA base. (Answer: (line) 121 V, 1.15 kV.)

5

Cables for Direct Current Transmission

5.1. INTRODUCTION

The use of d.c. transmission[1,2,3] has expanded at a modest rate, limited mainly by the high cost of conversion equipment traditionally based on the mercury-arc valve. However, the widespread development work on solid state controlled rectifiers and their use for high voltages is now producing designs which may to some extent change the economic picture. D.C. transmission with overhead lines has considerable technical advantages. With underground cables there are also several advantages, e.g. absence of dielectric loss and a.c. charging current, no requirement for reactive compensation, higher breakdown strengths and reduced partial discharge activity.[4,5]

With underwater crossings there is no alternative to the cable. Because of this nearly all major d.c. cable connections have been for underwater links. The solid mass-impregnated paper–oil type which on d.c. has better dielectric properties because of the much reduced corona (partial) discharge activity has been widely used. The only major land circuit to date is the British Kingsnorth–Willesden (London) connection using ±266 kV self-contained oil-filled cables.

An example of the saving in space by using d.c. is as follows. Six self-contained, 275 kV, 2000 mm^2 cables in two groups of three in horizontal formation (trench width 5.2 m) in soil of resistivity 1.2 deg-cm/W has an a.c. capacity of 1520 MVA. Two cables at ±500 kV d.c. have a capacity of 1600 MW in a trench of width 0.68 m. Table 5.1 lists the principal schemes installed or at an advanced stage of planning for which cable has been used for the main transmission circuits and gives details of the type of cable used.

Comparison with A.C. Cables

The extensive knowledge available for a.c. cables results in any consideration of d.c. cable requirements beginning with a comparison between the two systems. Instead of the familiar three-conductor system used for a.c., the d.c. transmission system requires two single-core cables respectively for double-pole systems, sometimes with

Table 5.1 Some high voltage d.c. cable systems

Location	Type of cable	Cable voltage	Approx. route length (km)	Date installed	Rating (A)	Conductor area (mm^2)
Sweden Mainland to Gotland	Submarine Solid type	100 kV	100	1954	200	90
England/France Cross channel link	Submarine Solid type	100 kV	64	1961	800	344
New Zealand Cook Strait	Submarine High pressure Gasfilled (Nitrogen)	250 kV	41	1965	1200	520 & 805 3 cables
Sweden/Denmark Konti-Skan	Submarine Solid type (Swedish side) Flat Pressure (Danish side)	250 kV	74	1965	1000	625
Sardinia/Italy	Submarine Solid type	200 kV	118	1968	1000	420
England Kingsnorth/Willesden	Land oil-filled	266 kV	81	1973	1200	800
Vancouver Is. to British Columbia, Canada	Submarine Solid type	300 kV	36	1976	1200	800

a lightly-installed neutral cable. Each of these cables must withstand half the total voltage between poles of the d.c. system, plus a ripple voltage. A change in the direction of power flow is achieved by a reversal of the polarity of the d.c. voltage.

If d.c. cables are connected to an overhead line, surge voltages of atmospheric origin may reach the cables in the same way as for a.c. cables and the impulse strength of the cables must be considered. Since no designs of switchgear capable of interrupting high voltage d.c. circuits exist as yet, d.c. cables are not subjected to switching surges in the same way as a.c. However, maloperation of the converter equipment can give rise to overvoltages of a form peculiar to d.c. transmission systems; these are referred to later in connection with testing.[1,2,6]

The losses in a d.c. and a.c. cables are in some respects very different. With the d.c. cable the additional losses in the conductor due to skin and proximity effects, and in the sheath due to circulating current and eddy loss, will be extremely small and are caused by the small ripple component of the load current. There is therefore no necessity for special constructions such as the Milliken conductor used to reduce skin and proximity effects in a.c. cables. Also, the loss in the dielectric of a d.c. cable will also be very much reduced. This loss in an a.c. cable is largely due to the oscillation of particles and polar elements in the dielectric, due to the alternating electric field, and is absent when the applied voltage is constant when the only power loss is that due to the small leakage current in the dielectric. The most fundamental difference is in the steady-state stress distribution in the dielectric, and this will be discussed in some detail.

In an a.c. cable the stress is dictated by the capacitance of the dielectric, whereas in a d.c. cable the electric stress is only dependent on the electrical conductivity of the dielectric. A theory for paper–oil insulation consistent with experimental results suggests that the conduction current flows mainly in the oil in the channels that permeate the structure of impregnated paper. The free carrier density in these channels is controlled by a trapping process, the trapping centres being associated with the cellulose dipolar nature of the material.[7,8,9]

The stress dependence of conductivity is explained by assuming that the conduction along channels limited by barriers representing (on a microscopic scale) the low conductivity of cellulose relative to oil. The process of conduction is limited by the rate at which carriers succeed in penetrating this barrier by tunnel effect and this is controlled by the applied stress **E**.

The d.c. conductivity (σ) of paper/oil appears to obey the following law:

$$\sigma = A_0 \exp\left(-\frac{W}{kT} + b\mathbf{E}\right). \tag{5.1}$$

A_0 and b are constants dependent on the impregnant (b varies from 0.35×10^{-6} m/V for fully impregnated paper to 0.08×10^{-6} m/V for unimpregnated paper). Also, k = Boltzmann constant, T = absolute temperature (K), W = the activation energy characteristic of cellulose ($\simeq$1 eV).

5.2 DISTRIBUTION OF STRESS – STEADY-STATE OPERATION[6,7,9,10,11,12,13,14,15]

In a coaxial a.c. cable the stress is determined by capacitance and at radius x,

$$\mathbf{E}_x = \frac{\mathbf{V}}{x \ln(R/r)} . \tag{5.2}$$

The distribution of stress in a d.c. cable with dielectric at uniform temperature (i.e. on no load) is also given by this relationship. Consider the situation for a coaxial cable of radii r (over conductor) and R (over insulation or screen) with a direct voltage V applied. The stress is determined by resistivity only. The electrical conductivity of paper/oil dielectrics (see Figure 5.1) may be written as

$$\sigma = \sigma_0 \, e^{(-a/T + bE)} \tag{5.3}$$

where E is stress and T is the absolute temperature (K), σ_0 is the electrical conductivity at reference temperature and at low stress. σ_0 varies for different impregnants and papers but a and b are reasonably constant for a type of insulation.

The temperature across the dielectric is given by

$$\frac{gq}{2\pi} \ln\left(\frac{R}{r}\right) \tag{5.4}$$

where g = thermal resistivity (deg C-m/W), q = conductor loss (W/m).

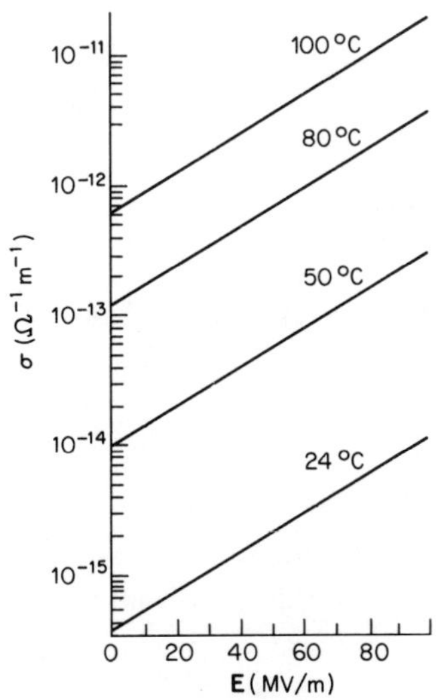

Figure 5.1 Electrical conductivity of paper/oil dielectric as a function of stress and temperature.[7] (Reproduced by permission of IEEE.)

A simpler equation for σ which is valid over the temperature range to be expected is as follows:

$$\sigma = \sigma_0 \exp(\alpha\theta + b\mathbf{E}) \tag{5.5}$$

where, θ is the temperature (°C). Typical values for paper/oil are, $\alpha = 0.105$ (deg C)$^{-1}$ and $\sigma_0 = 0.37 \times 10^{-16}\ \Omega^{-1}\ \text{m}^{-1}$.

Consider the flow (i) of direct current through a ring of dielectric of radius x and thickness dx (see Figure 5.2).

$$i = 2\pi x\sigma \frac{\delta V}{\delta x} = \text{constant (steady-state operation).}$$

Hence, using (5.5),

$$x\frac{\delta V}{\delta x} = D\text{e}^{-\alpha\theta}\text{e}^{-bE} \quad \text{and} \quad \frac{\delta V}{\delta x} = E.$$

At the conductor ($x = r$),

$$E_r = \frac{D}{r}\text{e}^{-\alpha\theta_r}\text{e}^{-bE_r}.$$

Therefore,

$$\frac{E_x}{E_r} = \frac{r}{x}\text{e}^{\alpha(\theta_r - \theta_x)}\ \text{e}^{-b(E_x - E_r)}. \tag{5.6}$$

The temperature difference between radius x and radius r is

$$\theta_r - \theta_x = \frac{gq}{2\pi}\ln\left(\frac{x}{r}\right).$$

Hence,

$$\left(\text{e}^{b(E_x - E_r)}, \frac{E_x}{E_r}\right) = f\left(\frac{x}{r}\right). \tag{5.7}$$

The value of E_x may be obtained as follows. Choose a value for E_r and substitute

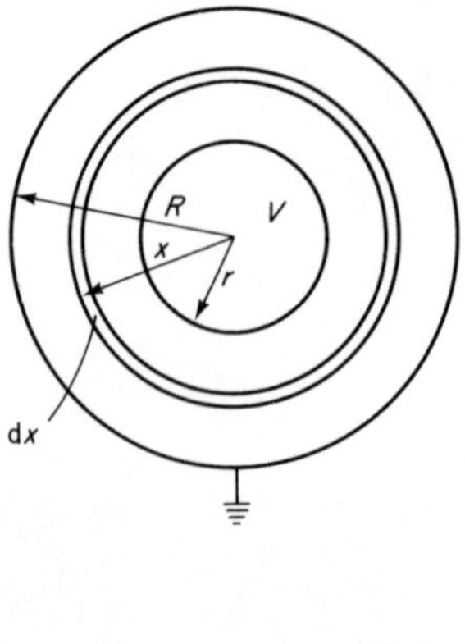

Figure 5.2 Coaxial cable with direct voltage applied.

in equation (5.6), thus getting a relationship between E_x and x which can be solved by iteration.

$$V = \int_r^R \frac{dV}{dx}\, dx = \int_r^R E_x\, dx$$

and values of V are found. E_r is systematically varied until the correct value of V is obtained.

Simplification is obtained if the variation in dielectric conductivity with electric stress is assumed negligible, i.e.

$$\sigma = \sigma_0\, e^{\alpha\theta}.$$

The temperature difference across the insulation from radius x to R,

$$\Delta\theta_x = \frac{I^2 \rho_c}{\pi r^2} \frac{\ln(R/x)}{2\pi g}$$

where I = load current and ρ_c = electrical resistivity of the conductor.

The ratio of the conductivities at any radius x and sheath (R) is

$$\frac{\sigma_x}{\sigma_R} = \frac{\sigma_0\, e^{\alpha\theta_x}}{\sigma_0\, e^{\alpha\theta_R}} = e^{\alpha\Delta\theta_x}$$

$$= \exp\left(\frac{\alpha I^2 \rho_c}{\pi r^2} \frac{\ln(R/x)}{2\pi g}\right).$$

Let

$$n = \frac{\alpha\rho_c}{\pi^2 2g}\left(\frac{I^2}{r^2}\right).$$

Then

$$\frac{\sigma_R}{\sigma_x} = \left(\frac{x}{R}\right)^n.$$

Stress ratio,

$$\frac{E_x}{E_R} = \frac{\sigma_R}{\sigma_x}\frac{R}{x} = \left(\frac{x}{R}\right)^{n-1}.$$

Thus

$$V = \int_r^R E\, dx = \int_r^R E_R \frac{x^{n-1}}{R^{n-1}}\, dx = \frac{E_R}{n} R\left[1 - \left(\frac{r}{R}\right)^n\right]. \qquad (5.8)$$

5.3 DISTRIBUTION OF STRESS – TRANSIENT OPERATION

(*a*) A step voltage of same polarity as the d.c.

At the instant of arrival of the transient, the stress in the dielectric,

$$E_x = E_{\text{dc}} + E_{\text{transient}}$$

$$= E_{\text{dc}} + \frac{(V_{\text{new level}} - V_{\text{dc}})}{x \ln(R/r)}. \tag{5.9}$$

$E_{\text{transient}}$ is determined by capacitive distribution because it is time dependent.

(*b*) A step voltage of reverse polarity such that the external voltage is V_r.

$$E_x = E_{\text{transient}} + E_{\text{dc}}$$

where E_{dc} is a portion of initial d.c. stress due to voltage V_{d} which cannot be discharged because it does not have capacitive distribution.

$$E_x = \frac{V_r}{x \ln(R/r)} + \frac{V_{\text{d}} - (E_{\text{d}}\, x \ln(R/r)}{x \ln(R/r)}$$

$$= \frac{V_r + V_{\text{d}}}{x \ln(R/r)} - E_{\text{d}}. \tag{5.10}$$

5.4 APPLICATION OF STRESS CALCULATION

Figure 5.3 shows the results of stress calculations made by this method. It should be noted that when the dielectric is at a constant temperature throughout, if the dependance of conductivity on stress is ignored, then the stress distribution is identical to that calculated for an a.c. cable. When the dependence of σ on E is included in the calculation the conductor stress is somewhat reduced, as shown. When the cable is heated the stress distribution is completely changed and the maximum stress appears at the screen, the value of this stress increasing as the temperature difference across the dielectric increases. Hence for a d.c. cable it is essential to specify this temperature difference in order to ensure that the operating stress stays within the design limits. At first sight it seems that when the cable passes from one environment to another having differing thermal properties the differences in temperature between longitudinally separated points at the same radius would give rise to differences in voltage at these points and hence to longitudinal stresses. Fortunately this is not so because the voltage distribution is controlled by the temperature difference *across the dielectric* which remains approximately constant for a given current even though the actual temperatures may change.

When considering the application of a d.c. voltage to an impregnated paper sample the initial surge of current caused by the charging of the capacitance of the sample is followed by a period in which the current steadily decreases over some hours until the steady value is reached. This dielectric absorption current can be

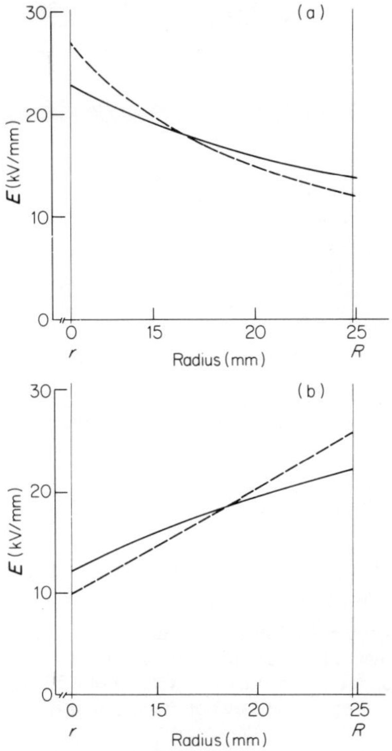

Figure 5.3 Stress distribution through paper/oil dielectric: (a) uniform temperature; (b) cable on load.
——— Stress dependence included.
— — — Stress dependence neglected.[7]
(Reproduced by permission of IEEE.)

explained by the assumption of a stratified form of dielectric and the redistribution of charge between the layers until the current is steady. Thus in the dielectric the voltage distribution is initially determined by the capacitances of the layers and finally by their resistances, with a transient condition intervening.

Figure 5.4 shows the results of stress calculations obtained for the case of the sudden application of voltage to a heated cable, showing the very long times, of up to 24 hours, needed before the stress distribution reaches its final form. When the polarity is suddenly reversed the stresses in the cable may be up to twice the normal maximum stress, slowly decreasing over some hours. Thermal transients will produce similar effects.

With a polarity reversal the charges in the conductor reverse immediately, whereas in the dielectric the change is very slow. After the reversal the charges in the dielectric and conductor have opposite signs and an overstress appears near the conductor. The polarity reversal may be considered as the superposition on the voltage V of a step function of magnitude $-2V$. Hence the gradient E is the superposition of the gradient E_v which existed with V, and $-E_t$ corresponding to

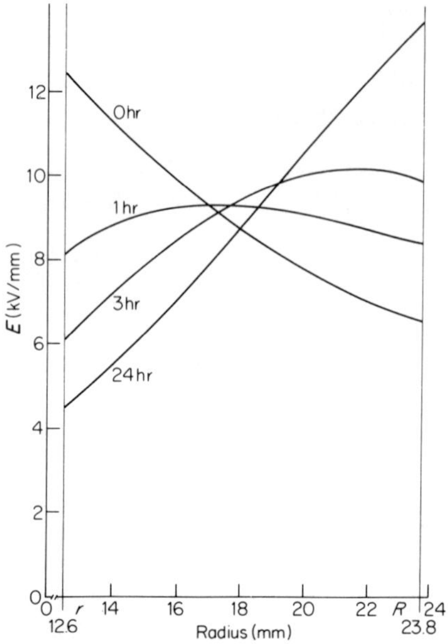

Figure 5.4 Stress distribution at various times after application of direct voltage 100 kV to a heated cable ($\Delta\theta = 20\,^\circ$C).[7] (Reproduced by permission of IEEE.)

($-2V$), i.e.

$$E = E_v - E_t$$

where,

$$-E_t = \frac{-2V}{r \ln(R/r)} \quad .$$

In Figure 5.5 the variations of electric stress for a polarity reversal are shown. A more refined model to represent the transient performance of a dielectric is shown in Figure 5.6. For this network the values of resistance and capacitance are calculated using the appropriate values of the permittivity at high frequencies and the d.c. conductivity. The application of this more refined method to dielectric design is described in Reference 7.

Summary of Design Criteria

Steady state – no load,

$$E_{r0} = \frac{V}{r \ln(R/r)} < E_d \text{ (steady state design maximum stress).}$$

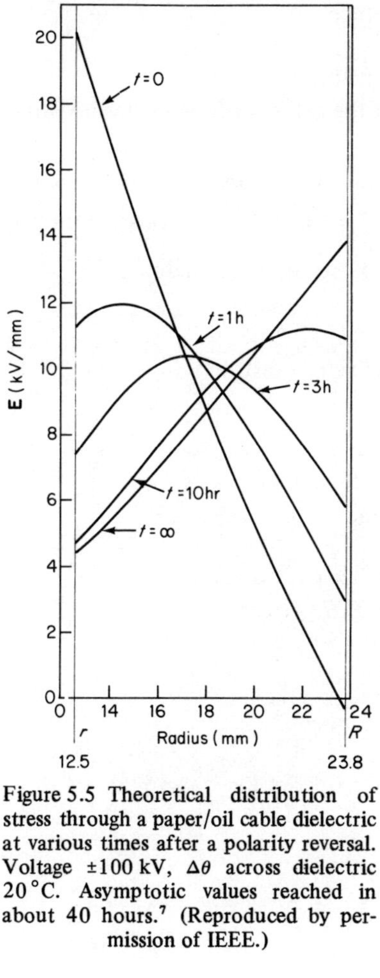

Figure 5.5 Theoretical distribution of stress through a paper/oil cable dielectric at various times after a polarity reversal. Voltage ±100 kV, $\Delta\theta$ across dielectric 20 °C. Asymptotic values reached in about 40 hours.[7] (Reproduced by permission of IEEE.)

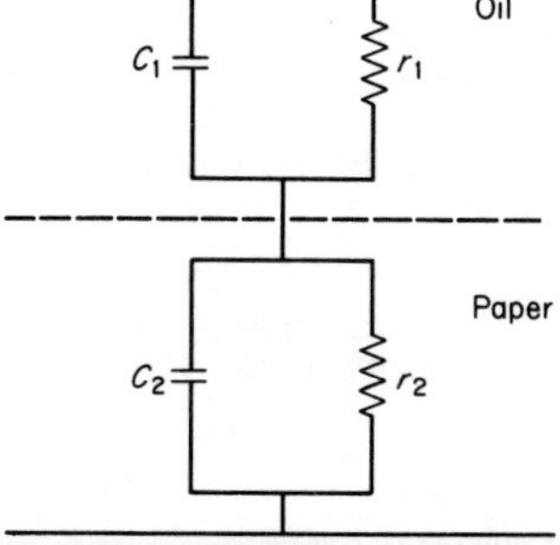

Figure 5.6 Equivalent circuit of a dielectric with layers of paper and oil.

Steady state – nominal load

E_{R0} (at sheath) $< E_d$.

Temperature drop across dielectric must be such as to produce this criterion.

Polarity reversal,

$$E = E_v - E_t = (\text{Steady state stress at conductor on load} - 2E_{r0})$$
$$< E' \text{ (transient breakdown strength} > E_d).$$

Surge of magnitude 3 V, stress at conductor $< E'$.

Example 5.1

Determine the steady-state stress distribution in the ±250 kV, d.c. cable specified as follows:

Conductor radius 17.75 mm
Radius over dielectric 28.05 mm
Conductor area 809 mm^2 (1.6 x 10^6 c mil)
Power = 600 MW
$\alpha = 0.1$ (1/deg C), $b = 3 \times 10^{-3}$ (cm/kV)
Conductor I^2R loss = 0.278 W/cm.

Solution

From Equation 5.6,

$$\ln E_x - \ln E_r + b(E_x - E_r) = \left(\frac{\alpha q g}{2\pi} - 1\right) \ln\left(\frac{x}{r}\right).$$

Hence,

$$bE_x + \ln E_x = bE_r + (\ln E_r) + \left(\frac{\alpha q g}{2\pi} - 1\right) \ln\left(\frac{x}{r}\right).$$

For a chosen value of E_r the variation of E_x with x is obtained by computer subroutine and the area under the $(E_x - x)$ curve obtained. This area is equal to

$$\int_r^R E_x \, dx = V = 250 \text{ kV}.$$

The area may be obtained graphically or by a computer subroutine.

Procedure

1. Let $E_r = 160$ kV/cm, $bE_x + \ln E_x = 5.555 + 2.1 \ln(x/r)$

E_x(kV/cm)	160	170	180	190	200	220	250	300
x (cm)	1.775	1.85	1.93	2.05	2.095	2.255	2.495	2.92

Total area (see Figure 5.7 = 231.1 kV.

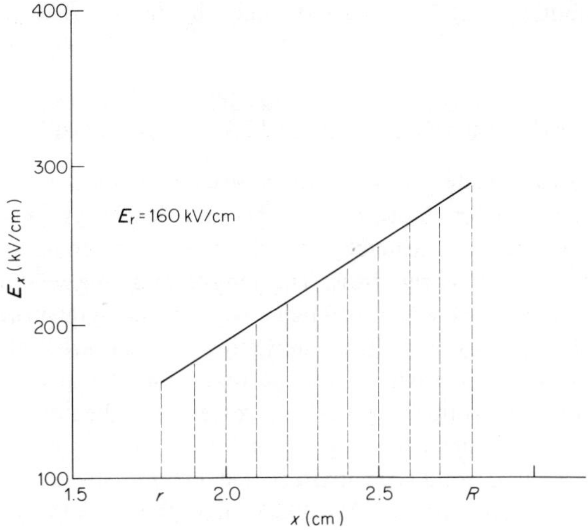

Figure 5.7 Example 5.1 Area under E_x–x curve for E_r = 160 kV/cm.

2. Let E_r = 180 kV/cm, $bE_x + \ln E_x = 5.73 + 2.1 \ln(x/r)$

E_x(kV/cm)	180	190	200	220	250	270	300	320
x (cm)	1.775	1.85	1.92	2.075	2.305	2.485	2.7	2.86

Total area = V = 252.9 kV.

3. E_r = 200 kV/cm, $bE_x + \ln E_x = 5.9 + 2.1 \ln(x/r)$.
Area = V = 277.78 kV.

4. E_r = 240 kV/cm, V = 314.9 kV.

5. E_r = 250 kV/cm, V = 322.87 kV.

6. E_r = 260 kV/cm, V = 351 kV.

7. E_r = 280 kV/cm, V = 377.76 kV.

8. E_r = 310 kV/cm, V = 406.45 kV.

From the graph of $E_r - V$, for V to be 250 kV, E_r = 178 kV/cm and $bE_x + \ln E_x = 5.71 + 2.1 \ln(x/r)$ and the stress distribution is:

E_x(kV/cm)	190	200	210	220	230	240	250	270	300	350
x (cm)	1.87	1.95	2.05	2.085	2.165	2.245	2.35	2.48	2.725	3.125

The stress distribution may be readily obtained by means of a simple computer program.

5.5 BREAKDOWN PARAMETERS AND TEST REQUIREMENTS[16,17,18,19]

The short-term steady state d.c. breakdown strength is of the same order as the impulse strength which for impregnated paper–oil dielectrics (oil-filled and solid types) is about 100 kV/mm. Generally, the short-term d.c. strength is at least three times higher than the working stress. The long-term strength is obviously more important, and for this there is a difference between non-pressurized or gas-filled dielectrics and oil-filled cables, the latter inhibiting corona (partial discharges).

Tests performed on the partial discharge repetition rate in air-filled cavities in impregnated paper dielectrics (typical of a severely drained solid-type cable dielectric in practice) indicate that ageing due to discharges is a major factor in determining the life of solid-type impregnated cables. The upper limit for the service stress is of the order of 25–30 kV/mm. With gas-filled cables the gas pressure enables a higher operating stress to be used. With oil-filled cables service stresses of 35–40 kv/mm are acceptable.

The following test requirements have been proposed for d.c. cables. At least 50 metres of cable including the relevant accessories must be subjected to 100 daily heating cycles to a temperature of 5°C in excess of the specified maximum and with a temperature drop across the insulation at least equal to that specified. The first 20 cycles to be made at twice working voltage, positive polarity, the following 20 cycles at the same voltage, negative polarity. The final 50 cycles to be made at 1.5 times working voltage with a polarity reversal (in approximately 1 second) every 2 hours during the heating period. Following this test, 10 metre lengths are cut from the cable and subjected to an impulse test and a high-voltage d.c. test whilst hot, i.e. maximum working temperature plus 5 °C. There is also a need to prove that the cable system can withstand surges originating from faults in the converter plant, and a typical form of test is to superimpose an a.c. voltage onto half the normal d.c. voltage such that the peak voltage equals twice the normal d.c. voltage, for durations up to 1 second.

When impulses are applied to a d.c. cable as a result of atmospheric surges they will be superimposed upon the d.c. voltage already present. Investigations have shown that the behaviour of impregnated paper insulation when subjected to these superimposed surges is rather unexpected. Because of the much greater resistivity of the paper as compared with the impregant the d.c. stress appears almost entirely across the paper layers, whereas the transient voltage stresses the paper and oil in relation to their permittivities. If the applied impulse is of the same polarity as the d.c. voltage then the breakdown at ambient temperature is a function only of the resultant peak stress. If the polarities of the two components are opposite, the breakdown is controlled only by the surge component.

These inferences have been obtained from model tests rather than on actual cables, but they indicate that separate impulse and d.c. tests do not necessarily indicate the behaviour of the dielectric when subjected to combined voltages. This

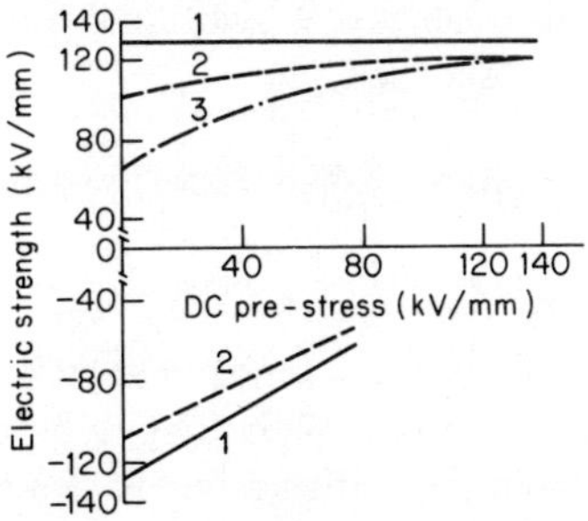

Figure 5.8 Electric strength of oil/paper insulation for surges superposed on direct voltage.[16] (1) Impulse on d.c. (2) Oscillatory surge on d.c. (3) a.c. on d.c. (Reproduced by permission of CIGRE from Report 21-03, 1974.)

is of special importance for systems in which a surge limiting device, such as a rod gap, is used. These gaps may be set to flashover at, say, twice the d.c. voltage; flashover occurs at much lower surge voltage than for opposite polarities. The cable must be designed, however, to withstand the maximum surge voltage, corresponding in this case to about 3 times the d.c. voltage. The electric strength of oil/paper for various overvoltage conditions is shown in Figure 5.8. The decrease in electric strength with increase in temperature of paper/oil is shown in Figure 5.9.

Permissible ratios of overvoltage to nominal direct voltage are as follows for paper/oil cables.[6]

Impulse (additive) on cold cable:

$$\frac{V_{\mathrm{imp}}}{V_{\mathrm{d}}} = 1.9 \text{ for oil-filled and 2.8 for solid-type.}$$

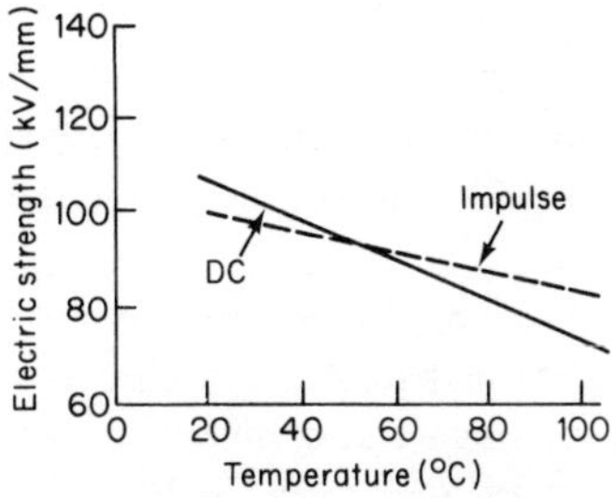

Figure 5.9 Electric strength dependence on temperature (paper/oil cable model 1.5 mm thick).[16] (Reproduced by permission of CIGRE from Report 21-03, 1974.)

Impulse (subtractive) to warm cable: 1.9 (oil-filled) and 2.8 (solid). Alternating voltage (power frequency) for 3 seconds,

$$\frac{V_{\mathrm{rms}}}{V_{\mathrm{d}}} = 1.3 \text{ (oil-filled) and } 1.5 \text{ (solid).}$$

The normal polarity reversal test uses $1.5\,V_d$ and gives rise to stresses of 2.5 x normal working stress. This is no problem for oil-filled cables but can be a design constraint for gas-filled cables. With long submarine cable the refracted portion of a lighting surge from an overhead line is reduced because of the cable's low surge impedance. When the cable is short, however, a voltage build-up occurs because of reflection at both ends. Generally a surge of subtractive polarity is more severe than one of the same polarity.

5.6 TYPES OF CABLE AND ACCESSORIES

'Solid' Type (Oil-impregnated Paper)[5,19]

Cables of the so-called 'solid' type have been successfully used for several d.c. systems, as seen in Table 5.1. Since this type of cable after heat cycling inevitably contains void spaces filled with gas, often below atmospheric pressure, it is of limited usefulness in a.c. systems because of the damaging effects of discharges in these voids. In a.c. cables discharges occur at rate of a least 100 per second, i.e. each half cycle, and the rate at which damage to the dielectric occurs depends upon this discharge rate. In a d.c. cable containing voids, repetitive discharges will still occur but their rate will be very much reduced and will depend upon the dimensions and surface resistivity of the void spaces. For solid cables the rate will certainly be several orders of magnitude less than for a.c. cables. The discharge rate is increased by stress reversals.

The solid type of cable has been shown to be prone to failure during the load cycle test and that failure almost always occurs between 1/2 hour and 2 hours after switching off the load current (cable cooling down). Increasing the rate of cooling increases the tendency to failure. This behaviour is thought to be due to over-stressing of the paper tapes brought about by the relative immobility of space charge in the dielectric. A correct choice of insulating papers and impregnant can greatly reduce this effect, the objective being to promote movement of charge within the dielectric.

This type of cable is attractive for submarine use because it is entirely self-contained, but for more general use the oil-filled cable is likely to be a more economic alternative. The solid cable also suffers from compound migration which reduces the electric strength of the cable. Solid cables for d.c. use have been employed up to 250 kV and with a stress of 25.0 kV/mm. These values probably represent the upper limit of their usefulness. A cross-section of such a cable is shown in Figure 5.10. The cross-section is generally circular for water depths of less than 90 m and is oval for greater depths to avoid creasing the lead sheath by the water pressure in thermal cycling. The electric strength[6] on polarity reversal is

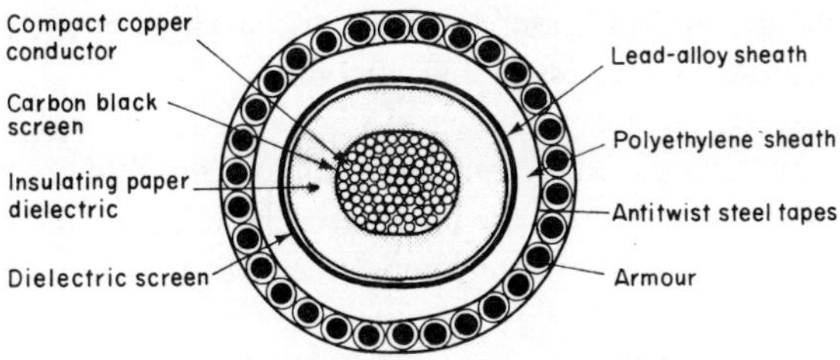

Figure 5.10 Cross-section of impregnated paper (non-pressurized) cable.

50 kV/mm and for normal direct voltage with additive or subtractive impulse is 90 kV/mm.

Oil-filled Cables[9,21,22]

These have superior electrical properties compared with solid cables. They do not suffer from internal discharges nor are they subject to overstressing during transient cooling conditions because of the fluidity of the oil which has a 'washing' action in the dielectric. Being liquid-filled this cable may be laid to water depths of 1000 m. The design is dependent upon the oil feeding lengths. However, by using the lowest viscosity oil (5 centistokes at 20 °C), an oil duct of diameter 25–30 mm, and a feed pressure of 25 atmospheres, a water crossing of length 60 km may be envisaged. The electric strength[6] on polarity reversal is 70 kV/mm, and for impulse on normal direct voltage is 95 kV/mm. The limit to conductor size for submarine cables at the present time is 1300 mm^2 copper. An advantage for this type is that the oil under pressure will flow out of a damaged cable, preventing the ingress of water. Conductor stresses of 40 kV/mm have been used.[21]

Gas-filled Cables

High pressure gas-filled cable has been used for d.c. in the 250 kV Cook-Strait Circuits.[24] Higher pressure gas-filling was adopted to ensure that in the event of a sheath failure water could not enter the cable even at the maximum depth. The gas

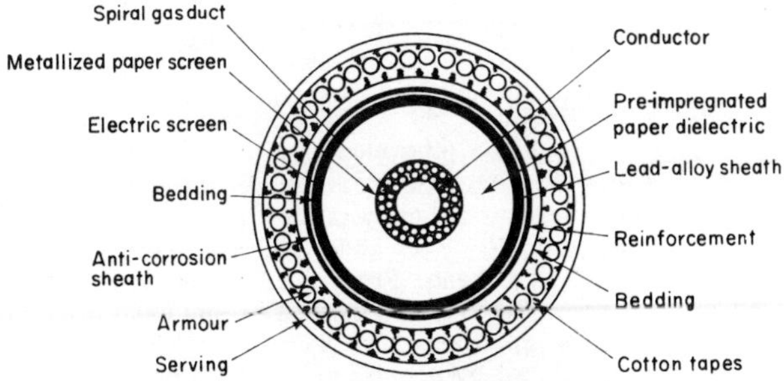

Figure 5.11 Cross-section of gas-filled cable.

also inhibits discharges and increases the dielectric strength of the cable. This cable was designed for a maximum stress of 25.0 kV/mm.

A typical cable cross-section is shown in Figure 5.11. The insulation consists of paper tapes pre-impregnated with petroleum compound. With damage, it is probable that water would ingress into the cable, especially if a gas duct was obstructed by the compound. The maximum laying depth seems to be 300 to 400 m.

Extruded Dielectric Cables[25,26]

Extruded polyethylene cables are inherently suitable for use with high voltage d.c. but the difficulties of ensuring consistently reliable manufacture, which have limited the usefulness of this cable in a.c. systems, apply to d.c. As yet no high voltage d.c. extruded cable has been used commercially. The advantages of polyethylene compared with oil-impregnated paper are much the same as for a.c. cables and include lower thermal resistivity (3.5 deg C-m/W). Also, discharge activity in cavities is reduced on d.c. The incorporation of fillers into polyethylene can control the electrical and thermal resistivities, although the long-term effects have yet to be ascertained.

The variation of electrical resistivity with stress at up to 80 °C and 30 kV/mm for a number of possible materials for extruded cables has been measured.[25] Although the temperature coefficient ***a*** appears well defined, ***b*** requires further investigation. The measured values of resistivity decrease with both stress and temperature. The electric stength–temperature characteristics of PVC, polyethylene, and EPR in sheet form are shown in Figure 5.12. Experimentation to obtain the variation of the conductivity of polyethylene with stress and temperature has shown that the conduction current takes a considerable time to reach a steady value, e.g. many days or even weeks, especially at near-ambient temperatures. The difficulty in obtaining true steady values has prompted the investigation of the effects of additives which may decrease the time to steady-state.

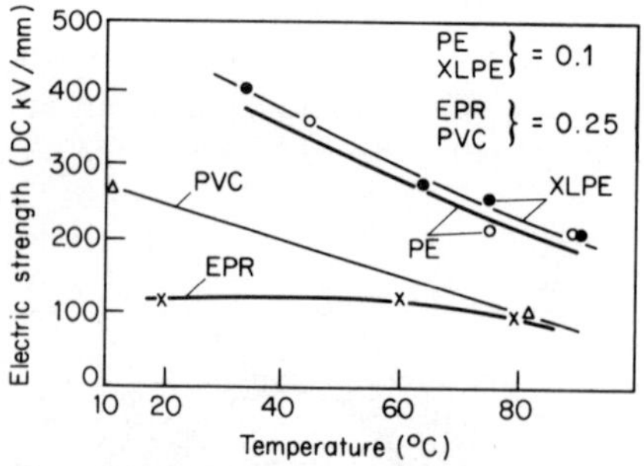

Figure 5.12 Variation of electric strength (fast-rise) with temperature for polyethylene (PE), PVC, EPR, and crosslinked PE (XPLE) sheets. PE and XLPE samples approximately 0.1 mm thick; EPR and PVC samples approximately 0.25 mm thick.[6] (Reproduced by permission of CIGRE from Report 21-03, 1974)

For cross-linked polyethylene insulation the value of the temperature coefficient of resistivity (a) is about 0.1 deg C^{-1}; a wide range of values of b have been reported, ranging from 0.1 kV/mm to 0.6 kV/mm. There appears to be a temperature dependence for b. The different values obtained from different reports probably stem from variations in additives and manufacturing methods.

A test cable for ±250 kV using crosslinked polyethylene has been reported.[16] The sample cables were designed for a standard impulse of 1170 kV at room temperature and a maximum (d.c.) working stress of 15 kV/mm. Tests on the cable resulted in many breakdowns and although the thin sample d.c. breakdown strength is more than 100 kV/mm the cables (wall thickness of 25 mm) broke down on long-term test at 20 kV/mm. This is probably caused by abnormally high stress which may be caused by accumulation of space charge or by non-uniformity of resistivity. Clearly, further development is required for this type of d.c. cable.

Accessories

It has been seen that the calculation of stress in a d.c. cable is more difficult than for a.c. Whilst the electric field in a cable is radial and amenable to calculation this is no longer true in an accessory in which the electric field is only calculable by field-plot methods. There are therefore complications in approaching the design of d.c. accessories by the normal process of calculating the field distribution. However, the iterative methods referred to in Chapter 3 for the calculation of temperature and electric stress may be readily adapted to this problem.

The approach so far has been to take established designs of accessories for a.c. systems and subject them to the type tests necessary for d.c. It has been found that accessories designed for a.c. also function satisfactorily on d.c. An exception is the case of outdoor terminations, for which the creepage distance necessary for d.c. is considerably greater than for a.c. in seaside or heavily polluted conditions. The design of a thorough joint for a proposed ±600 kV HPOF pipe-type cable is shown in Figure 5.13.

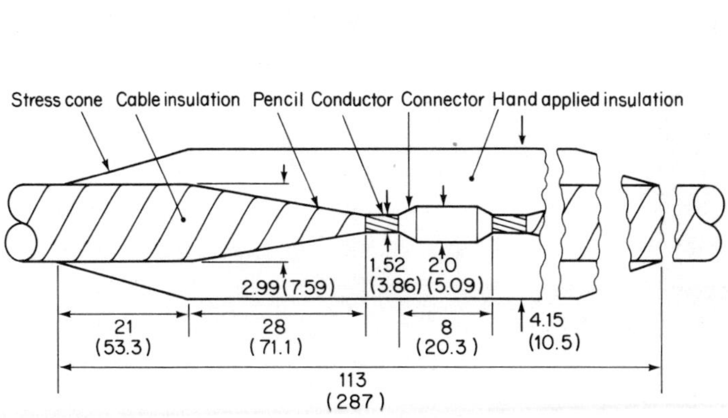

Figure 5.13 Schematic of joint for HPOF pipe-type cable rated at ±600 kV. Dimensions are in inches (cm).[30] (Reproduced by permission of IEEE.)

5.7 SUBMARINE CABLES[10,19,20,22,23,27,28,29]

This section applies equally to a.c. cables, but as the major application of d.c. cables is for underwater links it is included in this chapter. A summary of site preparation, earth work, and general environmental aspects related to river crossings is given in Section 1.9. If possible the cable is buried below the river bed to avoid problems from ships' anchors, etc., although for wide crossings, e.g. more than a kilometre, the cable tends to be laid on the bottom with perhaps partial burial in shallow, near-shore areas. Methods for cable burial are:

(*a*) Cut and cover – cable placed in trench and covered.
(*b*) Jetting – cable placed on bottom and underlying bed loosened by water jet allowing the cable to sink.

Several installation methods are used, such as pulling the cable from shore to shore, floating the cable from shore to shore and then removing buoyancy, and finally laying the cable from a barge or ship. The pipes for pipe-type cables are naturally buoyant and must be weighted down with concrete before cable installation.

The problems concerning submarine cables are mainly mechanical.[21,22] When the armoured cable is coiled into the factory or ship tanks it requires a 360° twist in the armour for each loop. If the armour wires are laid-up at an angle of lay (α) the above twist results in an elongation of $(r \tan \alpha)/R$ per unit cable length, where r and R are the cable and coil radii. If r is large, damage to the dielectric and joints may result. Also, 'birdcaging' may result at a point where the outer covering is weak enough to be torn by the pressure of the armour wires.

When on the drum several turns of cable are necessary to hold the tension during the laying operation. This produces a movement of the conductor relative to the armour wires, and possible dielectric damage. During laying, the cable is anchored at the drum (on the ship) and on the shore. Between the drum and the point at which the cable touches the sea bed the cable is subjected to a force **F** determined by the cable weight. This gives rise to a torque **T** (see Figure 5.14) due to the helically spiralled armour wires, given by **F** $r \tan \alpha$. The armour increases its lay on

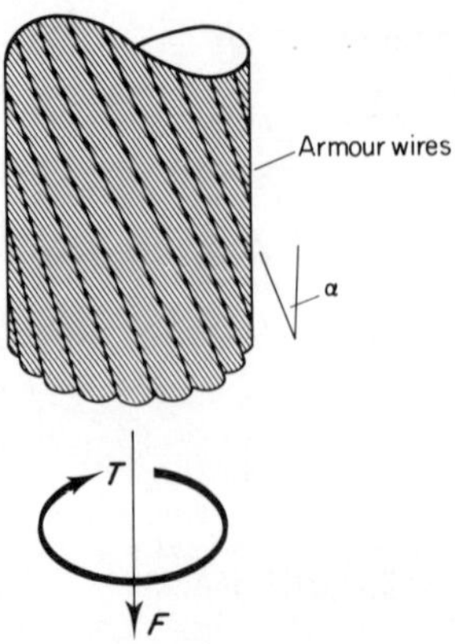

Figure 5.14 Torque induced in armour wires along a longitudinal force in a submarine cable.

the upper part of the suspended cable but returns to its normal lay on the lower part, resulting in torsion and elongation with possible dielectric deformation. This can be limited by the use of tapes wound helically underneath the armouring and in the opposite direction.

If the mechanical tension absorbed by the conductor is $\mathbf{T}_1$ and R the minimum radius of the reels of the laying machine, then the force of the conductor on the dielectric[22] is $(\mathbf{T}_1/R)$. For example for a depth of 1000 m and a cable whose weight per metre in water is 20 kg, $\mathbf{T}_1$ can reach (during recovery) a value of 20×10^3 kg. Hence with a 3 m diameter reel the force is about 13 tonnes per m. However, paper/oil is well able to withstand this pressure without damage to the tapes.

Summarizing, the cable must de designed to withstand both laying and recovery forces without damage. It is possible for the design to be such that the conductor takes not more than 40% of the total tension. This makes possible the use of heavy cables to depths of 1000 m without damage.

Manufacturing facilities exist for the manufacture of continuous (joint-free) lengths of single-core cable up to 64 km in length. Such long lengths tend to give less trouble in service than conventional shorter lengths with many joints. However, the extra cost of the manufacturing plant required must be weighed against this extra reliability. An interesting use for high power submarine cables a.c. or d.c. in the future could be the connection of off-shore floating nuclear plants to the mainland system.

Current loadings

The value of thermal resistivity to be used for the cable environment is often difficult to determine. Although the cable outer surface could be in contact with the water and hence at that temperature often the cable becomes partially or wholly embedded in sand or mud which respresents an appreciable thermal resistance. Often a value of 0.3 deg C-m/W is used and as a rough estimate the rating for the underwater portion of a cable link is increased by 40% over the land cable.[27]

5.8 LEAKAGE CURRENTS

The leakage current density $\mathbf{J}$ in a d.c. cable is given by (stress/resistivity). The order of magnitude of this current is obtained as follows.

Assume the resistivity to be 10^{16} Ω-cm at 20 °C, $\mathbf{E}$ to be 25 kV/mm, and α (paper/oil) to be 0.1 deg C^{-1}. Then

$$\rho = \rho_{20} e^{-\alpha(\theta_1 - 20)}$$

if the stress dependency is neglected. Consider a sheath temperature of 50 °C for a buried land cable. Then

$$\rho = 10^{16} e^{-3} = 0.05 \times 10^{16} \quad \text{and} \quad \mathbf{J} = 0.5 \times 10^{-12}\ \text{A/cm}^2.$$

For a paper–oil underwater cable (sheath temperature ⊁15 °C)

$$\rho = 10^{16}\ e^{+0.5} \quad \text{and} \quad J = 0.015 \times 10^{-12}\ \text{A/cm}^2.$$

These currents are very low and should not constitute an electrolytic hazard to the sheath material as the current enters the surrounding medium.

In designs where the sheath is insulated from the armouring the leakage current flows out at both ends. A fault in this insulating layer can be caused by the passage of short-circuit current through the sheath, setting up large longitudinal voltage drops. A local puncture in the sheath/armour insulating layer can result in a concentration of leakage current which may be sufficient for electrolysis to occur. To avoid this, the sheath and armour are electrically connected at intervals in the order of 2 km, which limits the voltage to about 5 kV for a surge of 1.7 V_{dc}.

The voltage difference between the sheath and the sea during transient conditions, e.g. a flashover at the cable terminals, is given by

$$V \frac{Z_2}{Z_1 + Z_2} [1 - e^{xR/2Z_2}]$$

where

V = transient voltage, conductor to sheath
Z_1 = surge impedance, conductor to sheath
Z_2 = surge impedance, sheath to armour
R = resistance of lead sheath
x = distance from grounded terminal of sheath.

5.9 SUMMARY

For solid (oil-impregnated) type cables the long-term working stress is about 25 kV/mm with a maximum temperature not exceeding 50 to 60 °C, to avoid undue displacement of the insulating compound. The maximum working voltage is limited to 250–300 kV. Internally generated overvoltages (i.e. in the d.c. system) limit the nominal stress to 25 kV/mm. This type of cable is suitable for long submarine lengths, but the conductor size is limited to about 1300 mm^2. The thermal resistivity of the insulation is relatively high, and the capability of a ±250 kV bipolar circuit limited to about 600 MW.

With gas-filled cables the nominal stress is limited to 25 kV/mm by impulse and polarity reversals. The long-term working stress is marginally higher than this. This type is of interest mainly for long submarine links. Oil-filled cables have a long-term working stress of 35 kV/mm. With impulse withstand of up to three times the working voltage the nominal stress is 35–40 kV/mm. Internally generated overvoltages and polarity reversals do not constitute a limitation. This cable may be used for land and submarine links up to a maximum length of 60 km. With a maximum service voltage of the order of ±750 kV power ratings of 2 to 3 GW could be achieved. A design for an oil-filled pipe type cable is reported in reference 30.

The specifications of three different types of d.c. cable are as follows:

(*a*) *Kingsnorth–Willesden, U.K.*
LPOF ±266 kV; 1200 A (all year) – depth 1 m in stabilized backfill, g_s of 1.2 deg C-m/W.
Conductor – temperature 85 °C, 800 mm² cross-section, diameter over screen 55.5 mm.
$\Delta\theta$ across dielectric 14 deg C on full load, dielectric thickness 10.3 mm.
$\mathbf{E}_{max}$ 32 kV/mm (withstands 800 kV impulse).

(*b*) *Vancouver Island to British Columbia (Canada)*
Solid-type paper/oil, 300 kV, 1200 A.
Conductor – sea cable 800 mm², land cable 1800 mm², temperature 45 °C.
Overall diameter 99 mm (sea), 117.2 mm (land).
$g_{sea\ bed}$ 0.40 deg C-m/W g_{mud} 0.6 deg C-m/W.
$\mathbf{E}_{nominal}$ 28 kV/mm, $\Delta\theta$ (dielectric) 20 deg C on full load.

(*c*) *Cook Strait (New Zealand)*
Gas-filled cable, ±250 kV, 1200 A, $\mathbf{E}_{max}$ 25 kV/mm.
Conductor section (sea 515 mm² (copper), 60 °C maximum temperature.
Pre-impregnated paper insulation – nitrogen filled at 2.95 MN/m².

REFERENCES

1. C. ADAMSON and N. G. HINGORANI, *High Voltage Direct Current Power Transmission*, Garraway, London, 1960.
2. E. W. KIMBARK, *D.C. TRANSMISSION*, Vol. 1, Wiley, New York, 1972.
3. B. M. WEEDY, *Electric Power Systems*, Third Ed., Wiley, Chichester, 1979.
4. ROGERS, E. C. and SKIPPER, D. J., 'Gaseous Discharge Phenomena in High Voltage D.C. Cable Dielectrics', *Proc. I.E.E.*, **107**, 241–251 (1960).
5. GAZZANA PRIAROGGIA, P., MORELLO, A., and PALMIERI, N., 'Research on the electric strength of fully impregnated paper dielectrics for h.v.d.c. cables', *CIGRE* Paper No. 205 (1962).
6. UHLMAN, E., *Power Transmission by Direct Current*, Springer-Verlag, Berlin, 1975.
7. OCCHINI, E. and MASCHIO, G., 'Electrical characteristics of oil-impregnated paper as insulation for h.v.d.c cables', *I.E.E.E. Trans. PAS-86*, No. 3, pp. 312–326 (March 1967).
8. OCCHINI, E. and MASCHIO, G., 'Transient and steady-state electrical conductivity of paper impregnated with synthetic oil as insulation for e.h.v. d.c. oil-filled cables', *1967 Annual Report, Conference on electrical insulation and dielectric phenomena*, National Academy of Sciences, Washington, pp. 178–186.
9. GAZZANA PRIAROGGIA, P., PALANDRI, G. and PALMIERI, N., 'The use of oil-filled cables for high voltage d.c. transmission systems', *CIGRE*, Paper No. 206 (1966).
10. OUDIN, J. M. *et al.*, 'Design and development of d.c. cables', *I.E.E.E. Trans. PAS-86*, pp. 304–310 (1967).

11. CHU, S. C., 'Design Stresses and Current Ratings of Impregnated Paper Insulated Cables for H.V.D.C.', *I.E.E.E. Trans. PAS-86*, pp. 1029–1036 (1967).
12. OUDIN, J. M. and THEVENON, H., 'Theory of d.c. cables – Calculation of gradient and its correlation with breakdown gradient', *CIGRE*, Paper No. 208 (1966).
13. HUGHES, K. E. L. and NICHOLSON, J. A., 'Design features of high voltage d.c. land cables', *I.E.E. Conference on high voltage d.c. transmission, London, 1966*, Paper No. 65, pp. 315–317.
14. OUDIN, J. M. and FALLOU, M., 'Some considerations on the design of d.c. cables', *I.E.E. Conference on high voltage d.c. transmission, London, 1966*, Paper No. 66, pp. 318–321.
15. NYBERG, B. R. *et al.*, 'Numerical Methods for Calculation of Electrical Stresses in H.V.D.C. Cables with Special Application to the Skagerrak Cable', *I.E.E.E. PAS-94*, pp. 491–497 (March/April 1975).
16. YOSHIDA, K. *et al.*, 'Research and Development of H.V.D.C. Cables in Japan', *CIGRE*, Paper No. 21–03 (1974).
17. C.I.G.R.E. Study Committee No. 21, 'Recommendations for tests on d.c. cables for a rated voltage up to 550 kV'. *CIGRE*, SC/CE 21–69/04, SC/CE 21–70/01, SC/CE–71/01.
18. BAHDER, G., GARCIA, F. G., and BROOKES, A. S., 'Insulation co-ordination in high voltage d.c. cables', *CIGRE*, Paper No. 21–03 (1972).
19. HANSSON, B., 'A submarine cable for 100 kV d.c. power transmission', *I.E.E.E. Trans. PAS-73*, Part III-A, pp. 599–610 (June 1954).
20. GAZZANA PRIAROGGIA, P. and MASCHIO, G., 'Continuous long length a.c. and d.c. submarine h.v. power cables – The present state of the art', *I.E.E.E.*, Paper No. 773, pp. 127–8 (1973).
21. LAST, G. H., GAZZANA PRIAROGGIA, P., and MIRANDA, F. J., 'The Underground h.v.d.c link for the transmission of bulk power from the Thames estuary to the centre of London', *I.E.E.E. Trans. PAS-90*, No. 4, pp. 1893–1901 (July/August 1971).
22. MASCHIO, G. and OCCHINI, E., High Voltage Direct Current Cables – The State of the Art', *CIGRE*, Paper No. 21–10 (1974).
23. OUDIN, J. M., 'Le câble sous-marin à courent continu Sardiagne–Italie', *Revue Generale de L'Electricité*, **75**, No. 12, 1391–1398 (1966).
24. WILLIAMS, A. L., DAVEY, E. L., and GIBSON, J. N., 'The 250 kV d.c. submarine power cable interconnection between the North and South Islands of New Zealand', *Proc. I.E.E.*, **113**, No. 1, 121–133 (January 1966).
25. WEEDY, B. M. *et al.*, 'Parameters of Extruded Insulation for H.V.D.C. Cables', *Proc. I.E.E.E. Conference Electrical Insulation, 1978*, pp. 46–49.
26. KIKUCHI, K. *et al.*, 'Characteristics of d.c. Dielectric Breakdown of Plastic Insulation', *Proc. I.E.E.E. Conference Electrical Insulation, 1978*, pp. 327–332.
27. BARNES, C. C., *Submarine telecommunication and power cables*, I.E.E. Monograph Series 20, 1977.
28. OUDIN, J. M., EYRAUD, I., and CONSTANTIN, L., 'Some mechanical problems of submarine cables', *CIGRE*, Paper No. 21–08 (1972).
29. CIGRE Study Committee, No. 21, 'Recommendations for mechanical tests on submarine cables', *Electra*, pp. 990–103 (October 1972).
30. BAHDER G. *et al.*, 'Development of ±400 kV/±600 kV High and Medium Pressure Oil Filled Paper Insulated D.C. Power Cable Systems', *Trans. I.E.E.E. PAS-97*, pp. 2045–2056 (1978).

PROBLEMS

Problem 5.1

Prove that for a coaxial cable with dielectric of constant electrical resistivity ρ and permittivity ϵ that the insulation resistance (d.c.) is given by

$$\frac{\rho}{2\pi}\ln\left(\frac{R}{r}\right) \text{ per metre length}$$

where R and r are the radii to the screen over the dielectric and to conductor respectively. Show that the product of capacitance and insulation resistance is independent of cable dimensions. (Answer: $R \times C = \rho\epsilon$, where $\epsilon = \epsilon_r \epsilon_0$.)

Problem 5.2

A d.c. cable has an extruded dielectric of relative permittivity 2.2 and electrical resistivity at 20 °C of 10^{16} Ω-cm. Determine the time for the cable to lose half of its initial voltage when isolated from the d.c. source and short circuited. (Answer: 1350 seconds.)

Problem 5.3

The expression for the conductivity of a d.c. cable dielectric is given by

$$\sigma = \sigma_0 \exp(\alpha\theta + b\text{E}).$$

Determine the value of temperature coefficient α such that the electric stress through the insulation of a coaxial-type cable is constant. Assume conductor loss is q w/m and dielectric thermal resistivity to be g_d deg C-m/W. For a cable of conductor loss per m of 30 W with a g_d of 5 deg C-m/W, calculate the value of α for uniform radial stress. (Answer $\alpha = 2\pi/g_d q = 0.042$/deg C.)

Problem 5.4

The parameters in the conductivity equation for the polyethylene insulation in a single-core extruded cable are as follows:

$$\alpha = 0.07 \text{ deg C}^{-1} \quad \text{and} \quad b = 0.015 \text{ cm/kV}$$

$$\sigma_0 = 10^{-17}\ (\Omega\text{-cm})^{-1}.$$

The radius over the conductor is 28.4 mm and over the dielectric 54.4 mm. The thermal resistivity of the polyethylene is 3.5 deg C-m/W and the circuit rating is 315 MVA for a system of voltage ±100 kV. If the cable is buried in soil the effective thermal resistance is 1.6 deg C/W per metre length. Calculate the stress at the sheath on full-load. (Answer: 5.2 kV/mm.)

Problem 5.5

Design a 100 kV d.c cable having the following parameters and calculate the current rating for installation underwater ($\theta_{sheath} \not> 15\ ^\circ C$) and underground ($\theta_{sheath} \not> 50\ ^\circ C$). Parameters:

Area conductor 1000 mm^2 (effective)
Space factor 0.8
Design working stress 25 kV/mm
Insulation – solid mass impregnated paper/oil
$\alpha = 0.1$ deg C^{-1} $\quad \sigma_{\substack{50^\circ C \\ E=0}} = 10^{-14}\ \Omega^{-1}\ m^{-1}$

$(1/g_d) = 0.0017$ W/deg C-cm.

Assume ratio, radius over dielectric to radius over conductor to be 2 and ignore stress dependence of conductivity. (Answer: $r = 20$ mm. No load, **E** on conductor 7.23 kV/mm. Underwater, $\mathbf{I} = 2160$ A and $\mathbf{E}_{sheath} = 21.8$ kV/mm (Full load). Stress at conductor on polarity reversal 14.3 kV/mm. Buried: $\mathbf{I} = 1070$ A, $\mathbf{E}_{sheath}$ (full load) 6.4 kV/mm, $\mathbf{E}_{conductor}$ on polarity reversal 12.5 kV/mm.

Problem 5.6

The resistivity of a single-core, high voltage, d.c. cable insulation may be assumed to vary with the square of the radius measured from the centre of the cable. The diameter over the conductor is 25 mm and over the insulation 70 mm. Determine the electric stress at the conductor and sheath if the conductor voltage is 100 kV. (Answer 1.17 kV/mm, 3.28 kV/mm).

Problem 5.7

A self-contained oil-filled single core cable has a 800 mm^2 copper conductor of diameter (over screen) 35.5 mm. The dielectric thickness is 10.3 mm. The cable is installed in ground of g 1.2 deg C-m/W at a depth of 1 m with natural cooling.

Determine approximately the power capability per cable core for the following schemes: (i) ±266 kV d.c. two-pole scheme, cables spaced 10 cm apart; (ii) 3-phase a.c. scheme with 3 cables in horizontal formation with 10 cm spacing, $\mathbf{E}_{ac}$ at conductor $\not> 12$ kV/mm and $\mathbf{E}_{impulse}$ at conductor $\not> 90$ kV/mm.

Problem 5.8

A ±266 kV d.c. oil-filled cable is used for an underwater crossing. Between the lead sheath and steel wire armouring there exists a sheathing of polyethylene. At intervals of 3 km the lead sheath and armouring are connected electrically. The basic dimensions of the cable are as follows:

Diameter over conductor	= 24 mm
Diameter over insulation	= 62 mm
Diameter over lead sheath	= 68 mm

ϵ_r paper/oil = 3.5
ϵ_r polyethylene = 2.2
Diameter of polyethylene sheath = 74 mm
Resistivity of lead sheath = 23×10^{-6} Ω cm at working temperature.

Determine the maximum surge voltage that occurs between the lead sheath and armour when an overvoltage of 1.7 x operating voltage occurs at a terminal of the cable. (Answer: 8.23 kV).

6

Compressed Gas Insulated (CGI) Cables

6.1 INTRODUCTION

The application of the electronegative gas sulphur-hexafluoride (SF_6) for electrical insulation purposes on a large scale occurred in the 1960s, initially for circuit breakers. This led to the gas-insulated metal clad substation with the environmental advantage of much smaller physical size and less visual impact. The design of gas-spacer transmission line has evolved from the substation philosophy and consists of three isolated phases. The rigid conductor is concentrically supported in the outer tube as shown in Figure 6.1 by insulator-spacers, often constructed of epoxy. Cable sections are fabricated in lengths of the order of 13 m, this being limited by transportation. The gas pressure is 3 atmospheres (50 psig) at room temperature. Development work is in progress to establish the design of a cable with three conductors in one containing pipe or sheath; this will be discussed because of its potential advantages.

Both forms of rigid spacer cable have the following advantages compared with traditional oil-filled systems.

(*a*) Large cross-sectional area conductor tubes are possible thereby decreasing the conductor loss for a given power.

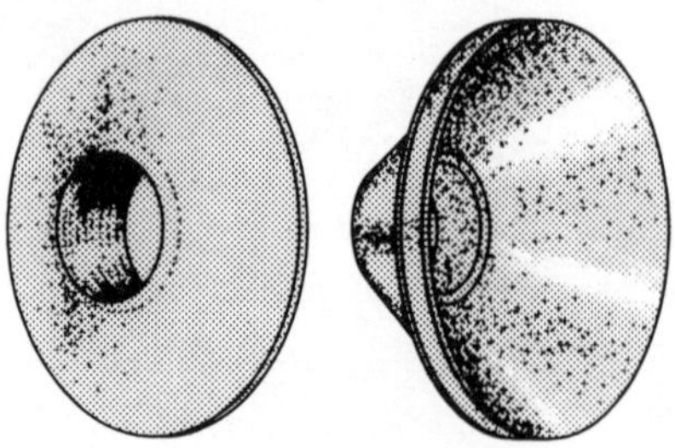

(a) (b)

Figure 6.1 Spacer-insulators: (*a*) disc; (*b*) conical.

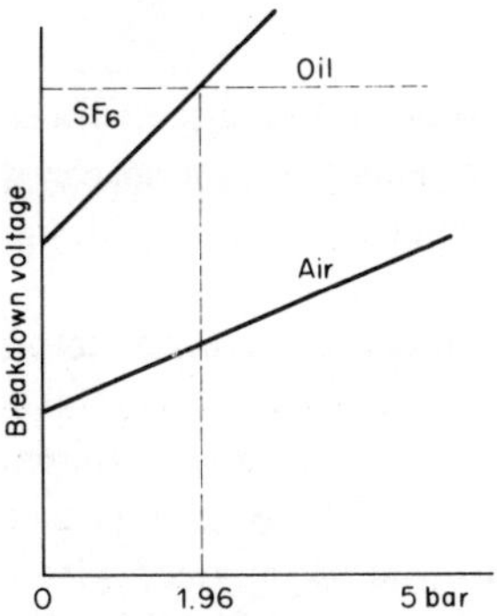

Figure 6.2 Comparison of SF_6, air, and oil. Variation of breakdown voltage with pressure.

(*b*) Small dielectric loss and much reduced reactive compensation needed owing to the low value of the relative permittivity of the gas.
(*c*) More efficient heat transfer from the conductor to the sheath.
(*d*) Simpler terminations because of the isotropic nature of the gas insulation.

The electrical advantage of SF_6 over air is shown in Figure 6.2; further it is non-toxic, chemically stable, and nonflammable. On the debit side it is expensive (£2.2/kg, 1977 prices) and at 0 °C it liquefies at a pressure of 22 bars, which precludes its use at high pressures unless heaters are used. The density of SF_6 is 6.13 g/litre compared with 1.22 for air. The basic components of a typical isolated phase system are shown in Figure 6.3, and installations may be divided into subsections by means of stop-joints which isolate the SF_6 between sections. The stop-joints have bypass valves which are left open in normal operation to allow gas flow from one section to another. Corrosion protection is required for the outer pipe when buried, and coatings such as epoxy or bitumen are applied with cathodic protection if considered necessary.

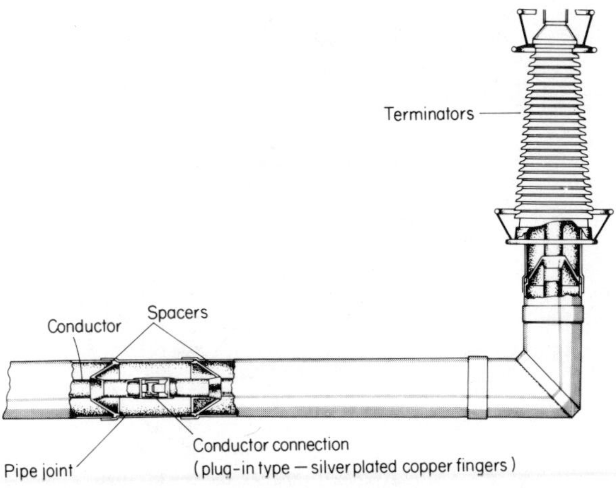

Figure 6.3 Components of SF_6 gas-insulated transmission line. (Reproduced by permission of Gould-Brown Boveri, Power Systems Division, Greenburg, Pennsylvania, USA.)

Flexible CGI Cable

The advantages of a flexible form of spacer cable with 100–200 m lengths transported on a reel are considerable. Initial studies[1] have shown that a spacer is required every half-metre length (single-core cable) and that a satisfactory design using disc-type or similar spacers is unlikely.

A development[2] which shows promise is the use of two or more subconductors per phase as shown in Figure 6.4, the subconductors being at the same potential. The ratio of gas stress to spacer-surface stress can be varied by using different number and diameter of subconductors. A cable working at 13.5 atms (g) requires a stress ratio of 2 to be tolerant of contamination and this can be obtained with a trefoil spacer as shown in Figure 6.5. Work on this form of cable is still in the research stage. Development of a cable with lapped polyethylene tapes pressurized with SF_6 has been reported.[3,4]

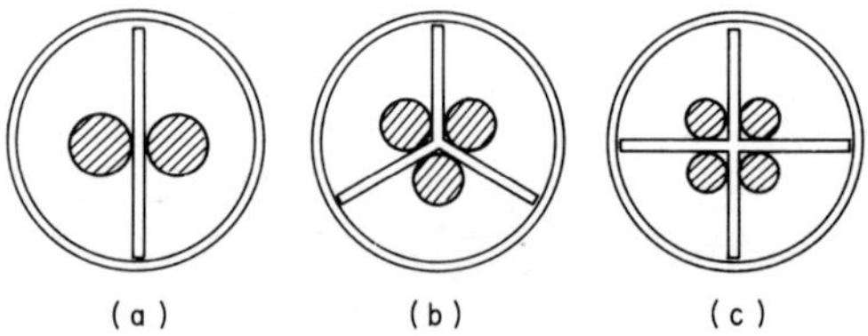

Figure 6.4 Flexible SF_6 spacer cable: cross sections of cables with multiblade spacers. (Reproduced by permission of IEE.)

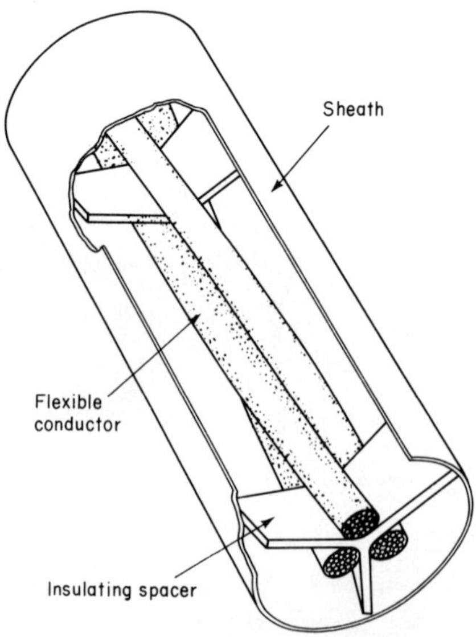

Figure 6.5 Flexible SF_6 spacer cable: arrangement of conductors in 3 conductor trefoil arrangement. (Reproduced by permission of IEE.)

6.2 BREAKDOWN CHARACTERISTICS

Although the gas dielectric in the system is normally SF_6, compressed air has been considered and breakdown parameters of both air and SF_6 will be discussed. One of the major factors influencing the breakdown strength of compressed gas is the insulator/spacer design and the presence of conducting particles in the gap. Before these are discussed a general approach to breakdown values will be summarized. Theoretical approaches, e.g. based on the streamer model, do not yield breakdown values which are consistent with experimental results and empirical results must be used. The breakdown or flashover strength (E_b) characteristics are conveniently expressed in terms of $r_1 E_b$ and $r_1 p$, where r_1 (mm) is the radius of the smaller electrode and p the gas pressure (Bar). The relationship for air and SF_6 for 50 Hz flashover[5] is shown in Figure 6.6 and for positive and negative lightning impulses in Figure 6.7. These graphs are based on the results taken from several published

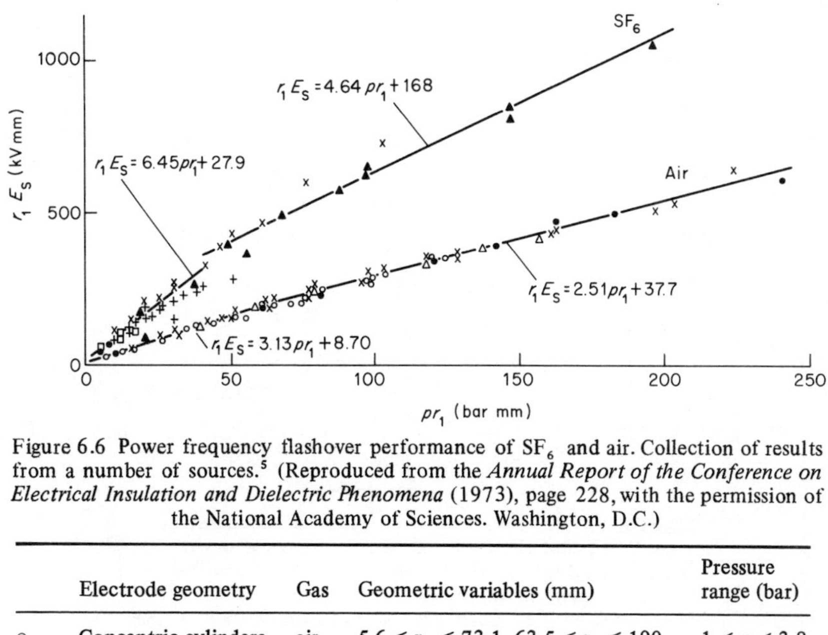

Figure 6.6 Power frequency flashover performance of SF_6 and air. Collection of results from a number of sources.[5] (Reproduced from the *Annual Report of the Conference on Electrical Insulation and Dielectric Phenomena* (1973), page 228, with the permission of the National Academy of Sciences. Washington, D.C.)

	Electrode geometry	Gas	Geometric variables (mm)	Pressure range (bar)
○	Concentric cylinders	air	$5.6 \leqslant r_1 \leqslant 73.1, 63.5 \leqslant r_2 \leqslant 190$	$1 < p < 3.8$
×	Concentric sphere/ Hemisphere	air SF_6	$10 \leqslant r_1 \leqslant 25, r_2 = 38.1$	$1 < p < 7.9$ $1 < p < 4.1$
△	Concentric cylinders	air	$r_1 = 20, r_2 = 32$	$1 < p < 7.8$
●	Concentric cylinders	air	$59.6 \leqslant r_1 \leqslant 90, r_2 = 99.7$	$1 < p < 2$
+	Concentric cylinders	SF_6	$5 \leqslant r_1 \leqslant 13.3, 15 \leqslant r_2 \leqslant 40$	$1 < p < 4.1$
▲	Concentric cylinders	SF_6	$19 \leqslant r_1 \leqslant 50, 100 \leqslant r_2 \leqslant 270$	$1 < p < 4.1$
□	Concentric cylinders	SF_6	$5 \leqslant r_1 \leqslant 7.5, r_2 = 15$	$1 < p < 2.2$

papers but are over a limited range of dimensions. Breakdown stresses for concentric cylindrical electrodes in SF_6 for impulse and switching urges are given in Figure 6.8.

In practice flashovers occur at lower voltages than model tests in the laboratory would suggest. The reason is the presence of particles (conducting and non-conducting) which, influenced by electrostatic forces, move in the gap. Particles are the byproduct of manufacturing and installation and may consist of spheroids (200 μm diameter), carbon fibres, say 200 μm in length. It is virtually impossible to ensure absolute cleanliness in such structures as pipe lines. One method overcoming this problem is the provision of particle traps along the bottom of the outer tube which may consist of perforations in the outer tube through which the particles pass and are retained. A further type is illustrated in Figure 6.9. The path of a particle is shown in Figure 6.10, it eventually entering the field-free region. The effects of the particles on breakdown strength appear to increase with particle size,

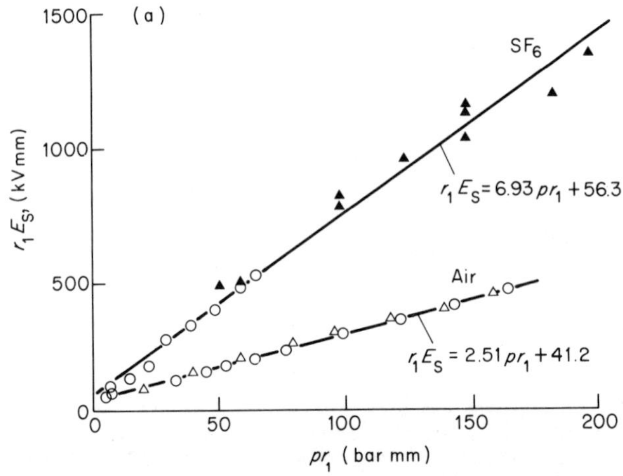

Figure 6.7(a) Negative lightning impulse flashover performance of SF_6 and air. Collected data from several sources for concentric cylinders.[5] (Reproduced from the *Annual Report of the Conference on Electrical Insulation and Dielectric Phenomena* (1973), page 229, with the permission of the National Academy of Sciences, Washington, D.C.)

	Gas	Geometric variables (mm)	Pressure range, bar
○	air	$31.8 \leqslant r_1 \leqslant 64, 89 \leqslant r_2 \leqslant 23.8$	$1 < p < 5.9$
	SF_6	$7.1 \leqslant r_1 \leqslant 28.6, 18.1 \leqslant r_2 \leqslant 77.9$	$1 < p < 3.1$
△	air	$r_1 = 20, r_2 = 32$	$1 < p < 7.8$
▲	SF_6	$19 \leqslant r_1, \leqslant 100, 100 \leqslant r_2 \leqslant 270$	$1 < p < 4$
□	SF_6	$S \leqslant r_1 \leqslant 7.5, r_2 = 15$	$1 < p < 2.2$

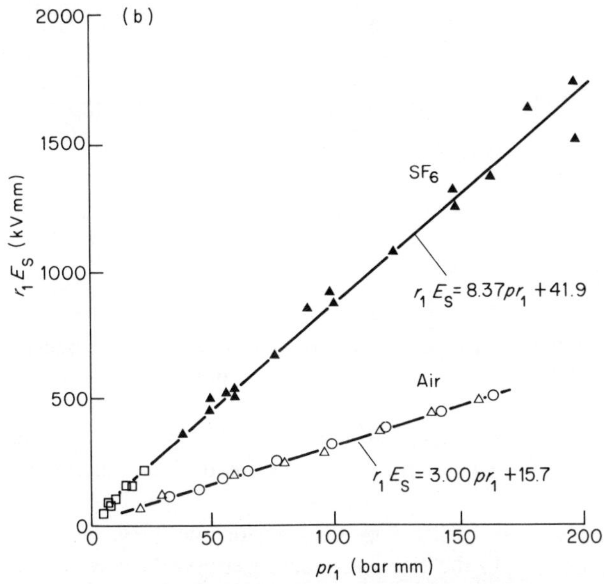

Figure 6.7(b) Positive lightning impulse flashover performance of SF_6 and air. Collected data from several sources (Concentric Cylinders).[5] (Reproduced from the *Annual Report of the Conference on Electrical Insulation and Dielectric Phenomena* (1973), page 230, with the permission of the National Academy of Sciences, Washington, D.C.)

gas pressure, and with positive polarity. Several hypotheses have been advanced to explain the reduction in breakdown voltage with particle size. Particles become charged and oscillate from the electrodes in the gas due to the a.c. field.[10] When a particle hits an electrode it loses its charge through a gas micro-discharge and this may cause main breakdown. On the other hand it has been proposed that the

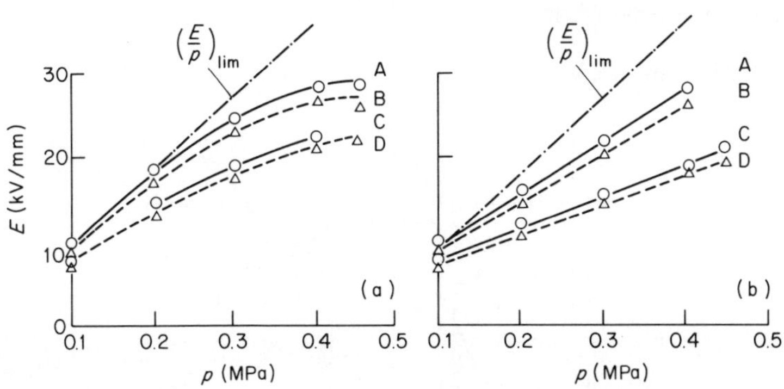

Figure 6.8 Variation of 50% breakdown stress and critical withstand stress (E_W) with pressure of SF_6. Concentric cylindrical aluminium electrodes, radii 70 and 190 mm². Stressed area 0.21 m². (a) Positive polarity, (b) Negative polarity. A,B lightning impulse; C,D switching impulse. —○— E_{50} (A and C), —△— E_W (B and D), —·—·—·— $(E/p) = 89$ kV mm^{-1} Pa^{-1}.[6] (Reproduced by permission of CIGRE from Report 15-01, 1978.)

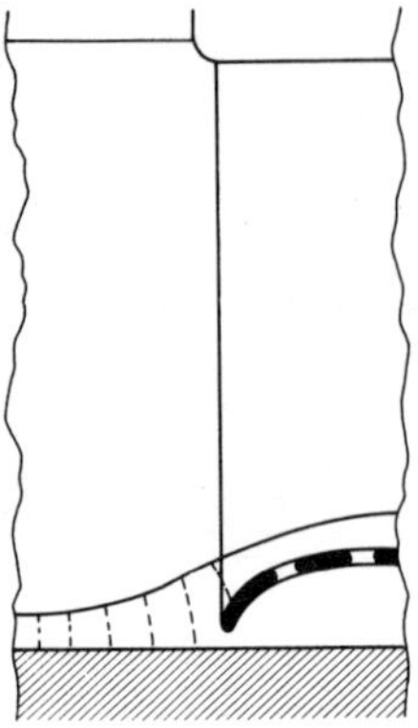

Figure 6.9 Field associated with trap. (Reproduced by permission of IEEE.)

reduction in breakdown voltage is due to field emission from a particle, or that the conducting particle on an electrode increases the local field intensity which precipitates breakdown.

On a.c., particles can be lifted by the electrostatic force, and wire-type particles stand perpendicularly on the electrode causing stress enhancement and breakdown. Consider the cylindrical conducting particle shown in Figure 6.11. Assuming that the lifting forces act on the upper end, neglecting the field distortion, the field strength normal to the face of the cylinder is[13]

$$E_n = 2E_0 \cos\theta \text{ V/m}.$$

The electrostatic force acting on the upper half of the metallic cylinder

$$F = \int_0^{\pi/2} \tfrac{1}{2}\epsilon_0 E_n^2\, 2a\, d\theta l \cos\theta$$

$$= \tfrac{8}{3}\epsilon_0\, a l E_0^2 \text{ (Newtons)}.$$

Gravity force downwards is

$$G = \pi a^2 l(\rho_p - \rho_g)g \text{ (N)}$$

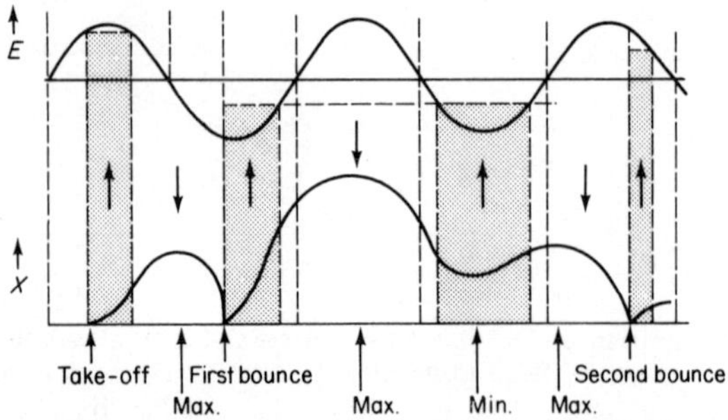

Figure 6.10 Motion and forces acting on a bouncing conducting particle in an alternating electric field. The hatched areas indicate the intervals of upward acceleration.

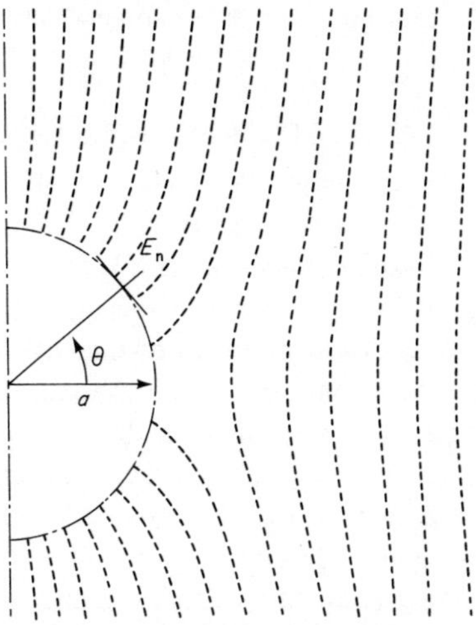

Figure 6.11 Conducting particle of cylindrical form in electric field in conductor-sheath gap. E_0 = uniform electric field strength (V/m), a = radius of cylinder (m), l = length of cylinder (m). (Reproduced by permission of IEEE.)

where ρ_p and ρ_g are the specific gravities of the particle and gas respectively and g the acceleration due to gravity. Hence the cylinder is just lifted when $F = G$, i.e.

$$E_0 = \frac{1}{2}\left[\frac{3}{2}\frac{\pi}{\epsilon_0}ag(\rho_p - \rho_g)\right]^{1/2} \text{ V/m.} \tag{6.1}$$

Example 6.1

Calculate the a.c. field strength to lift required for aluminium particles of diameter 1.6 mm and length 15 mm.

$$\rho_p = 2.7 \times 10^3 \text{ kg/m}^3\text{; neglect } \rho_g.$$

$$E_0 = \frac{1}{2}\left[\frac{3 \times 3.14 \times 0.8 \times 10^{-3} \times 9.8 \times 2.7 \times 10^3}{2 \times 8.855 \times 10^{-12}}\right]^{1/2}$$

$$= 1.67 \text{ kV Peak/mm}$$

$$= 1.185 \text{ kV (r.m.s.)/mm.}$$

For spherical conducting particles the lift-off electric field at the sheath,

$$E_L = \frac{1}{\pi}\left[2g(\rho_p - \rho_g)a/\epsilon_0\right]^{1/2} \tag{6.2}$$

where a = the sphere radius.

The average field in a coaxial system giving a certain transit time τ for the spherical particle to travel from the outer to the inner electrodes is as follows:

$$E_c = \left[\frac{2a}{\pi^2 \epsilon_0} (\rho_p - \rho_g) \left(g + \frac{2(R - r)}{\tau^2} \right) \right]^{1/2} \tag{6.3}$$

where R and r are the outer and inner electrode radii. For a coaxial system with $R = 305$ mm and $r = 104$ mm and a particle of radius 0.4 mm, the lift-off voltage is 46 kV and the voltage for a transit time τ of 5 ms is 800 kV. For practical systems it is unlikely that many particles will traverse the gap, although most will lift off. They will then follow paths as they move from high to low stress areas. As the particles approach a spacer they come under the influence of the spacer modified field and move under the shield or trap into a field-free region where they remain.

Bortnik[8] compares the results of several workers relating flashover voltage and pressure for laboratory rigs with SF_6. He concludes that below a stress of 15 kV/mm (uniform field) good agreement is found between published results relating breakdown stress and gas pressure, which lie close to a straight line of slope, 89 kV-cm/kgf. Above this stress, results from different authors diverge due to the surface area of the electrodes and the material and condition of the electrodes. The difference becomes more pronounced with large-diameter coaxial systems. The electric strength (a.c. and impulse) of the gas depends only upon its density. Also the moisture content of the gas has no effect on the flashover voltage if the dew point is below the gas temperature.

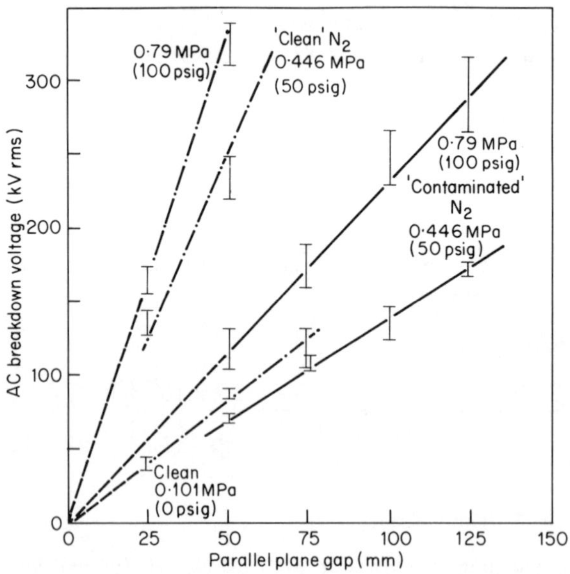

Figure 6.12 Effect of aluminium wires 0.43 mm (0.017 in) diameter and 1.58 mm ($\frac{1}{16}$ in) long on a.c. breakdown in nitrogen.[9] (Reproduced from the *Annual Report of the Conference on Electrical Insulation and Dielectric Phenomena* (1973), page 238, with the permission of the National Academy of Sciences, Washington, D.C.)

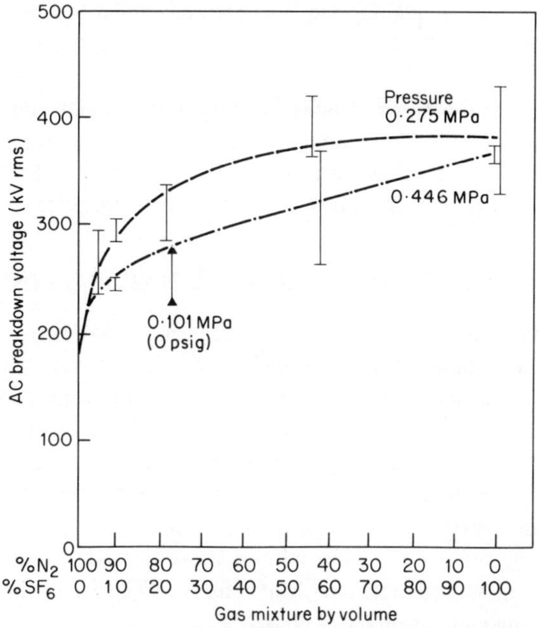

Figure 6.13 A.C. Breakdown in nitrogen–SF_6 mixtures in a 127 mm (5 in) plane gap with 0.45 mm (0.018 in) diameter and 6.3 mm (¼ in) long copper particles.[9] (Reproduced from the *Annual Report of the Conference on Electrical Insulation and Dielectric Phenomena* (1973), page 239, with the permission of the National Academy of Sciences, Washington, D.C.)

Bortnik[7] also summarizes published data on flashover in SF_6 with *impulse* voltages with the following general conclusions.

(i) All results, including large-area electrodes, are closer to the relation $E_b = 89$ p than the a.c. data; the spread is $E_b = (0.9$ to $1.1)89$ p; the largest deviations occur with the large electrode areas.

(ii) The flashover stress (E_b) is higher for positive impulses than for negative, but the difference is within 10%.

(iii) The impulse ratio (E_b impulse/E_b a.c.) for standard impulses depends on the area and pressure and for practical conditions with a gas pressure of 3 to 4 kgf/cm^2 (2.94×10^5 to 3.92×10^5 N/m^2) it may be taken as 1–1.1 for small-area electrodes and 1.3–1.4 for large areas.

(iv) For switching surges the impulse ratio is close to unity (often less than 1.05 for the negative impulses).

Breakdown in Nitrogen–SF_6 Mixtures

The flashover voltage dependence on electrode gap for alternating voltage across parallel plane gaps (uniform field) for nitrogen[9] is shown in Figure 6.12. Also indicated is the effect of conducting particle contamination in the form of aluminium wires. In Figure 6.13 the dependence of the a.c. flashover voltage on the SF_6/N_2 mixture is indicated;[9] contamination is present in a controlled manner in the form of copper particles. Under contaminated conditions the increase in breakdown voltage of SF_6 with increase in pressure, as found under clean conditions, no longer holds. However, with nitrogen the breakdown voltage still increases with pressure, certainly up to 8×10^5 N/m^2 (100 psig).

6.3 SPACER DESIGN AND BREAKDOWN IN SF_6 CABLES[11,12,13,14]

It will be evident from the preceding section on breakdown in gases that the working voltage and cable dimensions will be influenced greatly by the degree of contamination by particles present. Similarly, when discussing spacer design not only must the shape of profile be considered but also the influence of particles. One problem in testing spacers is the change in surface condition caused by the discharge. Tests on plain annular discs cut from 6 mm thick Perspex sheet, with silver coated glass sphere in the range of diameter 35–210 μm acting as contamination, have been reported.[9] The spheres were placed on the spacer surface adjacent to the centre conductor with a density of 20 particles/mm^2. From the results the following conclusions may be made.

(i) Conducting particles have a greater effect with positive than with negative impulses.
(ii) Particles of a given size reduce the flashover stress more at high pressure than at low pressure; this is true of both polarities, e.g. particles (125–210 μm range) reduce E_b by 55% at 1.5 MN/m^2 but only by 13% at 0.47 MN/m^2.
(iii) The effect of particles increases with their size, especially at the higher pressures. Tests with unsilvered glass spheres (i.e. nonconducting) revealed no reduction in E_b compared with clean spacer-discs.
(iv) Conducting particles up to 200 μm diameter will not greatly reduce the flashover voltage of SF_6 cables for gas pressures up to 0.5 MN/m^2. At higher pressures even very small particles (10's μm) have a marked effect.

The probability (P) of a length of cable failing by spacer flashover is given by:

$$P = 1 - (1 - p)^n$$

where p is the probability of a spacer failing under the same voltage when tested in isolation, and n is the number of identical spacers in the cable, say 200/km for a rigid cable and 1000/km for a flexible gas cable. If n is taken to be 500 and if p is 0.1%, then $P = 39\%$. Similarly, if $p = 0.001\%$, $P = 0.5\%$. Hence p must be less than 0.01% for an installation to have a good chance of passing its acceptance tests.

Spacers are usually shaped to reduce the stress at the electrodes and to give a more uniform surface stress. Apart from plain discs other basic shapes are the corrugated disc and the cone. With a plain disc the voltage distribution is the same as in the gas but the gas in the layer between the disc and electrode is critical, having a

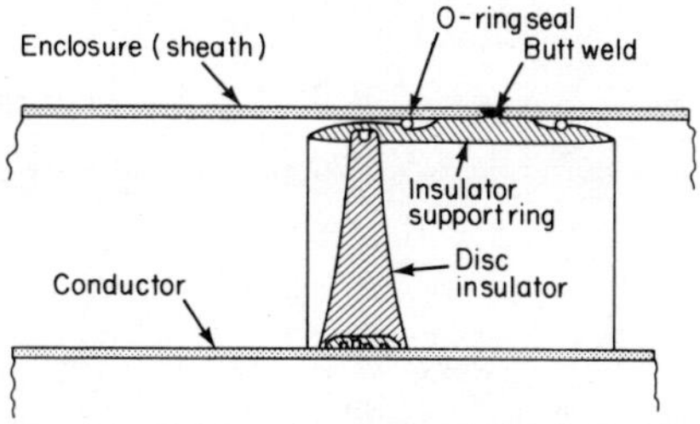

Figure 6.14 Joints in isolated-phase cables (Reproduced by permission of Gould-Brown Boveri, Power Systems Division, Greensburg, Pennsylvania, USA.)

stress $E = \epsilon_r E_0$ where ϵ_r = relative permittivity of the spacer and E_0 = stress at conductor surface between insulators.

One solution to this problem is to provide a conducting surface on the underside of the insulator and connect it electrically to the conductor. In practice the disc is often shaped to give a uniform stress distribution along its surface, when it theoretically becomes stronger than the gas. However, this ignores contamination and in practice a corrugated or conical shape may have some advantages, e.g. the former may have a large creepage distance as in outdoor insulators and the latter will have a relatively small amount of solid in series with the spacer–electrode gas gap, eliminating the need for the electric connection mentioned above.

Basic spacer types are shown in Figure 6.1; the construction can be of Teflon or Epoxy. The breakdown characteristics of spacers are still (at the time of writing) the subject of extensive research. Much work has been performed on small laboratory systems, and its direct relevance to large practical systems is in doubt. A comparison of two published impulse results[13,11] indicates little difference for breakdown voltage between disc and cone spacers (in the order of 18.5 to 20.5 kV/mm). An example of field jointing of the pipe-spacer sections is shown in Figure 6.14.

6.4 THREE-CONDUCTOR CABLES

There is obvious interest in the possibilities of having the three conductors in one containing pipe or sheath, resulting in a considerable reduction in both cable and trenching costs. A suggested[15] conductor support arrangement is shown in Figure 6.15. In this the spacers are mounted on a common ring which permits movement to take place relative to the sheath. This has the advantage that only free gas exists between the conductors. In the isolated-phase case relative movement is possible with the conductor sliding in an aluminium ring. It is possible to modify the field in the vicinity of the spacers by arrangements of metal inserts in the spacers.

Overvoltage surge tests have been reported[12] on a test model of a 500 kV, 3-core CGI cable. The inter-phase strength has been determined for the disc-type insulator (conductors pass through holes in a disc sliding in the outer pipe). The

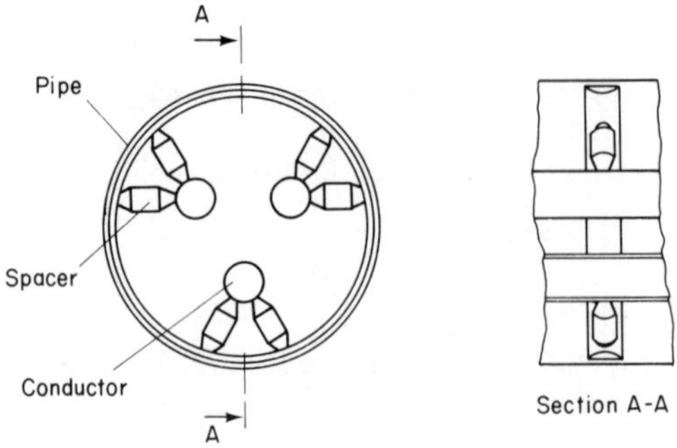

Figure 6.15 3 conductor rigid spacer cable arrangement.[15] (Reproduced by permission of CIGRE.)

breakdown voltage (crest value) as a function of gas pressure is shown in Figure 6.16 for two wavefront durations (200 and 400 μs). The general conclusions from this work are as follows.

(*a*) The switching surge breakdown strength in SF_6 gas is proportional to (absolute gas pressure).$^{0.7}$
(*b*) The breakdown voltage is not affected by time differences between two opposite polarity switching surges.
(*c*) Insulation designs can be accomplished by considering the maximum stress in the gas related to published breakdown figures.

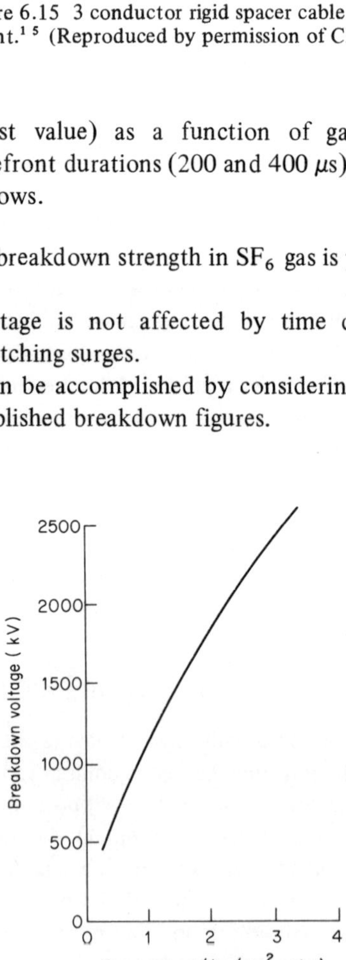

Figure 6.16 Interphase dielectric characteristics of 500 kV, 3-core SF_6 cables. Disc spacer. Surge on one conductor.[12] (Reproduced by permission of IEEE.)

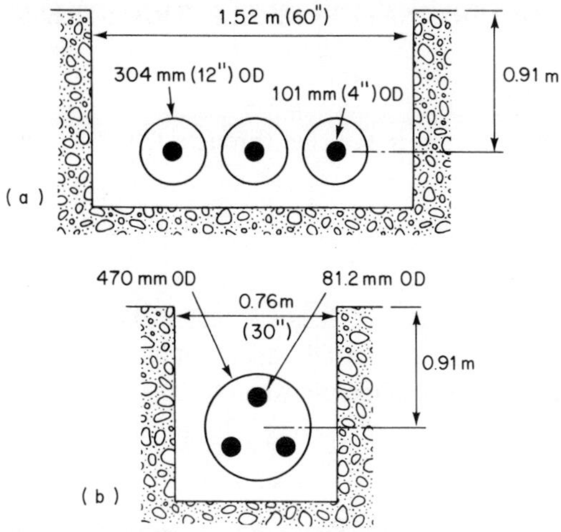

Figure 6.17 Comparison of trench sizes for (a) 230 kV isolated-phase system rated at 620 MVA, and (b) a 230 kV three conductor cable rated at 590 MVA.[23] (Reproduced by permission of IEEE.)

Advantages of the 3-core design are as follows: much less sheath material, lower sheath losses, fewer welds in the field, narrower trenches, and lower installed costs; also the current-carrying capacity is little reduced. This is illustrated in Figure 6.17 for a 230 kV system in which the 3 cores (590 MVA) have a pipe outer diameter of 0.47 m (18.6 in) and the isolated phase (620 MVA) of almost comparable rating has a pipe diameter of 0.304 m (12 in). This assumes an earth thermal resistivity of 0.9 deg C-m/W and a sheath temperature rise of 35 deg C above a 25 °C ambient.

A comparison of isolated phase and three-conductor CGI cables is given in Table 6.1 (adapted from Reference 16).

Table 6.1

	230 kV Isolated Phase	230 kV 3-core	345 kV 3-core	345 kV Isolated Phase
Sheath material	Al	Al	Al	Al
Sheath outer diameter (mm)	304	472	762	360
Sheath wall (mm)	7.6	7.6	12.7	12.7
Conductor wall (mm)	12.7	12.7	12.7	12.7
Conductor outer diameter (mm)	103	81	127	103
Phase–ground clearance (mm)	94	71	107	–
Phase–phase spacing (mm)	500	115	173	710
Current rating (A)	1550	1480	2000	2000
MVA rating	620	590	1200	1200
Loss (W/m)	125	111.5	158	142
BIL (kV)	900	900	1175	1175
r.m.s. 60 Hz/1 minute withstand (kV)	400	400	550	550
Max stress in gas gap (kV(r.m.s.)/mm)	2.5	2.67	2.5	3.15

6.5 DETERMINATION OF ELECTRIC STRESSES

Isolated-phase Systems

The variation of electric stress radially through the dielectric is given by the expression

$$E_x = \frac{V_n}{x \ln(R/r)}$$

where V_n is the conductor phase-voltage, R and r the outer and inner electrode radii, and x the radial distance from the inner conductor axis. As for conventional single-core cables the stress is a minimum when $(R/r) = 2.718$.

3-core Cables

The stress between the phase conductors is $\sqrt{3}$ times the phase to ground value. Also, one conductor could receive an impulse voltage whilst a neighbouring conductor could be at a peak voltage of opposite polarity. The disposition of the three conductors and sheath are shown in Figure 6.18 along with the stress distributions.[15]

For reasonable design, $R_1 = er = 2.718r, R_2 = 2er = 5.436r$.

The stress distribution (E_2) from the conductors to sheath is almost the same as in the isolated phase construction. With the conductors in equilateral formation the distance between conductor axes is $(\sqrt{3})re$ and the maximum stress between conductors is of the same order as that for conductor to sheath.

Consider an impulse voltage of peak value V_i on one conductor whose neighbour is at its instantaneous peak (see Figure 6.19). If the system line to line voltage is V_L (r.m.s.), then the maximum instantaneous voltage difference between these con-

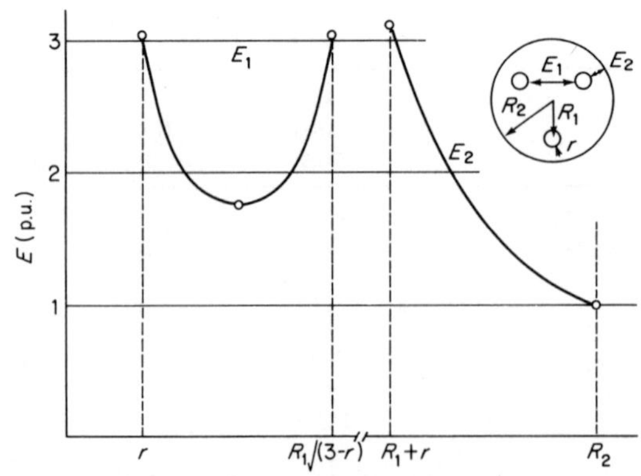

Figure 6.18 Variation of electric stress between conductors (E_1) and between conductors to ground (E_2) for a conductor CGI cable.[15] (Reproduced by permission of CIGRE from Report 21-11, 1974.)

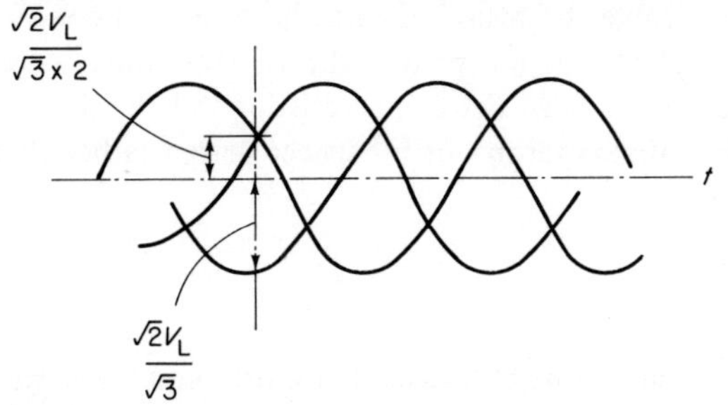

Figure 6.19 Maximum a.c. voltage between conductors.

ductors will be

$$\sqrt{2}\left(\frac{V_L}{\sqrt{3}}+\frac{V_L}{2\sqrt{3}}\right)+V_i=\sqrt{\frac{3}{2}}V_L+V_i.$$

If the other conductor is at V_i alone then the voltage across them is

$$2\frac{V_L}{\sqrt{3}}+V_i.$$

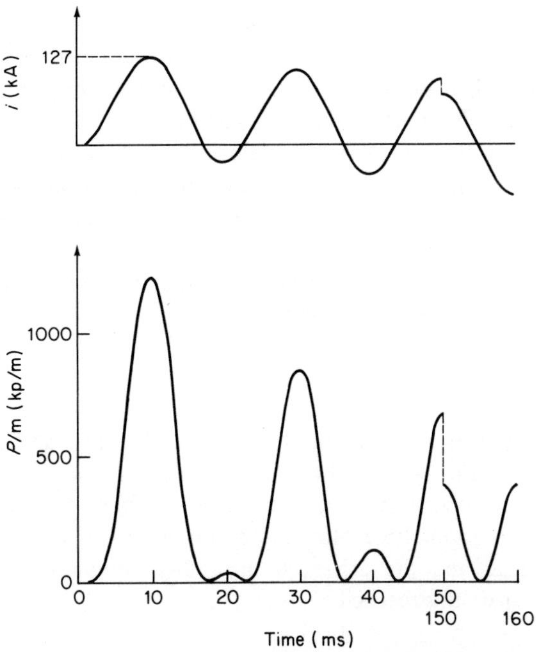

Figure 6.20 Variation of fault current and resultant force with time for a phase to phase short circuit (non-conducting sheath): – 3 conductor cable. Distance between conductor areas 265 mm.[15] (Reproduced by permission of CIGRE from Report 21-11, 1974.)

Hence for 3-core cable the overall sheath diameter is about 80% higher than the isolated-phase equivalent for one phase. The dimensioning suggested, however, is not critical as the optima have flat characteristics and the dimensions finally used may be decided by current ratings or by spacer design rather than by the stress in the undisturbed gas.

Forces

With full return current through the sheath a short circuit does not create forces on the conductor of the isolated-phase system. In the three-core cable system, forces are a major consideration, although if some current returns in the metallic sheath a small reduction (10–25%) in force occurs compared with a nonconducting sheath. In Figure 6.20 the waveform of an assumed short-circuit between two conductors of a 3-core system is shown[15]

6.6 INSTALLATION AND THERMAL CONSIDERATIONS

Installation and corresponding thermal conditions have many aspects in common with conventional cable systems. Again the thermal resistivity of the soil is very important. One great advantage of gas insulation is its low thermal resistance and the low dielectric losses.

The present limitation to current capacity is the temperature at the duct–soil interface due to moisture migration; this gives a maximum sheath or duct temperature of not more than 60 °C. With forced cooling of the sheath the next limit would be caused by softening of the corrosion mastic on the sheath at 80–100 °C. Epoxy spacers limit the conductor temperature to about 150 °C and limits may also be due to the push fitting on conductor connections or joints. As with conventional cables the current capability may be increased by forced cooling. It is interesting to note that it is the temperature at the duct–soil interface that is the limiting factor for current rating and not the conductor temperature. A discussion of possible forced cooling schemes follows.

(i) Movement of the gas

The gas may be blown along in the duct and passed through heat exchangers at intervals along the route. Because of the low specific heat and density of SF_6 the gas flow required for a reasonable temperature rise (i.e. inlet 35 °C, outlet 85 °C) over a kilometre or so is appreciable. More important, however, is the movement of free conducting particles and the enhanced possibility of breakdown. These factors seriously detract from this method.

(ii) Water cooling of the duct or sheath

Because of the relatively good heat-transfer from the conductor to sheath the difficulties inherent with direct conductor-cooling are best avoided and cooling

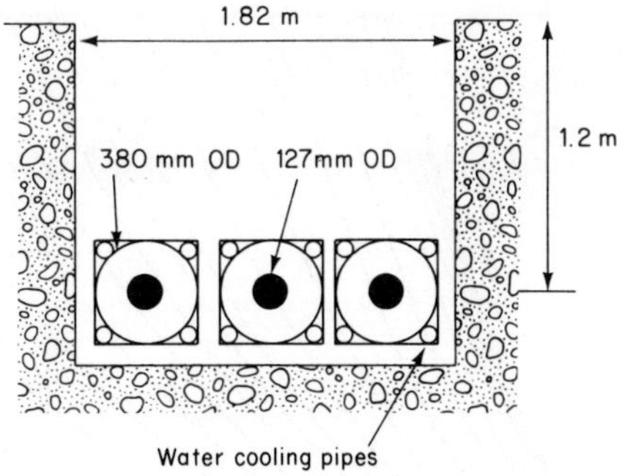

Figure 6.21 Proposed 345 kV isolated phase SF_6 cable rated at 2.8 GVA incorporating sheath cooling with water pipes. (The equivalent 3-conductor cable would require a trench width of 1.2 m.)[23] (Reproduced by permission of IEEE.)

concentrated at the sheath. A typical example of external water cooling is shown[16] in Figure 6.21. Small aluminium cooling pipes may be welded to the sheath in a linear or helical manner, during manufacture or plastic water pipes used. The water-return can either be via a separate plastic pipe in the trench or the cooling pipes of a separate parallel cable. The advantages of sheath cooling are as follows:

(*a*) Probably most economic as water can be used.
(*b*) No modifications to inside of cable required.
(*c*) Where adequate water make-up facilities are available once-through water cooling may be used, requiring no heat exchangers.
(*d*) The contamination and particle problem in the gas is not worsened.

The basic methods available for heat removal in the water-cooled scheme (500–1000 kW heat load) are the use of wet cooling and dry-type heat exchangers. The wet cooling arrangement is more economic, smaller, and less noisy than the dry-type, although objectionable vapour is created. The physical size of a water-to-air heat exchanger of capacity 500 kW is roughly 2.5 m x 2 m x 2.5 m and it would require an air-flow of 24.5 m^3/s (52×10^3 cu. ft/min).

Conductor Losses[17]

Full analytical treatments[18] to determine the a.c. resistance of a tubular conductor are reported in several texts. The skin depth of aluminium (50 Hz) is about 10 mm, and this gives an indication of conductor wall thickness to be usefully employed. Studies indicate a minimum value of R_{ac} at 60 Hz for copper and aluminium at wall thickness between 16 mm and 22 mm. The relationship between R_{ac} and electrical resistivity for various values of conductor inner radius (r_i) is given in Figure 6.22.

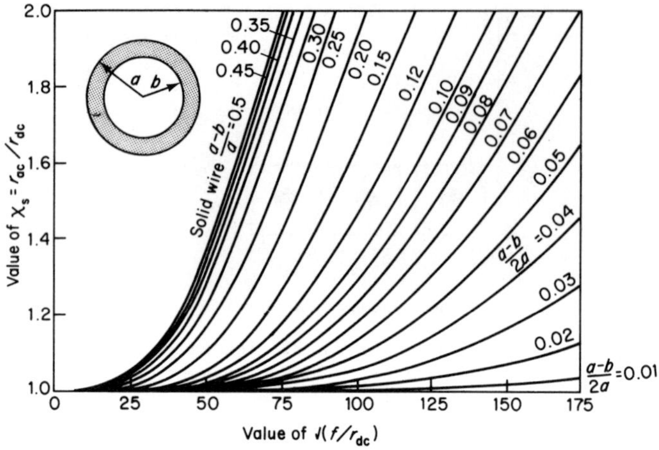

Figure 6.22 Variation of R_{ac} with electrical resistivity (ρ) for annular conductors. f = frequency r_{dc} Ω/mile (From *Trans. IEEE*, 41, 140 (1922). Reproduced by permission of IEEE.)

Losses in Sheath or Duct

As in conventional cable systems, increased current capacity may be achieved by crossbonding. This is done by interrupting the ducts of the isolated-phase system, say every 500 m, by the insertion of an insulation ring and crossconnecting the phases to neutralize the longitudinal induced voltage. The ducts must be insulated from the ground and this insulation is combined with corrosion protection.

The electromagnetically induced losses in ducts of nonmagnetic material may be calculated in the following manner using a formula due to Dwight[19] for three conductors equilaterally spaced in a duct. The loss,

$$Q_d = 12I^2R_{dc} \sum_{m=1} \left(\frac{c}{r}\right)^{2m} \frac{b^4}{m^2 + b^4} \sin\left(\frac{m\pi}{3}\right) \text{ W/m} \tag{6.4}$$

where

$$b^2 = \frac{2.37rt_d}{\rho_s 10^4}$$

r = radius of duct (m)
t_d = thickness of duct wall (m)
ρ_s = electrical resistivity of pipe material (Ω-m)
c = conductor axis to duct axis spacing (m).

More accurate solutions taking account of magnetic permeability in the case of steel pipes have been reported.[20] Because of the difficulty in calculating the duct loss calculations have been made assuming a duct loss equal to the total conductor loss. A comparison of effective electrical resistances for a certain design[15] is given in Table 6.2.

Table 6.2

Aluminium conductor 5000 mm² cross section	Single-phase ducts ($\mu\Omega$/m)	Three-phases in single sheath ($\mu\Omega$/m)
R_{dc} at 50 °C	3 x 6.5 = 19.5	3 x 6.5 = 19.5
Additional resistance (skin and proximity effect)	3 x 1.0 = 3.0	3 x 1.1 = 3.3
Sheath resistance (effective, i.e. eddy + hysteresis)	3 x 2.5 = 7.5	0 to 17.5*
Total	30.0	22.86 to 40.3

*The lower value applies to nonconducting materials (and approximately for stainless steel), the higher value for magnetic steels.

6.7 CALCULATION OF CURRENT RATING

The equation for radial heat conduction (Q) in an annular cylinder on a per unit length basis is

$$Q = \frac{2\pi k(T_i - T_o)}{\ln(r_o/r_i)}$$

where r_i and r_o are the inner and outer radii of the annulus and T_i and T_o the corresponding temperatures, and k is the thermal conductivity of the gas.

This equation can be used to describe the *natural convection* in an annular gas-gap by use of an apparent thermal conductivity, k_e, in place of k. A typical relationship[21] between k_e and k_g (the true thermal conductivity of the fluid) is

$$\frac{k_e}{k_g} = M(\mathrm{GrPr})^p \qquad (6.5)$$

where for

$$6000 < \mathrm{GrPr} < 10^6, M = 0.11, p = 0.29$$

and for

$$10^6 < \mathrm{GrPr} < 10^8, M = 0.40, p = 0.20.$$

The Grashof number (Gr) is based upon the thickness (δ) of the gas layer in the annular gap which is equal to $(r_o - r_i)$.

For SF_6 at atmospheric pressure and 20 °C, the thermal conductivity $k = 5.4 \times 10^{-5}$ W/m-K, viscosity $\mu = 5.3 \times 10^{-8}$ N-s/m², $c_p = 7 \times 10^3$ J/kgf-K, and $\rho = 6.2p293/T$ kg/m³.

For SF_6 at 4.45×10^5 (N/m²)(50 psig) and 70 °C, $\mathrm{GrPr} = 1.4 \times 10^6\,(r_o - r_i)^3\theta$, where $\theta = T_i - T_o$ (K). Also, $k = 2.68 \times 10^{-5}$ W/m-K. Hence the heat dissipated per

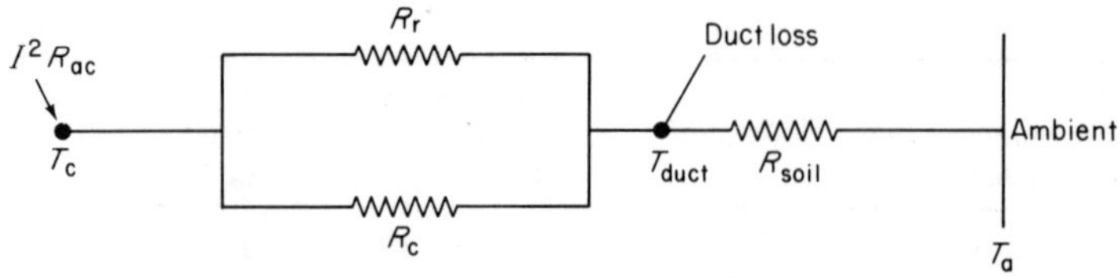

Figure 6.23 Thermal network representing heat transfer paths CGI cable. R_r = thermal resistance due to radiation; R_c = thermal resistance due to natural convection; R_s = thermal resistance of environment.

metre length by natural convection in SF_6 at 4.45×10^5 N/m^2 is

$$1.01 \times 10^{-3} \frac{(r_o - r_i)^{0.6}}{\ln(r_o/r_i)} \theta^{1.2} \text{ (W)} \tag{6.6}$$

where r_o and r_i are in mm.

Heat is also transferred by radiation, and the heat transfer per unit area,

$$Q_r = \frac{\sigma(T_o^4 - T_i^4)}{1/\epsilon_i + (1/\epsilon_o - 1)r_i/r_o} \text{ W/m}^2. \tag{6.7}$$

When ϵ_o approaches 1,

$$Q_r = \sigma\epsilon_i(T_o^4 - T_i^4) \text{ W/m}^2$$

where

T_o = duct temperature (K)
T_i = temperature of conductor (K)
ϵ_i = emissivity of conductor (usually about 0.9 to 0.95)
ϵ_o = emissivity of pipe
σ = Stefan Boltzmann Constant = 57×10^{-9} Wm^{-2} K^{-4}

It will be seen from Figure 6.23 that the temperature rise for a given conductor loss is not a straightforward calculation because both thermal resistances (i.e. convection and radiation) are nonlinear with temperature. The matter is further complicated by the loss in the duct and the temperature drop across the duct environment. The solution may be obtained by iterative methods, preferably by computer,[22] or use may be made of the graphs[23] shown in Figures 6.24 and 6.25.

Example 6.2

Calculate the rating of a 500 kV isolated-phase gas spacer cable assuming an ambient temperature of 25 °C and a soil resistivity of 0.9 deg C-m/W. The maximum allowable sheath temperature (to avoid soil drying) is 60 °C and the sheath loss is 30% of the conductor loss. The SF_6 pressure is 4.45×10^5 N/m^2 (50 psig). The depth of burial is 1.2 m (4 ft) and the axial separation 1.2 m. The radius over the conductor is 50 mm and over the pipe 150 mm. The thermal resistance of the

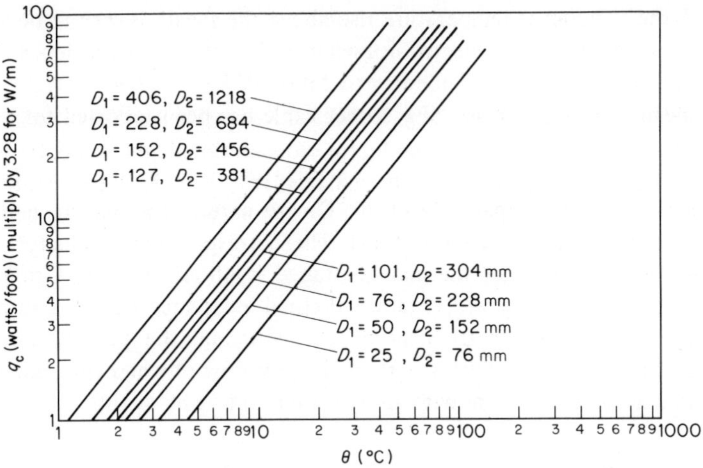

Figure 6.24 Heat transfer characteristics of SF_6 in a coaxial gap – temperature rise of conductor above sheath against heat flow. Natural convection at 3 bars pressure with sheath at 20 °C. Conductor diameter D_1, sheath diameter D_2 (mm).[23] (Reproduced by permission of IEEE.)

centre pipe is given by

$$\frac{g_s}{2\pi}\left[\ln\left(\frac{2h}{r_o}\right)+\ln\left(\frac{4h^2+s^2}{s^2}\right)\right]$$

$$=\frac{0.9}{2\pi}\left[\ln\left(\frac{2 \times 1.20}{0.5}\right)+\ln\left(\frac{4 \times 1.20^2+1.20^2}{1.20^2}\right)\right] \text{deg C/W per metre length.}$$

The temperature drop across the soil, $\theta_s = 60 - 25 = 35$ °C. The heat-flow corresponding to θ_s is $35/R_s = 63$ W/m per phase. As the sheath loss = 30% of the conductor loss, the allowable conductor loss = 63/1.3 = 48.5 W/m. From

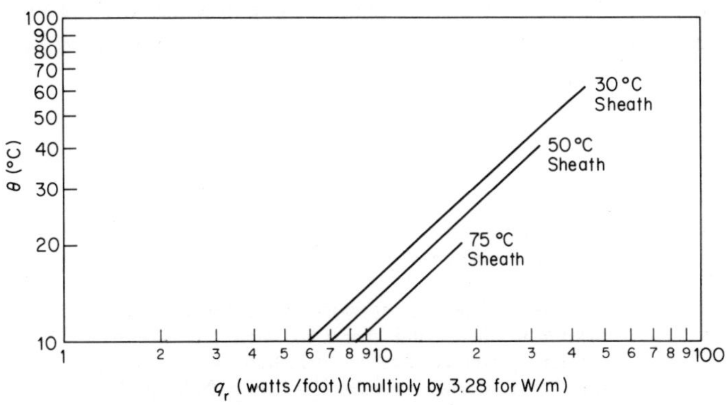

Figure 6.25 Radiation heat transfer from conductor (emissivity 0.95) for various sheath temperatures. Coaxial gap with SF_6 at 3 bars. Conductor diameter 101 mm. For other values of D_1 multiply q_r (for same θ) by $D_1/4$.[23] (Reproduced by permission of IEEE.)

Figure 6.25 the conductor temperature rise above the sheath is seen to be less than 10 °C. For the conductor the electrical resistance, R_{ac} = 6.0 μΩ/m. Hence, I^2R_{ac} = 48.5 W/m and I = 2850 A corresponding to 2500 MVA (3-phase).

A European[15] isolated-phase SF_6 spacer cable has been reported for a system voltage of 420 kV and aluminium conductor cross-sectional area of 5000 mm² (see Table 6.2). When buried in ground of thermal resistivity 1 m–K/W to a depth of 3 m with a specified sheath temperature of 40 °C with natural cooling, the permissible current (continuous) is 1.3 kA (50 W/m). The corresponding conductor temperature is 49 °C. The internal thermal resistance is about 0.25 deg C-m/W. With separate water-pipe cooling the external thermal resistance can be considerably reduced and the loss per metre rises to 150 W/m and the power transmission capacity increases by 170%. With sheath cooling by flow of water (integral cooling) the loss is increased to 347 W/m with an increase in rating by a factor of 260%.

Three-conductor Duct System

Assuming the limit to current capacity is at the duct–soil interface, ratings may be calculated without a detailed analysis of the heat transfer within the duct. The conductor effective a.c. resistance is increased by the proximity effect (dependent on the geometry but in the order of 10%). The total heat entering the soil is (3 x loss for one conductor) plus the duct loss. The product of this total loss and the effective thermal resistance of the soil gives the temperature rise of the duct above ambient. In calculating the soil thermal resistance the use of traditional formulae may be doubtful because of the large duct diameter relative to depth of burial. A field plot would yield a more reliable value.

Calculation of the conductor temperature rise presents a more difficult task, and to facilitate this considerable assumptions are made. A method which has been reported for an air-duct line[22] assumes natural convection from each conductor to a hypothetical cylinder of gas concentric with the duct, the thermal resistance of which is calculated. The complete steady-state thermal network is shown in Figure 6.26. Again, because of the temperature dependence of the convective and radiation heat-transfer resistances, an iterative procedure is used to calculate the conductor temperature.

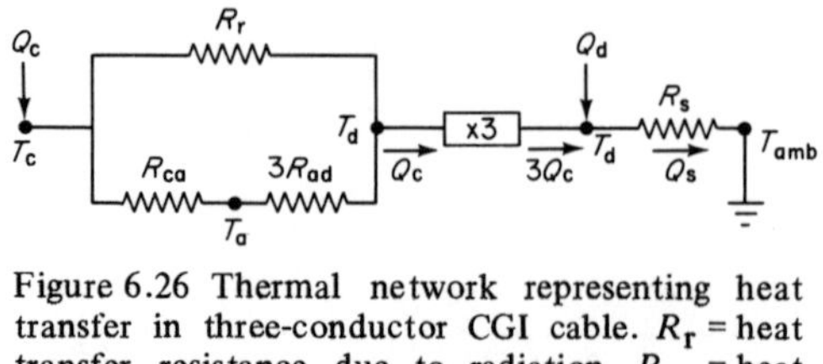

Figure 6.26 Thermal network representing heat transfer in three-conductor CGI cable. R_r = heat transfer resistance due to radiation. R_{ca} = heat transfer resistance conductor to hypothetical cylinder of gas; R_{ad} = heat transfer resistance from hypothetical cylinder of gas to duct; r_s = heat transfer resistance of soil. T_c, T_d = conductor and duct temperatures.

6.8 OTHER GASES

Two other common gases readily come to mind as alternatives to SF_6 for a spacer cable; these are nitrogen and air. For an impulse withstand stress of 10 kV/mm with the isolated-phase construction (radii 0.1 and 0.2 m) it would be possible to use SF_6 at 0.2×10^6 N/m^2 or nitrogen at 10^6 N/m^2. For an impulse strength of 40 kV/mm, SF_6 is just possible at 1.5×10^6 N/m^2 but nitrogen and air are not practical.[25] Although a smaller cable, or alternatively a lower pressure, is possible with SF_6 resulting in lower cost, this may be somewhat offset by lower gas replacement costs for nitrogen. Insufficient information regarding breakdown of compressed air with spacers is available to enable a judgement to be made at the time of writing.

Some development work has been carried out on an *air duct line*[22] at atmospheric pressure. For a system voltage of 230 kV using three conductors in one sheath arrangement, a sheath (duct) diameter of the order of at least 2.5 m is required. Of the alternatives for the sheath (duct) material i.e. concrete, plastic, or corrugated steel, the latter appears to be preferable on grounds of cost and the fact that it provides a return conductor for out of balance fault currents. The electromagnetic loss, however, is considerable in the duct and as the current-capability limit is the allowable temperature at the duct/soil interface, this loss is very important. The determination of current rating follows the general lines described for the SF_6 three-line phase. To achieve the high currents and powers to warrant such an air insulated system, water cooling of the buried duct would be required. The presence of particles, which has dominated the SF_6 spacer-cable developement, is of course a major problem here.

6.9 DESIGN AND ECONOMIC CONSIDERATIONS

(a) A.C. Systems

For a concentric isolated-phase arrangement for minimum stress E at the conductor, the ratio (R/r) is 2.718, whence E becomes $2.718\ V_p/R$, where V_p is the phase voltage. The mechanical stress[25] (Y) in a thin pipe is

$$Y = PR/t$$

where P is the pressure, t the pipe wall thickness, and Y the working stress (about 20 MNm^{-2} for aluminium). The weight of the pipe is $2\pi RT = KPR^2$ per m length, where K is a constant dependent on Y. Similarly, the weight of gas is $K'PR^2$.

As the costs are related to those weights, the cost of the gas and pipe is $(K + K')PV_p^2/E^2$. The cost initially falls with increase in pressure and reaches a minimum when increase in P no longer produces an increase in E^2 (occurs[25] at roughly $E = \frac{1}{2}E_m$). The maximum permissible conductor stress for air at atmospheric pressure is 0.37 kV/mm, and roughly extrapolated to 10 atmospheres is 3.7 kV/mm. Using SF_6 roughly doubles this value of stress, but the working pressure must be less.

For a 3-phase power transfer (W), the conductor loss for a single conductor/duct arrangement is

$$I^2 \frac{\rho}{2\pi rt} = \frac{W^2 \rho}{6\pi r V^2} \text{ watts per metre}$$

where t is the pipe wall thickness.

For the system with 3 ducts buried (as shown in Figure 6.17), the duct temperature rise above the ambient surface[26] is

$$q\, 0.367 L_f g_s \left[\log_{10}\left(\frac{2L}{R}\right) + \log_{10}\left(\frac{4L^2 + s^2}{s^2}\right)\right] \tag{6.8}$$

where

L_f = load factor (take as 1.0)
L = depth of burial (m)
s = duct separation (m)
q = total heat to ambient (W/m).

If it is assumed that $L = 2R$, $s = 4R$, and $g_s = 1.2$ deg C-m/W, then $\theta = 0.4q$. If θ is limited to, say, 40 °C then assuming the duct loss is equal to the conductor loss the upper limit for q is 100 W/m and for the conductor loss, 50 W/m. Hence the minimum lower limit to r is,[27]

$$r \geqslant \frac{W^2 \rho}{50t\, 6\pi V^2} \geqslant 3.2 \times 10^{-9}\, W^2/V^2.$$

(b) D.C. Systems

In d.c. systems there are no losses in the containing duct, and each duct contains one pole. Also, because of the higher working stress much larger power transfers are

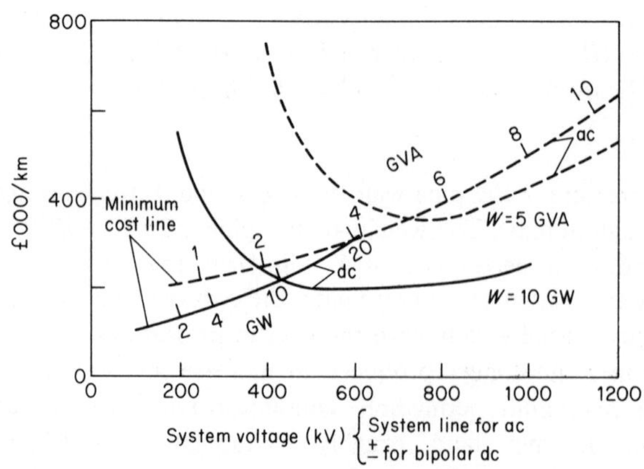

Figure 6.27 Comparative transmission costs a.c. and d.c. isolated phase CGI system (British figures, 1974 prices).

possible. The general approach is the same as used for a.c. for the relation between conductor radius and system voltage. 10 GW is transmittable at ±430 kV with an r of 230 mm and R of 660 mm giving a total trench width of 2 m. British cost[27] figures (based on 1974 prices) are shown in Figure 6.27. A study of future usage and economic assessment relative to pipe type cables in the USA has been reported.[28]

REFERENCES

1. HAMPTON, B. F. and FLEMING, S. P., 'Impulse Flashover of Particle-Contaminated Spacers in Compressed SF_6', *Proc. I.E.E.*, **120**, 514–518 (1973).
2. HAMPTOM, B. F. *et al.*, 'Outline of a flexible SF_6 Insulated E.H.V. Cable', *Proc. I.E.E.*, **123**, 159–165 (1976).
3. GIBBONS, J. A. M. and STANNETT, A. W., 'Present State of Development of a Gas Pressurized Lapped-Polythene Cable for E.H.V. Transmission', *Proc. I.E.E.*, **120**, 433–439 (1973).
4. GIBBONS, J. A. M. *et al.*, Gas Pressurized Lapped-Polythene Dielectric for Extra-High Voltage Power-Cable Systems, *Proc. I.E.E.*, **112**, 89–102 (1965).
5. MATTINGLEY, J. M. and RYAN, H. M., 'Correlation of Breakdown Gradients in Compressed Air and SF_6 for Non Uniform Fields', *Annual Report Conference on Electrical Insulation and Dielectric Phenomena', National Academy of Sciences, Washington D.C., 1973*, pp. 222–223.
6. RYAN H. M. and WATSON W. L., 'Impulse Breakdown Characteristics in SF_6 For Non Uniform Field Gaps', *CIGRE*, Report 15-01 (1978).
7. BORTNIK, I. M., 'Selecting Operating and Test Field Strengths for HV Gear with SF_6 Insulation', *Elektrichevstro*, No. 12, pp. 20–27 (1974). English Version: *Electric Technology, U.S.S.R.* pp. 97–113 (1974).
8. BORTNIK, I. M. *et al.*, 'Gas-Insulated Cables of Great Capacity', *Elektrichevstro*, No. 6, pp. 31–35 (1975).
9. COOKSON, A. H. and WOOLTON, R. E., 'Particle Initiated A.C. and D.C. Breakdown in Compressed Nitrogen, SF_6, and Nitrogen–SF_6 Mixtures', *Annual Report Conference on Electrical Insulation and Dielectric Phenomena, National Academy of Sciences, Washington D.C., 1973*, pp. 234–241.
10. COOKE, C. M. *et al.*, 'Influence of Particles on A.C. and D.C. Electrical Performance of Gas Insulated Systems at Extra High Voltage', *I.E.E.E. Trans. PAS-96*, pp. 768–777 (1977).
11. CRONIN, J. C. and PERRY, E. R., 'Optimization of Insulators for Gas Insulated Systems', *I.E.E.E. Trans. PAS-92*, pp. 558–562 (1973).
12. TAKAGI, T. *et al.*, 'Dielectric Strength of SF_6 Gas and 3-Core Type C.G.I. Cables Under Inter-Phase Switching Impulse Voltages', *I.E.E.E. Trans. PAS-93*, pp. 354–359 (1974).
13. MERIJU, S. *et al.*, 'Dielectric Breakdown of High Pressure SF_6 in Sphere and Coaxial Cylinder Gaps', *I.E.E.E. Trans. PAS-93*, pp. 1706–1712 (1974).
14. ITAKA, K. and IKEDA, G., 'Dielectric Characteristics of Compressed Gas Insulated Cables', *I.E.E.E. Trans. PAS-89*, pp. 1986–1994 (1970).
15. EIDINGER, A. and KLEMENT, W., 'Underground Power Transmission Using Compressed Gas Insulated Cables, *CIGRE*, Report 21-11 (1974).
16. DOEPKEN, H. C., 'Compressed-Gas-Insulated Cables with Increased Ampacity and Reduced Cost', *Proc. I.E.E.E. Conf. Underground Transmission and Distribution, 1974*, pp. 536–540.
17. KREUGER, F. H. and VAN DEVENTER, G. C., 'Spacer Cable', *CIGRE*, Report 21-04 (1974).

18. ZABORSZKY, J. T. and RITTENHOUSE, J. W., *Electric Power Transmission*, Rensselaer Bookstore, Troy, N.Y., 1969.
19. DWIGHT, M. B., *Electric Coils and Conductors*, McGraw Hill, New York, 1945.
20. EMANUEL, A. and DOEPKEN, H. C., 'Calculation of Losses in Steel Enclosures of Three Phase Bus or Cables', *I.E.E.E. Trans. PAS-93*, pp. 1758–1767 (1974).
21. GROBER, H., ERK, S., and GRIGULL, U., *Fundamentals of Heat Transfer*, 3rd Ed., McGraw Hill, New York, 1961.
22. WEEDY, B. M. and SCHWEIKART, D. L., 'Current Rating of an Airduct Transmission line', *Proc. I.E.E.*, **122**, 915–916 (1975).
23. DOEPKEN, H. C., 'Calculated Heat Transfer Characteristics of Air and SF_6', *I.E.E.E. Trans. PAS-89.*, 1979–1985 (1970).
24. DWIGHT, H. B., 'Proximity effects in Cable Sheaths', *A.I.E.E. Trans.*, pp. 993–998 (September 1931).
25. JONES, B., 'New Approaches to the Design and Economics of EHV Transmission Plant', Pergamon Press, Oxford, 1972.
26. GRAYBILL, H. W. and WILLIAMS, J. A., 'Underground Power Transmission with Isolated Phase Gas Insulated Conductors', *Trans. I.E.E.E. PAS-89*, pp. 17–23 (1970).
27. GARDNER, G. E. and KLEWE, R. C., 'Possible Designs and Applications for Compressed Gas Insulated Lines', *Proc. Universities Power Conference, 1976 (British)*, pp. 1–6.
28. KILAR, L. A. and CRONIN, J. C., 'A General Method for Predicting the Potential Usage of Specific Transmission Cables Applied to SF_6 Cables', *Proc. I.E.E.E. Conference Underground Transmission and Distribution, 1976*, pp. 437–441.
29. TAKAGI, T. *et al.*, 'Development of 500 kV Gas Spacer Cable: Parts I and II *I.E.E.E. Trans., PAS-97*, pp. 1732–1748 (1978).
30. Bibliography of Gas Insulated Substations (including Lines), *I.E.E.E. Trans. PAS-96*, 1280–1287 (1977).

PROBLEMS

Problem 6.1

Calculate (i) the critical lengths, and (ii) the var generation over a 20 km route, for the following types of transmission underground: (*a*) CGI – isolated phase – diameter over pipe 300 mm, over conductor 100 mm; (*b*) Pipe-type (HPOF) – diameter over dielectric 92 mm, diameter over conductor 53 mm. The voltage is 235 kV, 60 Hz, and the rating of scheme (*a*) is 600 MVA and of (*b*), 410 MVA. (Answers: (*a*) 578 km, 21.4 MVA; (*b*) 60 km, 147 MVA.)

Problem 6.2

Design CGI isolated phase systems for 500 kV and 750 kV and determine the MVA capability of each, the loss per km, and the critical length. Relevant parameters are as follows:

	500 kV	750 kV
Conductor OD (mm)	152	202
Peak gradient on conductor	4.45	4.57
Peak gradient on sheath (kV(p)/mm)	1.33	1.27
Phase spacing (mm)	1200	1500

(Answers: 500 kV: 2500 MVA, 189 kW/km, 220 km, sheath ID 508 mm; 750 kV: 4800 MVA, 235 kW/km, 209 km, sheath ID 760 mm.)

7

Cryogenic Cables

7.1 INTRODUCTION

As the end of the century approaches because of economic and environmental reasons and the availability of cooling water, generating stations will be increasingly concentrated in much large capacities than at present. Nuclear parks in the order of 10 GVA are anticipated, if possible on coastal sites, and often remote from load centres. Here underground circuits will be used to connect the stations or energy parks to the main overhead-line system. Large, e.g. 5 GVA and over, load transfers involving underground circuits will be required for the cross-flows across large urban conurbations, e.g. New York and London, where overhead circuits could not be used. The need for such power transfers has resulted in intense research and development activity in the areas of compressed gas insulated cables and low temperature (cryogenic) cables.

Ideas which theoretically are attractive frequently cannot be translated into an engineering entity because of lack of suitable materials or the fact that an existing process is cheaper. The continued use of paper/oil insulation up to the highest voltages over most of this century is a sobering lesson on the availability of materials and the importance of economics on engineering processes. The application of superconductivity to electric transmission was for many years a dream; the superconductors available lost their remarkable properties when associated with relatively low magnetic fields, being of little use in engineering applications. In 1961 niobium and later a compound of niobium and tin (Nb_3Sn) became available and remained superconducting at current densities of a suitable order, and their application to electric transmission and machinery became the object of great interest.[1,2,3,4,5]

An alternative to a superconducting cable is the use of the resistive operation of a conductor of conventional material (probably aluminium) at cryogenic temperatures. This utilizes the decrease in resistivity with decrease in temperature (see Figure 7.1) and uses a cryogenic fluid in the same manner as a coolant is used in internally cooled conventional cables; such a cable is known as cryogenic-resistive or cryo-resistive.[3,6] For both types, therefore, cryogenic liquid is required to cool the conductor to the appropriate working temperature. A significant cost of any cryogenic scheme will be the refrigerators, the size of which will be considerable

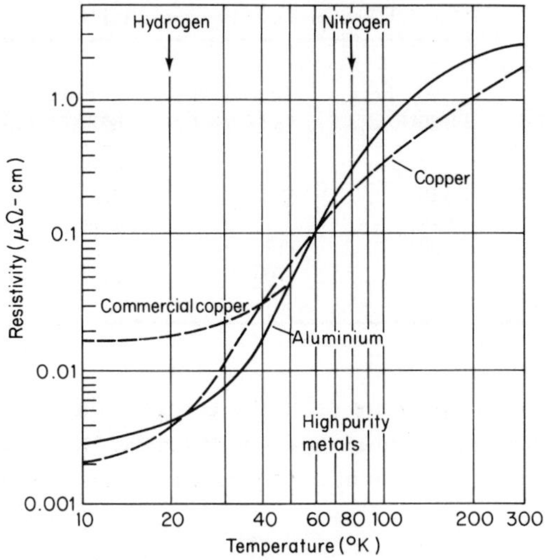

Figure 7.1 Electrical resistivity and temperature for copper and aluminium.

owing to their low efficiencies, especially at very low temperatures.[7,8] The basic parameters of coolants and refrigerators are summarized in Table 7.1. For resistive cables the optimum coolant, theoretically, is liquid hydrogen but the low cost and complete safety of nitrogen make it more attractive.[9,10,11] Whereas in conventional cables the major problem is the dissipation of the heat from the cable to the ambient, in cryogenic cables the reverse holds.

Although it would seem attractive to operate superconducting cables at low voltages, operating voltages in keeping with normal system usage are advantageous. For example, the provision of transformers at the low-temperature circuit terminations is much easier with higher cable voltages. Also the ability of circuit breakers to handle very high currents is limited. Some design aspects are common to both superconducting and resistive cables, e.g. electrical insulation, cooling systems, and the cryogenic envelopes to minimize heat entering the cable from the ambient (inleak).

The properties of the cryogenic liquids of interest are summarized in Table 7.2.

Table 7.1 Properties of metals and refrigerators.

Cable coolant	Resistivity of aluminium relative to ambient	Refrigerator ratio Watts in/Watts load Theoretical	Practical
Conventional oil	1	–	–
Liquid N_2	$^1/_{10}$	3	6–10
Liquid H_2	$^1/_{500}$	14	50–100
Liquid He	–	75	300–500

Table 7.2 Properties of cryogenic fluids.

Liquid	Boiling point (K)	Latent heat (J/gm)	Specific heat c_p (J/gm K)	Density (gm/cm^3)	Viscosity (milliNs/m^2)
Helium	4.2	21	4.5	0.13	0.004
Hydrogen	20.4	450	9.8	0.07	0.014
Nitrogen	77.3	198	1.1	0.74	0.16
Water	373	2300	4.2	1.0	–
Oil (Dodecylbenzene)	360	–	2.07	0.88	10.5

In all types of low-temperature cable the cryogenic liquid is pumped through conductor ducts absorbing the heat inputs in its passage from inlet to outlet. When a taped dielectric is used the system is very similar to the oil-filled pipe-type cable with forced flow of oil.

7.2 SUPERCONDUCTORS[7,12,13,14]

Below their critical temperature (T_c) super conductors exhibit perfect diamagnetism and the current flows on the conductor surface; above T_c they become resistive. Metals which switch state in this manner are called Type I and include pure metals (e.g. lead). Another variety, known as Type II superconductors, switch to a mixed state partly superconducting and partly resistive above the original critical field and do not become completely resistive until a higher magnetic field strength. When the Type II conductor is in the mixed state the current and flux penetrate the conductor. On alternating current, a loss occurs which at 50 Hz in strips of niobium or niobium–tin carrying 40 A r.m.s. per mm of circumference is less than 0.1 W/m^2. It is usual to express currents in terms of distance around the circumference as the current largely flows on the cylindrical conductor surface. Type II conductors comprise mainly brittle intermetallic compounds such as niobium–tin (Nb_3Sn), niobium–zirconium (NbZr), and niobium titanium (NbTi). Superconductivity in mercury, at a temperature of 4 K, was discovered by Onnes in 1911. In 1969 the Bell Laboratories developed a Type II superconductor with a critical temperature of 20.7 K. Over the years progress in developing new, higher temperature superconductors, may bring the possibility of liquid hydrogen as the coolant instead of liquid helium as at present, causing a marked change in the economic situation.

The generally accepted theory of superconductivity[15] is that of the 'Cooper Pair'. In this the force between two electrons is so modified by interaction with the lattice that two electrons (of opposite spin and momenta) effectively attract and form a Cooper Pair which cannot be scattered by the processes causing resistance in a normal conductor. In a super conductor the collision energy available at low temperatures is not sufficient to break the bond of a Cooper Pair and the 'Pairs' pass through the conductor without scattering, i.e. with zero resistance. In a superconducting material there are two electron types, the normal (higher energy) and the Cooper Pairs. At 0 (K) all electrons are superconducting but as the temperature rises collisions eventually break the pairs and create high energy electrons until T_c is reached, above which the conductor is normal. When a superconductor is subjected to an external magnetic field, surface supercurrents are induced, which screen the inside of the superconductor from the magnetic field. The exclusion of magnetic flux from the bulk of a superconductor is known as the Meissner effect. If the surface field is H, the field within the superconductor falls off rapidly so that, at depth x from the surface, the field is $H \exp(-x/\lambda)$ where λ is the penetration depth (typically 10–100 nm). If the applied magnetic field is too large, the superconductor will be driven normal. At any temperature below T_c there is a critical magnetic field H_c above which the metal is resistive. The relationship between the critical field and temperature is of the form

$$H_c = H_0 \left[1 - \left(\frac{T}{T_c} \right)^2 \right]$$

where T = absolute temperature, H_0 = critical field at absolute zero, and T_c = absolute temperature at which superconductivity occurs in zero magnetic field.

The current which a superconductor can carry is therefore limited by the magnetic field which it produces. The critical current I_c is that current which just generates the critical field at the surface of the wire. This is known as Silsbee's rule and is a good practical guide except for very thin conductors where critical currents are higher than predicted.

For a cable of conductor radius a, the critical magnetic field and critical current are related by:

$$H_c = \frac{I_c}{2\pi a} \text{ A/m.}$$

Comparing this expression with the critical field–temperature relationship yields a relationship between critical current and temperature (see Figure 7.2), namely

$$I_c = I_0 \left[1 - \left(\frac{T}{T_c} \right)^2 \right] \tag{7.1}$$

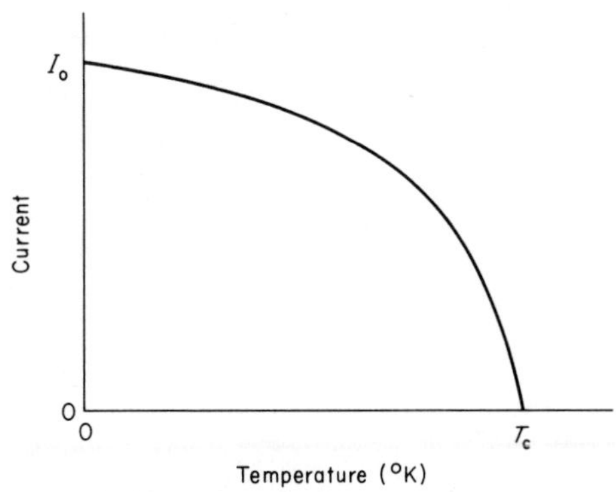

Figure 7.2 Variation of critical current with temperature.

Table 7.3 Type I and Type II superconductors.

Superconductor		Transition temperature (K)	Critical magnetic field at 4.2 K (T)
Type I	Pb	7.2	0.08 at 0 K
	Ta	4.5	0.08 at 0 K
Type II	Nb	9.3	0.15
	Nb–48% Ti	9.5	12
	Nb–25% Zr	10.8	7
	Nb_3Sn	18.4	22
	$Nb_3(Al_{0.8}Ge_{0.2})$	21.0	41
	Nb_3Ge	23.2	3 at 20 K

where I_0 is the critical current at absolute zero. In order to achieve economical current-carrying capacities it is therefore necessary to operate a superconducting cable at temperatures well below the critical temperature.

Type II superconductors are characterized by two critical fields. Below the lower critical field (H_{c1}) magnetic flux is expelled from the conductor as for a Type I material. Above H_{c1} magnetic flux penetrates the superconductor, which splits into normal and superconducting regions; this is known as the 'mixed state', The superconductor retains its property of zero direct current electrical resistivity up to the upper critical point (H_{c2}), but penetration of the superconductor by magnetic flux leads to a.c. losses at fields above H_{c1}.

Although many superconductors exist, types suitable for engineering purposes are quite few and are summarized in Table 7.3. For a.c. cable purposes, the losses of niobium–tin are acceptably low. However, the critical temperature of niobium is only 9 K, whereas niobium–tin (Nb_3Sn) has a higher critical temperature (18 K) and hence higher current-carrying capacity, but it must be made in thin strips to

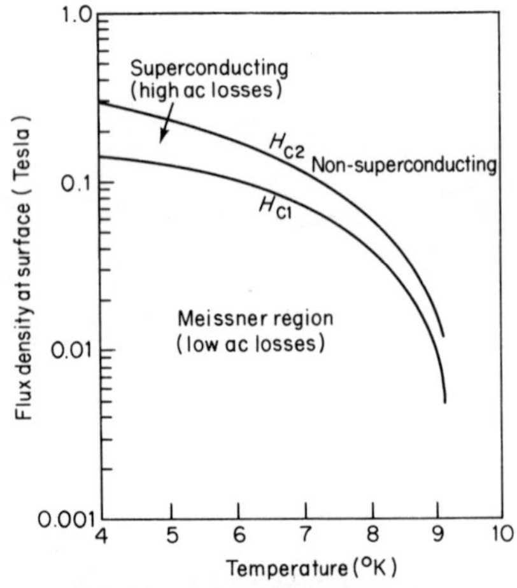

Figure 7.3 Magnetic field and temperature relation of niobium.

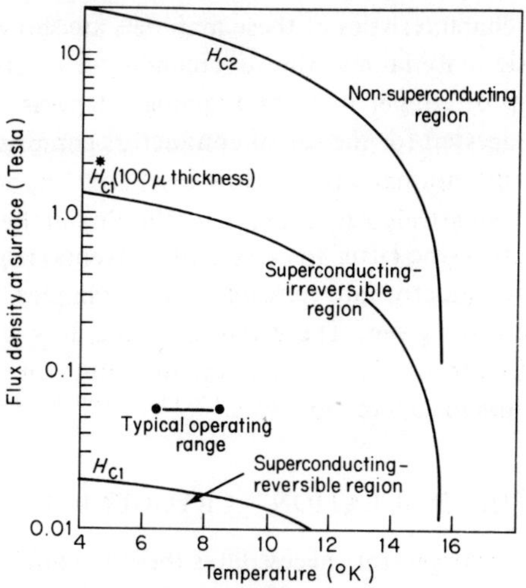

Figure 7.4 Magnetic field and temperature relation of niobium–tin ribbon.

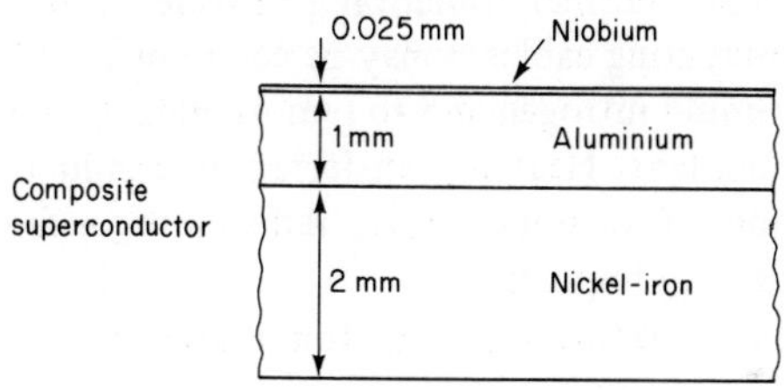

Figure 7.5 Composite superconductor.

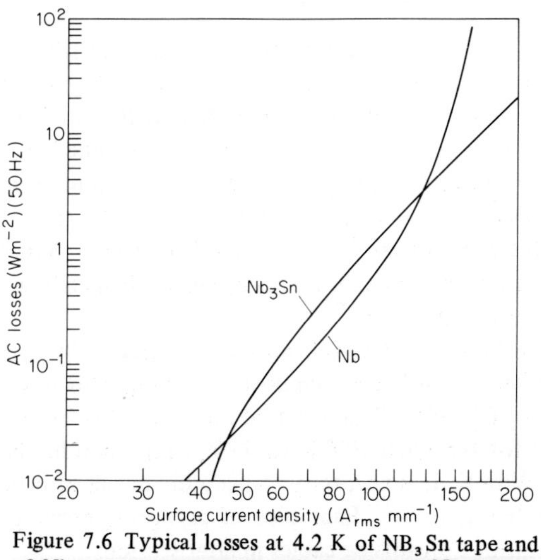

Figure 7.6 Typical losses at 4.2 K of NB_3Sn tape and of Nb–copper composite strip conductor.[49] (Reproduced by permission of CIGRE from Report 21-05, 1976.)

avoid high losses. The characteristics of these materials are shown in Figures 7.3 and 7.4. Nb_3Sn is a brittle material and the superconductor or strips would normally require a backing strip of copper to form a laminate tape as shown in Figure 7.5. Schemes have been suggested for the use of conductors comprised of two materials, e.g. niobium to carry the normal current (low loss) and Nb_3Sn to carry overloads (higher a.c. loss). Alternatively, niobium (or Nb_3Sn) etched on a backing of aluminium or copper strip (the latter to carry fault currents) may be used.

Several forms of conductor are possible and examples will be cited when particular cable designs are given. The variation of a.c. loss with the surface field strength is shown in Figure 7.6 for niobium and niobium–tin strip. Much work has been reported on a.c. loss in superconductors.[15,16,17,18,19,20]

7.3 THERMAL INSULATION – CRYOGENIC ENVELOPES

The low efficiencies of refrigerators necessitates the minimum possible temperature rise of the coolant. One major source of heat to the coolant is the heat transmitted from ambient (in the order of 300 K) to the conductor (say 5 K or 77 K). To minimize this heat 'inleak' thermal insulation is provided in the form of a cryogenic envelope. For superconducting cables it may be economical to have an intermediate heat shield cooled by liquid nitrogen and to remove most (about 90%) of the inleak via a liquid nitrogen coolant. Heat is transferred by conduction, convection, and radiation; the provision of vacuum greatly reduces the effects of the first two, making the latter mode very important.

In a vacuum the heat transmitted by radiation is given by

$$5.67 \times 10^{-8}(T_1^4 - T_2^4)\frac{\epsilon_1 \epsilon_2}{\epsilon_1 + \epsilon_2 - \epsilon_1 \epsilon_2}\ \mathrm{W/m^2} \tag{7.2}$$

where T_1, T_2 and ϵ_1, ϵ_2 are the temperatures (K) and emissivities of the surfaces containing the vacuum. ϵ_1 and ϵ_2 are both of the order of 0.03. Radiation is reduced by the use of several highly reflective shields. If n shields are used the heat transmitted is reduced by $1/(n + 1)$. Such shields are obtained by the use of layers of aluminium foil separated by thin glass-fibre spacers or a mesh of nylon, inserted into a vacuum space, and constitute 'superinsulation'. The effective thermal conductivity is of the order of 10^{-5} to 1.7×10^{-4} watt/m-K. Superinsulation is the most effective insulation; however, it is expensive, mechanically weak, and requires a vacuum of the order of 10^{-4} Torr.[7]

A less effective insulation, but one cheaper and mechanically suited to changes in shape is Aerogel powder, which also requires a vacuum. This has a conductivity in the region 6 to 12×10^{-4} W/m-K and is used down to 77 K from ambient. Finally, foams may be used for the range 300 K to 77 K with a conductivity in the region 260 to 360×10^{-4} W/m-K. It is necessary to keep the inleak into the helium part of a cable down to a few μW/cm^2. Examples of cryogenic envelopes are shown in illustrations of the various prototype cables in later sections.

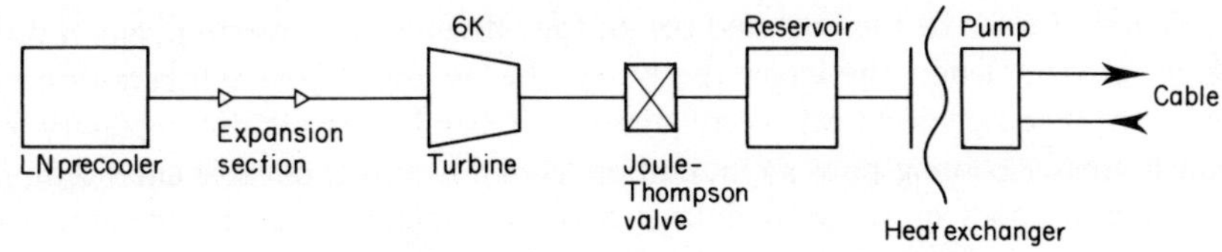

Figure 7.7 Schematic diagram of the Claude refrigeration system.

7.4 REFRIGERATION AND COOLING SYSTEMS

The refrigerators used are likely to work on the Claude system (see Figure 7.7) for a liquid helium version.[7,13,21] Turbo expanders have more life expectancy than reciprocating types, although the latter are more efficient. Efficiencies of refrigeration for various loadings are summarized in Figure 7.8[5] The cost of a refrigerator may be related to a known or standard cost (C_0) and size (H_0) as follows,

$$C = C_0 \left(\frac{H}{H_0} \right)^n \tag{7.3}$$

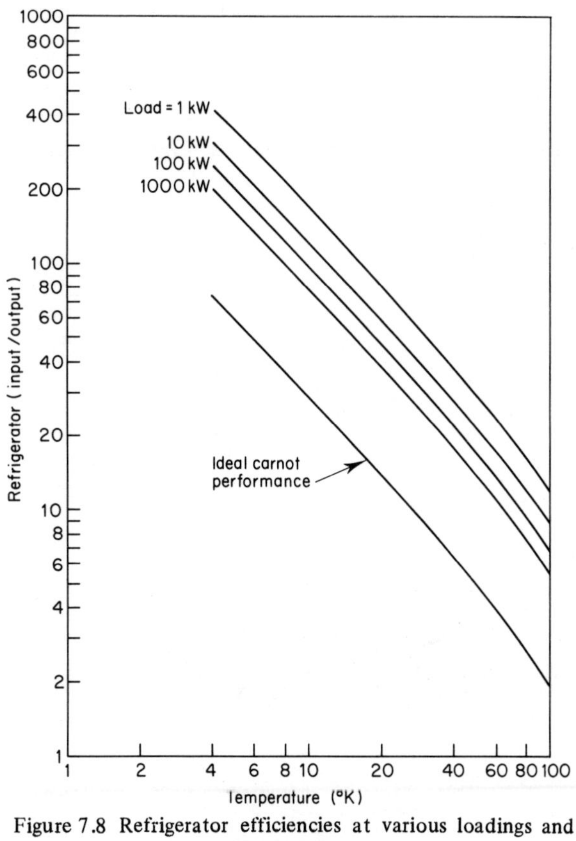

Figure 7.8 Refrigerator efficiencies at various loadings and temperatures.

where C and H refer to the cost and size of the refrigerator in question, and n is of the order of 0.6. Hence, the larger the system the less the relative refrigerator cost. Duplication of refrigerators will be required for reliability and is also necessary for the much greater cooling power required on 'cool down' before cable energization.

Two factors control the maximum length of cable between refrigerator and pumping stations: the maximum temperature rise and maximum pressure drop of the coolant. Two-phase flow, i.e. a mixture of liquid and gas, sets up high pressure drops in the cooling channels and should be avoided. Also this would affect the performance of electrical insulation where the liquid forms the impregnant. Coolants may be pumped through conductor ducts (Figure 7.9) and returned through pipes separate from the cable or returned through the conductor ducts of a second (parallel) cable. Alternatively the coolant could be returned in the spaces between the insulated cores and the containing pipe. For a single-phase flow (pressurized liquid or gas) the pressure drop (MKS units) in a pipe with turbulent flow is

$$\Delta p = \frac{0.092 L \rho^{0.8} \eta^{0.2} V^{1.8}}{D^{1.2}}$$

where, L = length, ρ = density, V = velocity, D = diameter of duct or channel, and η = viscosity. Also,

$$qL = G \frac{\pi D^2}{4} c_p \theta \tag{7.5}$$

where θ = temperature rise (K), c_p = specific heat, q = total loss per m, and G = mass flow.

$$\text{Pumping power} = \Delta p \frac{G \pi D^2}{4\rho}.$$

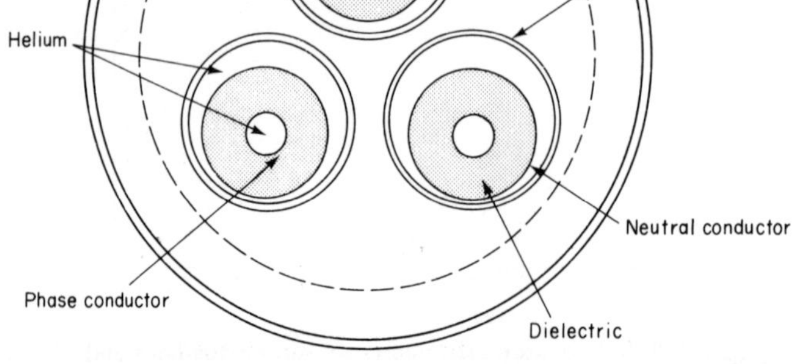

Figure 7.9 Schematic cross-section of typical superconducting cable.

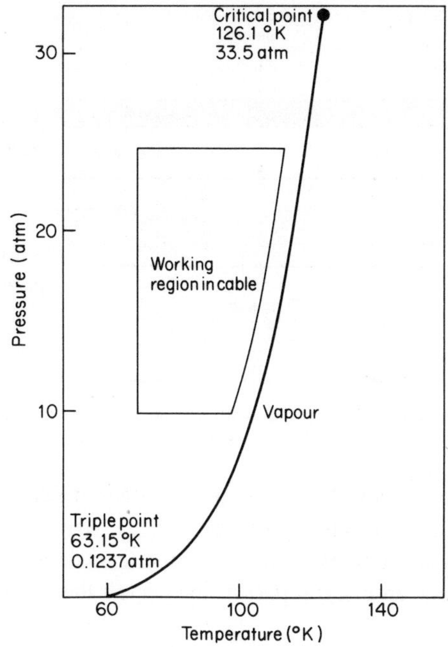

Figure 7.10 Pressure–temperature (liquid–vapour) curve for nitrogen.

From the above, it is seen that Δp is proportional to L^3 whereas the heat load varies with L. The total heat load, q, is the sum of conductor a.c. loss, heat inleak dielectric loss, and the heat evolved due to the turbulent flow of coolant.

To avoid two-phase operation of the coolant, a liquid must be pressurized, and the possible range of coolant operation is found from the pressure–temperature diagram; for liquid nitrogen this is shown in Figure 7.10. There is obviously a compromise to be made between the allowable pressure drop (Δp), inlet to outlet (which determines the flow rate), and the corresponding temperature rise.[22,23] In some flow configurations, for cables with flexible taped insulation the difference in temperature between the 'go' and 'return' streams (i.e. across the insulation) causes heat to leak between them, determined by the thermal resistivity of the insulation. The situation is illustrated in Figure 7.11 which shows an element of cable Δx in length; for simplicity only one conductor and idealized return flow channel are shown. The heat leakage, $q_L = (\theta_2 - \theta_1)/R_d$, where R_d is the thermal resistance of the dielectric over length Δx. The dielectric loss divides between the conductor and return coolant flows. The analysis of this system requires iterative techniques and a computer. Solutions have been obtained[22] in order to assess the effect of insulation thermal resistivity on the current rating of liquid nitrogen cables.

In superconducting systems[24,25] cooled with helium with synthetic tape dielectric, the thermal conductivity should be in the range 4.6×10^{-6} to

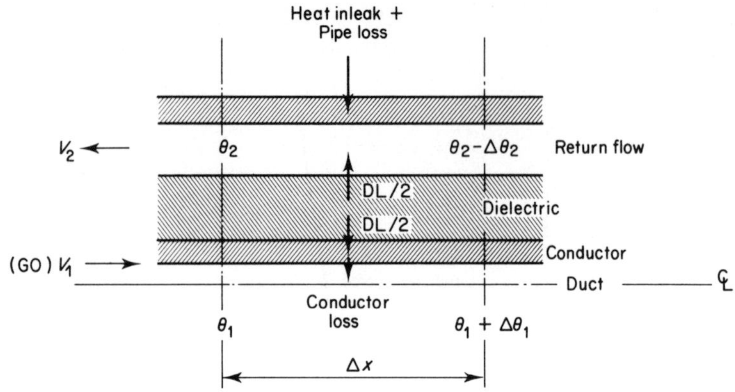

Figure 7.11 Heat sources in cable with go and return in same cryogenic envelope – heat leak across dielectric.

9×10^{-4} W cm/K. Less than the lower limit entails dielectric heating and local temperature excursions, and, above the upper value, significant leaks between the Go and Return streams. However, in one American design the thermal leakage between streams is made an integral part of the overall thermal design.[26,27]

7.5 DIELECTRICS

A number of possibilities exist for the electric insulation which will have to withstand relatively high voltages. As vacuum is needed for insulation it could be used as the dielectric as well.[28] Much work has been performed on the breakdown characteristics of vacuum, although in this case the presence of spacers to separate the annular surfaces reduces the breakdown voltage compared with that of plain vacuum gaps. The low-temperature environment will enhance the dielectric properties of vacuum, which include very low dielectric loss. The use of the cryogenic fluids themselves in conjunction with plastic tapes is attractive.

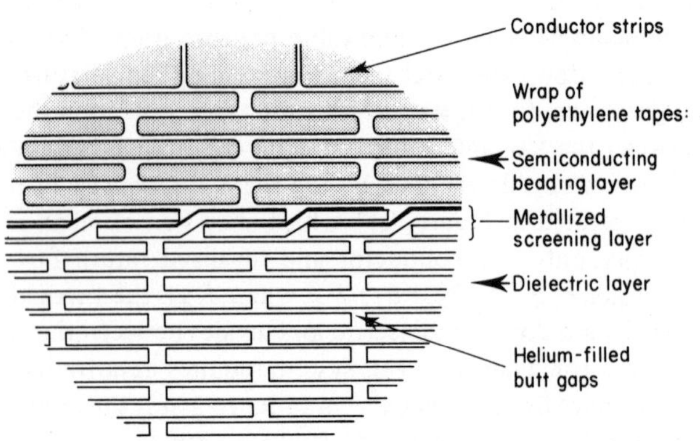

Figure 7.12 Section through wall of impregnated tape insulation, conductor strips, and semiconducting layers. (Reproduced by permission of CEGB.)

Both liquid nitrogen and hydrogen have breakdown and loss characteristics (measured between electrodes) superior to those of conventional paper/oil, while those of liquid helium are slightly worse although still usable.[9,29,12] A further advantage is the low permittivity of the cryogenic liquids compared with paper/oil giving lower capacitance and hence reducing system voltage-control problems. Typical dielectric loss angles ($\tan \delta$) are 2×10^{-7} hydrogen at 14 K, 2×10^{-6} for helium at 4.2 K, and 2×10^{-3} for paper/oil at 293 K. The insulation system most preferred comprises plastic tapes impregnated with the coolant, as shown in Figure 7.12 and for helium having $\tan \delta \leqslant 2 \times 10^{-5}$.

Vacuum Systems[12]

These imply spacers and probably a rigid assembly involving the making of many joints during installation because of the difficulty in transporting lengths greater than 20 m from the maker to the site. The great advantage of vacuum is that it could provide both electrical and thermal insulation, leading therefore to a compact design. Also, its relative permittivity is unity and dielectric loss small. In general, the lower the pressure the higher the breakdown strengths and with low temperature operation the gas molecules condense on the cold surfaces, further improving the vacuum. However, the lowest pressure attainable in a full-scale system will be limited by degassing of metal surfaces and leaks.

Much work has been done on the breakdown properties of vacuum gaps at room temperatures, and information is now available for low temperatures.[12] The break-

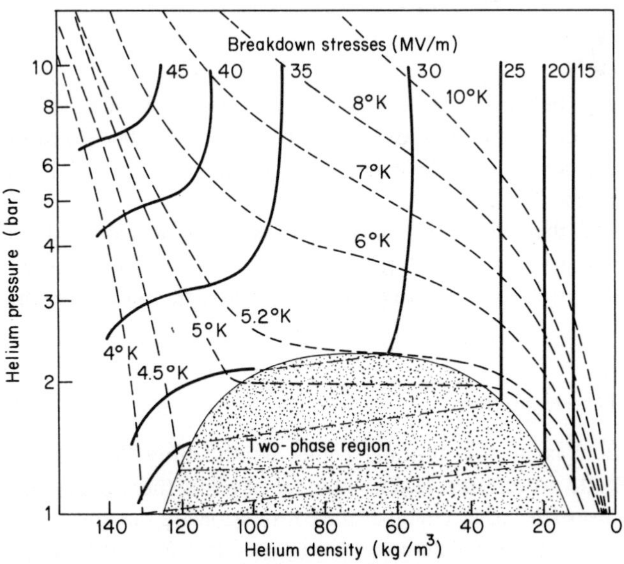

Figure 7.13 Contours of constant breakdown stress superposed on the isotherms of pressure–density diagram for helium. Results for direct voltage and uniform fields between brass electrodes. ——— Isotherms. (Reproduced by permission of CEGB.)

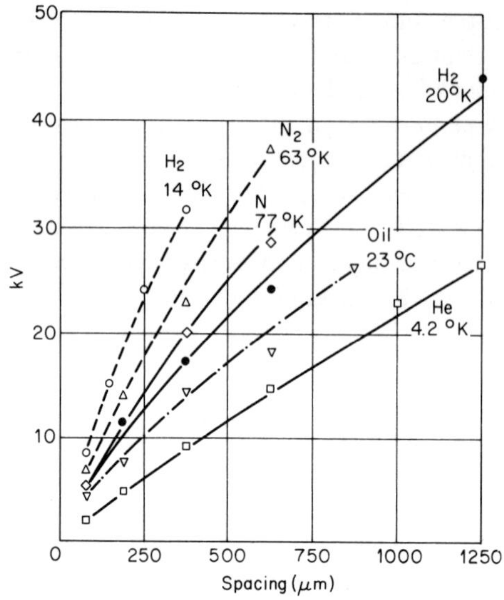

Figure 7.14 Breakdown in cryogenic fluids.[9]
(Reproduced by permission of IEEE.)

down strength is a complex function of several factors such as pressure, electrode separation and area, electrode material and surface finish, and temperature. The performance of vacuum as electrical insulant in a liquid nitrogen cable has been measured in full-scale tests.[30]

Cryogenic Liquids[9,31]

The relationship between the breakdown voltages of helium as a function of pressure and temperature is shown in Figure 7.13. Higher pressures lead to a notable improvement in the dielectric strength of helium, i.e. 70% higher at 10 atmospheres than at 1 atmosphere.

It is expected that with the insulation gaps in spacer cables the a.c. strength (r.m.s) of liquid helium (including spacers) will be in the range 7–10 kV/mm A.c. breakdown properties for liquid helium, nitrogen, and hydrogen are given in Figure 7.14.[9]

Impregnated Tape Insulation

Taped insulation as in conventional cables is inherently flexible and reasonable continuous lengths are obtainable. The number of joints therefore is relatively small. The essential properties of low-temperature insulation are the same as for dielectrics at ambient, i.e. breakdown voltage on a.c (power frequency) and impulse (1/50 μs and switching-surge waves), dielectric loss (tan δ), permittivity, internal partial (corona) discharge inception voltage, and, of course, cost. The basic considerations discussed in Chapter 2 are relevant, including the influence of butt gaps.

Table 7.4 Dielectric materials for superconducting cables.[32] (Reproduced by permission of Revue Generale Electricitie.)

Polymer type and trade name	Diel const	Diel loss	Elongation	Tensile strength	Tensile modulus	Cost
Polyethylene, low density	Exc.	Exc.	Poor	Not acc.	Not acc.	Exc.
Polyethylene, high density	Exc.	Exc.	Poor	Good	Poor	Exc.
Polyamide, Nylon 11, Rilsan-F	Good	Good	Good	Good	Poor	Good
Polysulfone, Udel	Good	Good	Good	Good	Good	Good
Polyimide, Kapton H	Not acc.	Not acc.	Good	Exc.	Exc.	Not acc.
Polyimide-Teflon, Kapton F	Good	Not acc.	Exc.	Good	Good	Not acc.
Polyester, Mylar A	Good	Not acc.	Good	Exc.	Exc.	Exc.
Polycarbonate, Makrofol KG	Poor	Poor	Exc.	Exc.	Exc.	Good

Exc. = Excellent Not acc. = Not acceptable

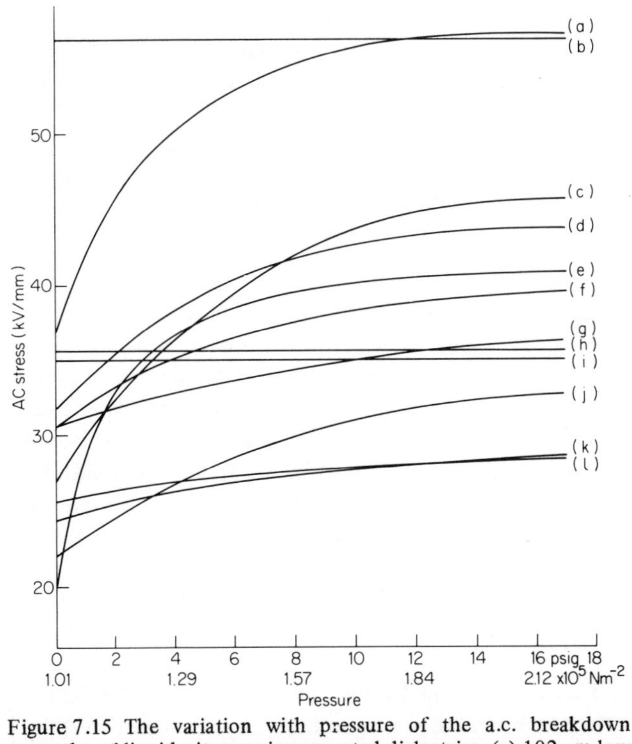

Figure 7.15 The variation with pressure of the a.c. breakdown strengths of liquid nitrogen impregnated dielectrics. (a) 102 μm low loss polypropylene–paper laminate, (b) 36 μm Melinex, (c) 104 μm Nomex, (d) 170 μm Nomex/Nylon, (e) 123 μm paper, (f) 140 μm FEP, (g) 125 μm Melinex, (h) 100 μm Nylon 11, (i) 98 μm embossed polycarbonate, (j) 132 μm Tyvek, (k) 158 μm polycarbonate, (l) 157 μm Tenax.[33] (Reproduced by permission of IEEE.)

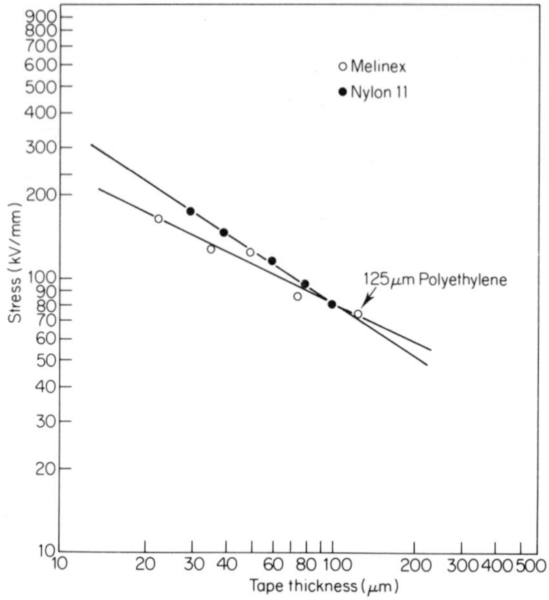

Figure 7.16 Impulse strength of liquid nitrogen impregnated taped insulation–variation with tape thickness. Pressure 1×10^5 N/m^2; temperature 77.3 K; 6 tape layers, negative surge (1/50 μs). (From Rigby, S. J. and Weedy, B. M., Vol. 16, p. 167 of *Cryogenics*, published by IPC Science and Technology Press Ltd., Guildford, Surrey, UK.)

At low temperatures the tape material must have mechanical integrity at the working temperature and also be able to be wrapped on the conductor at ambient. The large excursion in temperature involves large thermal contractions and the various materials contract by different amounts, e.g. polyethylene has a thermal contraction coefficient roughly ten times that of conductor materials. Several plastic materials, although suitable from the electrical standpoint, have mechanical properties which render them impossible to wind on conventional automated winding machines. A list of possible tape materials[32] is given in Table 7.4.

A summary of the electrical breakdown parameters[33,34,35,36,37] of plastic tapes impregnated with liquid nitrogen is given in Figures 7.15 and 7.16. Although designs are often decided by impulse requirements, the discharge inception voltage is very important. Considerable attention has been given to corona (partial) discharges in the butt gaps of taped insulation impregnated with liquid nitrogen. The variation of stress with discharge magnitude[38] is shown in Figure 7.17. A detailed discussion of the nature of the discharges appears in References 38, 39, 40, and 41.

Results[46] for tapes impregnated with liquid helium[42,43,44,45] are shown in Figure 7.18. However, because of the lower breakdown strength of helium gas, the partial discharge inception voltages are low and currently an a.c. working stress of 10 kV/mm is hoped for with high density polyethylene tapes.

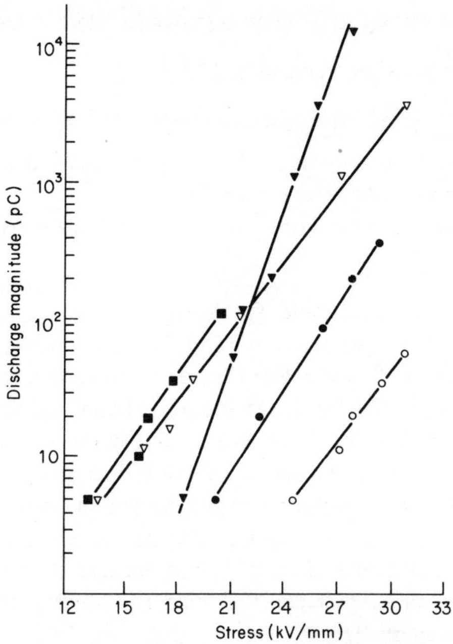

Figure 7.17 Effect of stress on discharge magnitude – liquid nitrogen impregnated taped insulation. • 98 μm embossed polycarbonate, ▾ 125 μm polyethylene, ▿ 140 μm Nomex, ▪ 157 μm Tenax, ○ 102 μm polypropylene/paper.[38] (Reproduced by permission of IEE.)

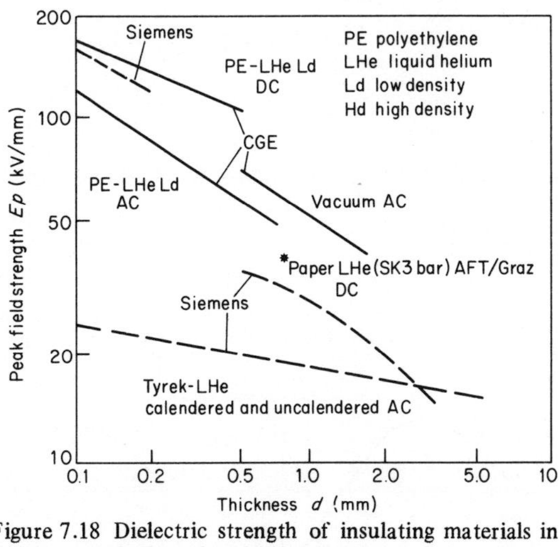

Figure 7.18 Dielectric strength of insulating materials in the form of single and multi-layered tapes as a function of total thickness (results from European laboratories). (From Bogner, G., Vol. 15, p. 79 of *Cryogenics*, published by IPC Science and Technology Press Ltd., Guildford, Surrey, UK.)

7.6 DESIGNS FOR SUPERCONDUCTING CABLES

The following designs have been considered,[46,47]

(*a*) rigid, liquid helium dielectric with spacers[48]
(*b*) flexible, wrapped tape impregnated with pressurized helium using a niobium conductor, maximum temperature 5 K.[42,49]
(*c*) as (*b*), but with niobium–tin conductor and supercritical helium up to 9 K.[43,14]

These designs are basically suitable for both a.c. and d.c. cables, although the absence of a.c. loss in the superconductor and the better properties of dielectrics on d.c. makes the d.c. design[50] easier. For the conductors niobium is possible in either straight or corrugated tubular form. On the other hand, segmental conductors may be constructed of wire or strip. To reduce costs a common cryogenic envelope for the three phases will be used. The use of a flexible corrugated cryogenic envelope is being investigated with the possibility of getting 150 m on to a cable drum.[51] The problem of thermal contraction requires the use of contraction joints. For the conductor, telescopic joints are difficult to use because of the high and somewhat capricious contact resistance between the fixed and moving parts. 'Ω' bends in the cable will be of considerable help.

Helically Wound Wires and Strips

Consider wire or strip laid up with an angle of lay α as shown in Figure 7.19. The following expressions[42] apply:

$$b^2 = L^2 + (2\pi r)^2$$

and

$$\tan \alpha = 2\pi r/L$$

where b is the actual length of tape. The change in the radial dimension, Δr, is

$$\Delta r = \frac{\Delta_b - \Delta_L \cos^2 \alpha}{\sin^2 \alpha}. \qquad (7.6)$$

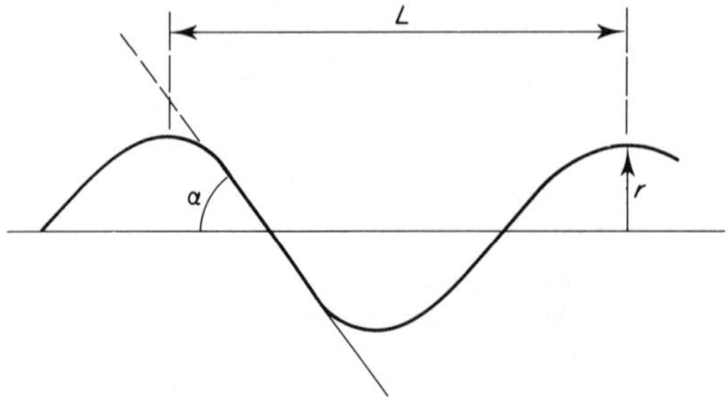

Figure 7.19 Helically wound wire or strip. Total length of one complete turn b m.

On cool-down the axial contraction causes a change in conductor diameter and, for a specified length and contraction change, the angle α is determined. The wire is anchored at its two extremities and L is fixed. As the wire contracts the change in r, Δr, modifies the difference in contraction between the dielectric and conductor, although the dielectric is still squeezed on cool-down which is beneficial to its performance. The thermal contraction coefficient of metals is about 0.3% and of taped dielectric about 3%.

A problem caused by this helical design is the axial flux set up by the conductor current.[52] This includes a circulating current in the surrounding duct. There is no outside radial flux because of the equal and opposite return current in the outside conductor or screen.

Basic Design Considerations

Consider a dielectric contained between the concentric arrangement of a conductor of radius r and dielectric screen of radius R, the voltage across the dielectric being V. It is readily shown that

$$R = r \exp\left(\frac{V}{rE}\right)$$

where E is the maximum working stress at the conductor, which must be considerably less than the breakdown value.

The field strength, $H < H_c$ (the critical field), $H = I/2\pi r$, where I is the conductor current. Also the power, P, is

$$P = VI$$

and

$$r = \frac{P}{2\pi HV}.$$

The optimum relationship[44] is obtained by taking

$$\frac{dR}{dV} = 0$$

when

$$\bar{V} = \left(\frac{PE_c}{4\pi H}\right)^{0.5} \qquad \bar{r} = \frac{2\bar{V}}{E_c} \qquad \bar{R} = \bar{r} \exp\left(\frac{\bar{V}}{\bar{r}E_c}\right). \tag{7.7}$$

Similarly, the thermal insulation design may be optimized.[44]

The conductor dissipation is somewhat less than 10 μW/cm^2 at 6 K and the value of H_c for niobium is less than 40 A/mm for a tubular construction. Strips have edge-losses, but by careful construction these may be reduced to about 20% of the main-conductor a.c. loss. The heat-leak from ambient is of the order of 2 to 3 times the total of other losses indicating the importance of cryogenic envelope design. The optimum length between refrigerators appears to be of the order of

10 km for liquid helium and the best operating voltage to be one quarter of that of a conventional cable for the same power level for the liquid helium and spacer insulation, and about one half for taped liquid helium insulation. For supercritical helium the distance between refrigerators may be considerably larger.[26]

7.7 PROPOSED SUPERCONDUCTING CABLES

In the USA, prototype a.c. designs have been developed by Union Carbide and the Brookhaven National Laboratory. The former uses helium as dielectric with spacers, the latter niobium–tin conductor and a taped polyethylene insulation impregnated with supercritical helium, as shown in Figure 7.20. A full-scale test underground superconducting cable using taped insulation is expected to be commissioned in the near future.

Brookhaven Design

The Brookhaven design uses supercritical helium with an inlet temperature of 6 K and maximum temperature of 8 K. The conductor is niobium–tin and the inner bore is pressurized with helium at 1.5 MN/m^2 (approximately 15 atmospheres). Discharge inception is critical and for the designs using polyethylene tapes a working maximum a.c. stress of 10 kV/mm is aimed at. Parameters of a cable[53] designed to convey a 4.8 GW from a nuclear complex to a substation 69 km (43 miles) away are given in Table 7.5. Each circuit is able to carry a continuous contingency loading of 4.8 GVA. One undesirable feature of this design is the high cost of the cryogenic envelope. A cheaper envelope could result in significant overall economy. In the original design study the cooling gas flowed through the cable and inner pipe of the cryostat for a sufficient distance to absorb enough heat to cause the maximum allowable temperature rise of 2 degrees K. At this point the cooling gas was removed from the cable and returned to the refrigerator using insulated pipes contained within the annulus between the inner and outer pipes of the cryostat. These pipes would be connected as each section of the crystat is installed, an expensive operation which also increases the likelihood of leaks in the numerous welds. A revised and improved version makes the cable and enclosure a heat-exchanger forming part of the refrigerator system. Expansion is required at the

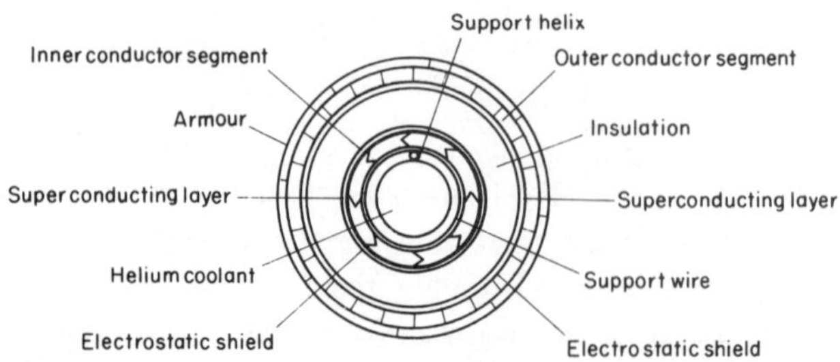

Figure 7.20 Cross-section of a flexible superconducting coaxial cable with taped insulation.[55] (Reproduced by permission of IEEE.)

Table 7.5 Parameters of Brookhaven Cable.[54] (Reproduced by permission of IEEE.)

Voltage kV	345
Length, km	68
Number of circuits	2
Cable diameter, mm	120
O.D. of cryogenic envelope, mm	500
Maximum electrical stress, MV/m	10
Mean electrical stress, MV/m	7.1
Maximum line current, continuous contingency rating, kA	8.0
Heat load at continuous contingency rating, W/km per circuit	262
Operating temperature range, °K	6.5 to 8.5
Number of refrigerator stations per circuit	6
Refrigerator size per station per circuit, H.P.	1700

end of the cooling loop. The increased conductance of the helium flow channels permits greater spacing between refrigerators, and smaller compressors are needed for the same heat load. The revised design is shown in Figure 7.21, in which the cables are surrounded by a sheath of corrugated stainless steel and form the high pressure Go-line. The Return line is formed by the space between the cables and the enclosure. The cables and envelope form a counterflow heat exchanger. At the far end of the cooling loop the gas is expanded and hence the cable is cooled from both ends.

European

In Europe research and development work has been carried out at several laboratories. Typical cable specifications are, CGE: Air Liquide (France), 125 kV/3 GW; Siemens (Germany), 120 kV/2.5 GW: CERL (Britain), 132 kV/1.4 GW. For d.c cables the preferred conductor material for CGE is NbTi multicore conductors embedded in high purity aluminium by an extrusion process, and Nb_3Sn ribbons for AEG. For a.c. cables niobium or niobium–tin in the form of thin layers on substrates of copper or aluminium is preferred.

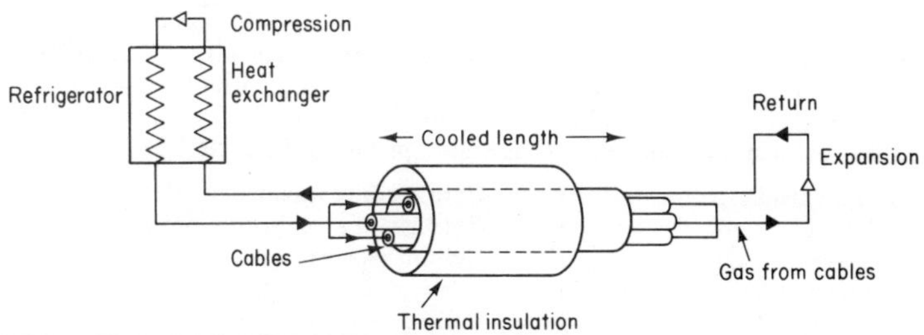

Figure 7.21 Superconducting cable cooling system – method of reducing cost by forming a counter-flow heat exchanger from the cable and cryostat.[55] (Reproduced by permission of IEEE.)

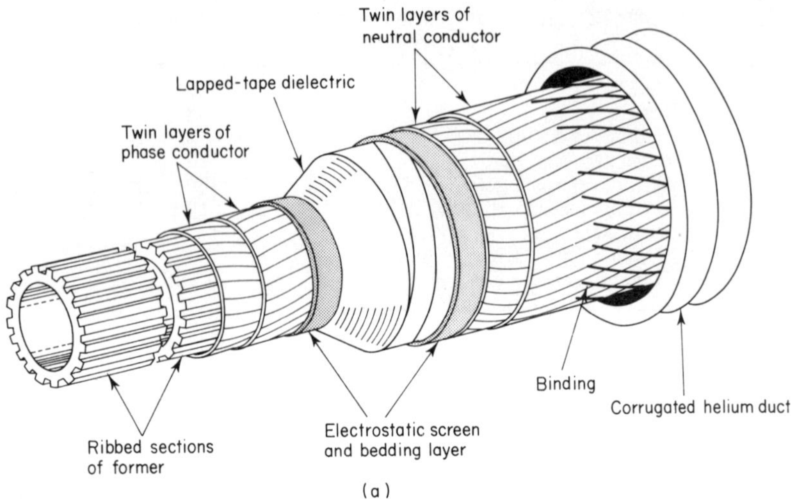

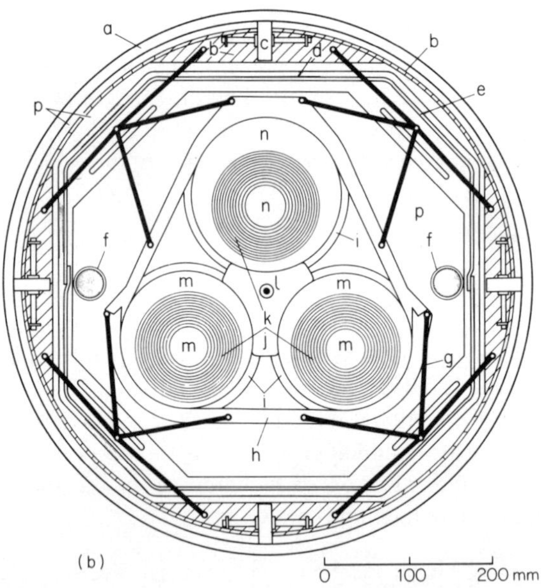

Figure 7.22 (a) Superconducting cable core. CERL design for 5 GVA at 400 kV. Helically wound coaxial conductors comprise twin-layers of niobium–copper composite strips. Overall diameter of core 130 mm. (Reproduced by permission of CEGB, UK.) (b) Cross-section of three-phase assembly of the CERL cable.[49]

a	Outer steel pipe	h	Core binding frame
b	Carriage	i	Corrugated steel helium pipes
c	Carriage wheels	j	Helium pipe spacer
d	Superinsulation	k	Cable cores
e	Heat shield	l	Pulling wire
f	Nitrogen pipes	m & n	Helium Go and Return flows
g	Support struts	p	Vaccuum

(Reproduced by permission of CIGRE from Report 21-05, 1976.)

The cross-section of the semi-flexible a.c. cable design of CERL is shown in Figure 7.22. An assessment[32] of tape materials for a superconducting cable is tabulated in Table 7.4. Polyethylene is available both as film or in spun bond (paper) form. Its glass-transition temperature is roughly 150 K (below which it is brittle) and as only 70% of the total thermal contraction has occurred by this temperature, the mechanical action of polyethylene is of concern. Nevertheless, the CERL design uses polyethylene tapes which have the following parameters: tan $\delta 6 \times 10^{-6}$, ϵ_r = 2.3 (at 4.2 K), thermal conductivity (polyethylene film) $(20–50) \times 10^{-3}$ W/m-K over the range 4–7 K, the corresponding value for helium being approximately 25×10^{-3} W/m-K. The design criteria are summarized in Table 7.6 for the CERL cable.

Basic system parameters[49] of proposed superconducting cables are compared with those for overhead lines and conventional self-contained cables in Table 7.7. Comparative costs for the CERL 5 GVA 400 kV design are given in Table 7.8 (mid-1975 prices). A rural route of at least 30 km has been assumed. Because of the high power transfers required to make superconducting cables economic relative to conventional systems it is not anticipated that in Europe a need for such cables will occur during this century. Accordingly, the research and development programmes have been greatly curtailed.

Japanese

Considerable research and development work is in progress in Japan. A proposed superconducting d.c. cable[47] is rated for 100 kV and 50 kA. The conductor consists of 17 layers of 7 μm NbTi ribbon on a substrate of a 5 cm diameter copper tube of wall thickness 7 mm. This is surrounded by a 12.5 mm thickness of polyamide tape insulation, the maximum stress being 9.5 kV/mm. The outer conductor comprises 12 layers of 7 μm NbTi ribbon on another copper cylinder. Supercritical helium at 5 K and 8 atm is used. The complete assembly is assembled in a steel pipe about 35 cm diameter, the cable being supported by nylon string reinforced by steel wire every 5 m.

Table 7.6 Dielectric design criteria and test conditions, British superconducting cable.[42]

Parameter	Value
Electrical stress, a.c.	10 MV m^{-1} (r.m.s.)
Electrical stress, impulse	82.5 MV m^{-1} (p)
Operating voltage (ϕ – E)	76 kV (r.m.s.)
Test voltage	100 kV (r.m.s.)
He pressure	4 bar
He temperature	5 K
Bending radius	2 m
Winding tension	2 MN m^{-2}

Table 7.7 Basic system parameters of British conventional and superconducting circuits.[49] (Reproduced by permission of CIGRE from Report 21-05, 1976.)

Voltage (kV)	Type of circuit	Series reactance (% km) (100 MVA base)	Shunt susceptance (% km) (100 MVA base)	Continuous rating (GVA)	Natural load (GW)
275	$2 \times 400\ mm^2$ overhead	0.040	0.27	1.0	0.26
	$2000\ mm^2$ water cooled underground	0.022	11	0.76	2.2
	superconducting underground	0.0055	4.4	3.0	2.9
400	$4 \times 400\ mm^2$ overhead	0.016	0.66	3.0	0.64
	$2600\ mm^2$ water cooled underground	0.011	20	3.0	4.3
	superconducting underground	0.0034	7.2	5.0	4.7

Table 7.8 Comparative British costs of 5 GVA 400 kV systems (1975).[49] (Reproduced by permission of CIGRE from Report 21-05, 1976.)

Item		Price (£m^{-1}) Superconducting cable	Conventional cable LPOF
(i)	Cable cores	250	540
(ii)	Cryogenic envelope or water cooling pipes	270	220
(iii)	Refrigeration or cooling stations	130	65
(iv)	Installation and civil works	160	250
(v)	Jointing	120	145
(vi)	Ancillaries and miscellaneous	90	95
(vii)	Helium	40	–
(viii)	Reactive compensation	10	55
(ix)	Running losses	130	250
	Total	1200	1620
	£MVA^{-1} km^{-1}	240	325

7.8 RESISTIVE CABLES (CRYORESISTIVE)

The options available are similar to those for superconducting cables, i.e. liquid/spacer insulated rigid cable and flexible tapes (impregnated). For the conductor either rigid tubes or conventional stranded conductors with a central duct in which the coolant flows (as in LPOF cables) are used. A design using liquid nitrogen with vacuum/spacer insulation[30] is shown in Figure 7.23. The conductors consist of extruded aluminium tubes through which the coolant is pumped. The cable is rated at 345 kV and the circuit comprises two separate three-phase cables laid one above the other in a trench 0.5 m wide. As there is refrigeration at one end only of the link the coolant flows through one cable and returns through the other. In the event of a failure in one cable, valves are incorporated to enable the coolant to

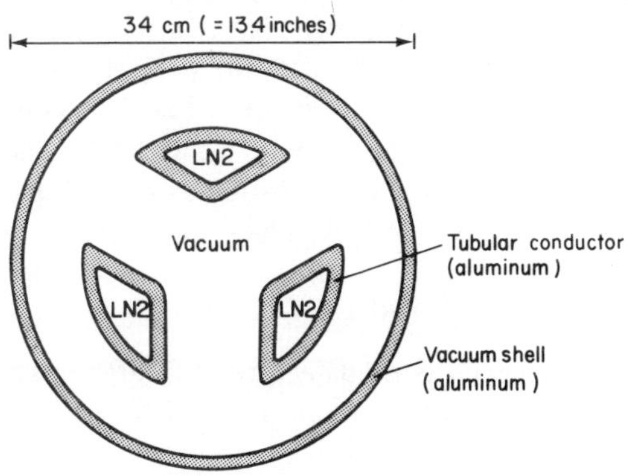

Figure 7.23 Cross-section through three-phase liquid nitrogen cooled, vacuum insulated rigid spacer cable – 138 kV.[30] (Reproduced by permission of IEEE.)

return through one of the three conductor channels of the operative cable. For a circuit of length 10 km, 1.6 MW of heat are removed and the rating is 3660 MVA. The segmented shape of the conductors is the result of design-analysis studies to reduce the a.c. resistance.

A major problem with resistive cables is the large influence of electro-magnetic effects, such as those of skin and proximity, on the conductor a.c. resistance; because of these some of the reduction in resistance at low temperatures is lost. The sum of the eddy and conduction losses per unit length in a stranded conductor is given by

$$\frac{I^2\rho}{A} + \frac{KAB_{\rm m}^2(\omega d)^2}{\rho} \tag{7.8}$$

where ρ = resistivity, d = strand diameter, ω = angular frequency, K = constant, $B_{\rm m}$ = maximum flux density, and A = conductor area. Hence an increase in area A and decrease in ρ decreases the conduction loss but increases the eddy loss. The effect may be reduced by using thin strands and transposing the strands.[3] Eddy-current losses also occur in the containing pipes, although these may be reduced by the use of low-resistivity shields around the conductors.

It is seen from Figure 7.1 that at liquid nitrogen temperatures the resistivity of aluminium is only marginally greater than that of copper and it would be profitable to use a stranded aluminium conductor. The crucial quantity with regard to eddy current loss is the skin depth (λ) of the material which is given by $(10^7\rho/f)^{1/2}/2\pi$, where ρ is the resistivity (Ω-m) and f the frequency (Hz). For aluminium of 99.8% purity λ is 3.8 mm at 77 K and the strands should be less than this in diameter. Practical investigation has shown that for a stranded transposed aluminium conductor at liquid nitrogen temperatures the ratio $R_{\rm ac}/R_{\rm dc}$ is approximately 1.6 when the individual strands are well insulated from each other and equal to 2.0 when this insulation is variable in quality, although it is hoped with improved design to reduce these values.

A design with a taped insulation closely approaches the oil-filled pipe-type conventional cable; high density polyethylene is a possible material for the tape wrapping.[22] Studies[22] have shown that the largest conductor area possible should be used, e.g. in the order of 2500 mm^2. Based on conventional manufacturing technology the maximum diameter of insulated conductor viable is about 150 mm diameter overall. With these constraints for a liquid nitrogen cable a throughput of the order of 5.0 GVA is possible.

7.9 FAULT CURRENTS

In a high voltage system, fault currents can reach fourteen times the normal load current in a cable conductor. The fault clearance times normally applicable are as follows:

100 ms – normal operation of protection
300 ms – operation of local back-up protection
up to 1 s – operation of remote back-up.

Apart from the heating aspect of the fault current, the mechanical forces set up on the conductors are very large (order of tons/cm^2) and constitute a major design problem. In a cryoresistive cable the conductors constitute a large thermal mass and the temperature rise is limited. The calculation of this rise is the same as previously described for conventional ambient cables. With a superconductor,[25] however, the thermal mass is low and the advent of a high fault current on a conductor designed for normal load current could be disastrous. This has already been referred to when the combination of Type I and II conductors was discussed.

A more popular solution is to place the superconductor designed only for normal current on a backing of copper sufficient in mass to absorb the short-circuit heating without causing enough temperature rise to drive the superconductor material 'normal'. Other remedies tackle the problem from a system standpoint and include the insertion of artificial inductance or the provision of special transformers to achieve this aim and the installation of resonant links.

7.10 TERMINATIONS

A critical part of the cable system exists where the conductor leaves the insulation to connect to an overhead line or transformer. This is accomplished by a sealing end (pot-head) as previously discussed for conventional cables. This already difficult task is considerably worsened by the transition from ambient to cryogenic temperatures taking place in the termination. Without careful design the heat inleak so caused can be seriously large. Further, the dielectric strength of helium gas at ambient temperatures is poor and so this cannot be used in the termination. Hence

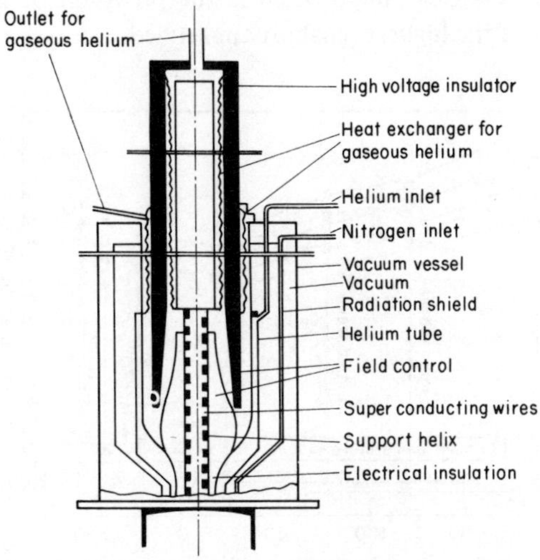

Figure 7.24 Coaxial terminal fitting for a 120 kV superconducting a.c cable (one phase). (From Bogner, G., Vol. 15, p. 79 of *Cryogenics*, published by IPC Science and Technology Press Ltd., Guildford, Surrey, UK.)

a vacuum or solid insulation will be required in the transition from 4 K to ambient (300 K). A coaxial terminal or pot-head for a German 120 kV superconducting a.c. cable[46] is shown in Figure 7.24.

7.11 OTHER BELOW-AMBIENT COOLANTS

Liquid oxygen will not be considered because of fire and explosion hazards, but above this temperature several coolants exist and are indicated in Figure 7.25.[52] Calculations[52] of possible current ratings for a given length and pressures sufficient to maintain single-phase flow indicate that liquid methane gives the highest current. However, the deciding features must be cost and considerations of fire hazards and safety. As operating temperatures decrease, refrigeration and thermal insulation costs increase. The most likely candidates are liquid ethane, liquid freon 22, and water plus 30% glycerol. In Figure 7.25[52] a comparison of unit capital costs for cables with these various coolants is given. The conductors are assumed to consist of transposed strands 1.59 mm in diameter and to have a smooth internal bore of 60 mm diameter. (R_{ac}/R_{dc}) varies from 1.0 at ambient to 1.3 at liquid nitrogen temperature. The spacing beteen the cooling stations is 10 km and the insulation is assumed to consist of a lapped polymer.

7.12 SUMMARY

The engineering problems associated with cryogenic cables, especially superconducting cables, are seen to be severe. Although emphasis here has been on electrical problems, the mechanical aspects are formidable, e.g. the accommodation of thermal contractions. Of great importance is the provision of pumps and large-capacity refrigerators of the highest reliability in service.

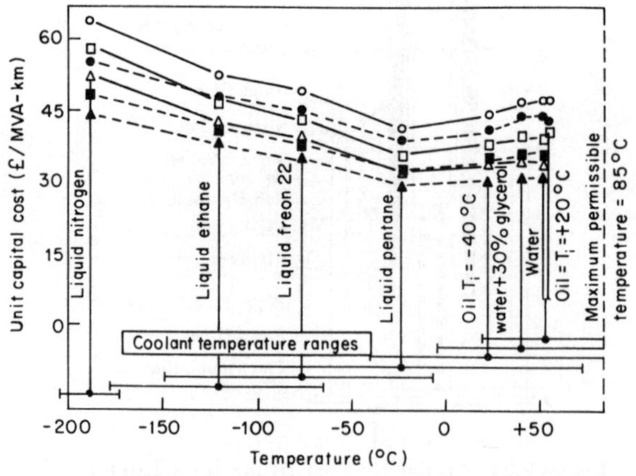

Figure 7.25 Variation of unit capital cost of 400 kV self-contained cables with conductor cooling duct with various below-ambient coolants and impregnants. The cooled temperature ranges are indicated. (Reproduced by permission of IEE.)

Basic system parameters for the British superconducting cable are summarized in Table 7.7. Superconducting cables are characterized by low reactances, high continuous ratings, and, in addition, by a minimum economic length of 5 to 10 km to avoid excessive refrigeration and termination costs. The last two characteristics make them unattractive for the undergrounding of short lengths in mainly overhead circuits. However, in a completely underground transmission system for a large conurbation, where the cable ratings need not necessarily be related to line ratings, superconducting cables deserve consideration.

The three major applications for superconducting cables would be:

(*a*) Connection of large blocks of generation into the main transmission system.
(*b*) Feeding power into a major conurbation.[49,53]
(*c*) Forming the basis of an underground transmission system utilizing the good transmission parameters of these cables. This application is the most uncertain and most distant in time.

A possible application[49] for (*b*) is the transmission of power into the Greater London area from a generation complex as shown in Figure 7.26.

The time scale before the commercial application of a superconducting cable is such that possibilities exist of major changes in its form, for example the use of niobium–germanium (Nb_3Ge) of critical temperature 23 K as the super conductor. This would bring the possibility of liquid hydrogen as the coolant at, say, 15 K which is electrically stronger and also cheaper than helium. However, Nb_3Ge is brittle and not yet available in quantity, and much development work is required. Innovations in the cryogenic envelope are possible in order to reduce cost; for example, the liquid nitrogen shield has been omitted in the Brookhaven design.

Economic studies over the last few years have indicated an increasing cost-advantage for the superconducting over the cryoresistive cable, although work on the latter is still being actively pursued, mainly in Japan. Estimated costs for the proposed British design are summarized in Table 7.8. The utmost care must be

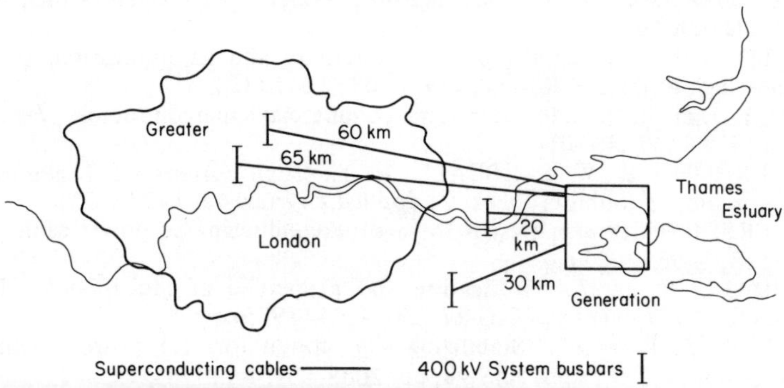

Figure 7.26 Possible British application of superconducting cables. Transmission or generation of Thames Estuary to London Area.[49] (Reproduced by permission of CIGRE from Report 21-05, 1976.)

taken to ensure the lowest possible failure rate. The average duration time of repairs to low-temperature cables will be much longer than for conventional cables because of the time required for warming to ambient and subsequent re-cooling. These repair times and attendant costs will therefore be considerable.

REFERENCES

1. McFEE, R., 'Application of superconductivity to the generation and distribution of electric power', *Elec. Engrg.*, **81**, 122–129 (February 1962).
2. KLAUDY, P., 'Elektrische Energieversorgungs- und Ubertragungs-einrichtungen mit tiefstgekuhlten Leitern', *Elektrotech. Maschinenbau*, **82**, 275–281 (1965).
3. WILKINSON, K. J. R., 'Prospect of employing conductors at low temperature in power cables and in power transformers', *Proc. IEE (London)*, **113**, 1509–1520 (September 1966).
4. GRANEAU, P. 'Low temperature conductors', *Electronics and Power*, **13**, 55 (February 1967).
5. *Underground power transmission*, Industry Advisory Committee, Federal Power Commission, Report, April 1966 (U.S.A.)
6. AFSHARTOUS, S. B., GRANEAU, P., and JEANMONOD, J., 'Economic assessment of a liquid-nitrogen-cooled cable', *IEEE Trans. PAS 1970*, pp. 1–8.
7. SCOTT, R. B., Cryogenic Engineering, Van Nostrand, New York, 1967.
8. LITTLE, Arthur D. (Inc.), *Underground Power Transmission*, Electric Research Council, New York, 1971.
9. MATHES, K. N., 'Dielectric Properties of Cryogenic Fluids' *IEEE Trans. Electrical Insulation EL-2, 1969*, pp. 24–32 (1969).
10. ROGERS, E. C., 'A comparison of super conducting and cryoresistive cables', *Proc. Conference on Low Temperature and Electric Power, March 1969*, British Cryogenic Council, London, pp. 83–86.
11. MINNICH, S. H. and FOX, G. R., 'Cryogenic Power Transmission', *Cryogenics*, **9**, 165–176 (1969).
12. RECHOWICZ, M., *Electric Power at Low Temperatures*, Oxford University Press, 1975.
13. HOARE, F. E., JACKSON, L. C., and KURTI, N., *Experimental Cryophysics*, Butterworths, London, 1961.
14. FORSYTH, E. B. (Ed.), *Underground Power Transmission by Superconducting Cable*, Brookhaven National Laboratory, Associated Universities Inc., Report 50325, March 1972.
15. HADLOW, M. F. G. *et al.*, 'Superconductivity and its applications to power engineering', *Proc. IEE Review*, **119**, 1003–1032 (1972).
16. CHANT, J. *et al.*, 'A.C. losses in some commercial superconductors', *Proc. IEE*, **117**, 1141–1147 (1970).
17. MAYERHOFF, R. W. and BEALL, W. T., 'High current a.c. losses in large superconducting niobium tubes', *J. Applied Physics*, **42**, 147 (1971).
18. ROGERS, E. C. *et al.*, 'Design for a superconducting ac power cable', *Proc. IEE*, **118**, 1493 (1971).
19. RANDALL, R. *et al.*, 'Fabrication and properties of multifilament Nb_3Sn conductors', *IEEE Trans. MAG-11*, 291–294 (1975).
20. HOFFER, J. K. *et al.*, 'Stabilizing superconductors for power engineering applications', *IEEE Trans. PAS-94*, pp. 2008–2014 (1975).
21. FORSYTH, E. B., 'Cryogenic Engineering for the Brookhaven Power Transmission Project', *Cryogenics*, **17**, 3–7 (1977).
22. WEEDY, B. M. and RIGBY, S. J., Thermal and Electrical Assessment of Flexible Cryoresistive Cables', *Cryogenics*, **17**, 453–459 (1977).

23. HELLER, I., 'Power losses in liquid nitrogen cooled high-current cables', *Cryogenics,* **15**, 243–295 (1975).
24. SWIFT, D. A., 'Optimum flow conditions for liquid coolants in superconducting power cables', *Cryogenics,* **8**, 238–293 (1968).
25. MELVILLE, P. H., 'Temperature rise in a superconducting cable during overload', *Cryogenic* **15** 265–269 (1975).
26. FORSYTH, E. B. *et al.*, 'The Technical and Economic Flexibility of superconducting power transmission – A case study', *IEEE Trans PAS-94*, pp. 161–167 (1975).
27. EDNEY, K., FOX, M., and GILBERT, G., 'Longitudinal temperature distribution in low-temperature cables with counterflow cooling', *Cryogenics,* **7**, 355–358, (1967).
28. GRANEAU, P., 'Economics of Underground Transmission with Cryogenic Cables', *IEEE Trans. 69 TP 94-PWR*.
29. FALLOU, B. and BOBOR, M., 'Electrical Properties of Insulating Materials at Cryogenic Temperatures', *Conference on Electrical Insulation and Electrical Insulation and Dielectric Phenomena, National Acad. Sciences, Washington, DC, 1973*, pp. 514–521.
30. GRANEAU, P. *et al.*, 'High Power Transmission Tests on Short LN2 Cryogenic Loop', *Proc. IEEE Conf. Underground Transmission and Distribution, 1974*, pp. 424–428.
31. MATHES, K. W., 'Cryogenic Cable Dielectrics', *IEEE Trans. EI-4, 1969.*
32. MULLER, A., 'Mechanical Properties of insulating tapes at cryogenic temperatures', *Revue Generale Electricitie (RGE)*, **84**, 568–572 (1975).
33. RIGBY, S. J. and WEEDY, B. M., 'Liquid Nitrogen-Impregnated Taped Insulation for Cryoresistive Cable', *IEEE Trans. on Electrical Insulation EI-10*, 1–9 (1975).
34. RIGBY, S. J. and WEEDY, B. M., 'The effects of semi-conducting screens on the dielectric loss of liquid nitrogen impregnated taped cable insulation', *Cryogenics,* **15** 465–474 (1975).
35. FUKASAWA, M. and NAGANO, H., 'Liquid Nitrogen impregnated insulation for cryogenic power cables', *Cryogenics,* **14**, 607–611 (1974).
36. BELANGER, B. C., 'Lighting Impulse and Switching Surge Breakdown of liquid nitrogen-impregnated wrapped dielectrics for cryogenic cables', *IEEE Trans. PAS 1972*, pp. 2616–2621.
37. RIGBY, S. J. and WEEDY, B. M., 'Impulse Strength of liquid nitrogen-impregnated taped cable insulation', *Cryogenics,* **16** 167–170 (1976).
38. RIGBY, S. J. and WEEDY, B. M., 'Partial discharges in liquid nitrogen impregnated taped insulation', *Proc IEE,* **123**, 165–170 (1976).
39. TAKAHASHI, Y., 'A review of ageing by Cryogenic Insulations by Partial Discharges', *Bulletin of Science and Engineering Faculty, Chuo University,* **19**, 143–156 (1976).
40. FUCHNIEWICA, J. R. *et al.*, 'Susceptibility of liquid nitrogen impregnated electrical insulation to partical discharges', *Cryogenics,* **15**, 419–421 (1975).
41. TAKAHASHI, Y. and OHTSUKA, K., 'Corona Discharges and bubbling in liquid nitrogen', *J. Phys.* D: *Appl. Physics*, **8**, 165–169 (1975).
42. SWIFT, D. A., 'Dielectric design for a superconducting cable with solid insulation', *Revue Generale de l'Electricitie*, No. 10, pp. 741–746 (1975).
43. FORSYTH, E. B. *et al.*, 'Flexible superconducting power cables', *IEEE Trans. PAS-92*, pp. 494–505 (1973).
44. SWIFT, D. A., 'Dielectric design criteria for solid insulated superconducting a.c. cables', *Proc. IEE,* **118**, 1237–1241 (1971).
45. SCHWENTERLY, S. W. *et al.*, 'A.C. Dielectric performance of Helium impregnated multi-layer plastic film insulation', *IEEE Trans. EI-12*, pp. 46–50 (1977).

46. BOGNER, G., 'Cryopower transmission studies in Europe', *Cryogenics*, **15**, 79–83 (1975).
47. HORIGOME, T., 'The present state of R & D for superconducting transmission in Japan', *Cryogenics*, **15** 91–95 (1975).
48. EIGENBROD, L. K. *et al.*, 'Conceptual design and economic analysis of a superconducting AC power cable system', *IEEE Trans' Paper 70T 168 PWR*, 1971.
49. MADDOCK, B. J. *et al.*, 'Superconducting power cables and their application', *CIGRE*, 21-05 (1976).
50. DEAN, J. W. and LEQUER, H. L., 'Conceptual design of a DC superconducting power transmission line', *Proc. IEEE Conference on Underground Transmission and Distribution, 1976*, pp. 417–423.
51. SUTTON, J., 'Induced circumferential currents and losses in flexible superconducting cables', *Cryogenics,* **15** 541–545 (1975).
52. ROGERS, E. C., 'Internally forced-cooled power cables: optimum working temperature in the range 70–358 K', *Proc. IEE,* **122**, 908–914 (1975).
53. ASTAKHOV, Y. U. N. *et al.*, 'Use of Low-cooled Power Transmission lines in large conurbations', *Elektrichestro*, No. 12, pp. 1–65 (1974).
54. FORSYTH, E. B. *et al.*, 'Long distance bulk transmission using helium-cooled cables', *Proc. IEEE Conference Underground Transmission and Distribution*, pp. 446–453 (1976).

PROBLEMS

Problem 7.1

Given that the maximum permissible pressure difference between inlet and outlet of the liquid nitrogen coolant is 10 atm (1.03×10^6 N/m^2) calculate the current rating of the cable (see Figure 7.27) assuming that the power loss in the conductor

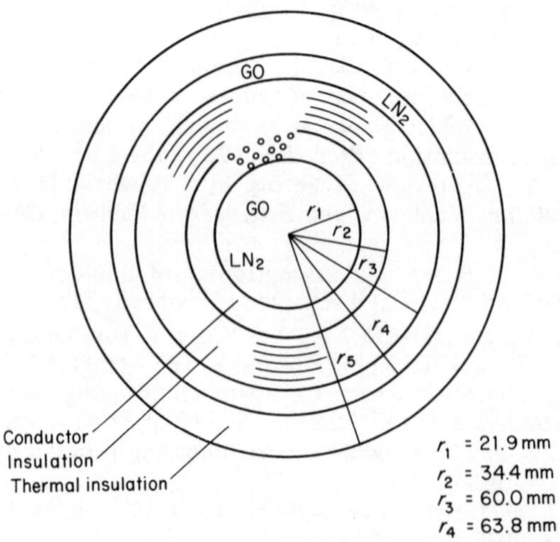

Figure 7.27 Cross-section of cable for Problem 7.1.

and insulation must be absorbed in coolant flow through the central duct. It assumed that the thermal inleak from ambient is absorbed by the flow of liquid nitrogen in the annular duct which may be considered as a separate study from the conductor duct flow and is not required here. The following data applies.

$$(\text{vel})^{1.8} = \Delta P \frac{d^{1.2}}{0.092 l \alpha^{0.8} \eta^{0.2}}$$

where

l	= length (m)
ΔP	= pressure drop (N/m^2)
d	= diam (m)
mean density α	= 740 kg/m^3
mean viscosity η	= 0.0001 Ns/m^2
mean specific heat	= 2100 J/kg K
inlet temperature	= 77 K } conductor duct
outlet temperature	= 106 K } conductor duct
conductor area	= 2200 mm^2 (Al.)

mean resistivity of aluminium between 77 and 106 K = 3×10^{-9} Ωm

capacitance C	= 214 pF/m
tan δ for insulation	= 3×10^{-4}
(R_{ac}/R_{dc})	= 2.0

(Answers: Velocity 0.91 m/s, 1200 A.)

Problem 7.2

The windings of a transformer are separated from the magnetic core by a cylinder of thermal superinsulation 2 cm in thickness. The core is at a temperature of 313 K and the windings are cooled by liquid hydrogen at 20 K.

Calculate the heat inleak per unit area into the hydrogen from the core. The superinsulation consists of 50 layers of aluminium foil ($\epsilon = 0.03$) in a vacuum. Give an approximate value for the refrigerator input power (per cm^2) that this loss would require. If the superinsulation is replaced by foam of thermal conductivity 10^{-3} W/cm K calculate the new loss and refrigerator input per cm^2 of surface. (Answers: 16 $\mu W/cm^2$, 8000 $\mu W/cm^2$, 0.15 W/cm^2, 7.3 W/cm^2.)

Problem 7.3

A long niobium conductor at 6 K is in the form of a thin cylinderical shell of radius 60 mm. If the current carried is 30 kA (d.c.) determine whether the conductor is operating in the Type I or II regions. (Answer: Marginal.)

Problem 7.4

In a superconducting cable each core comprises a tubular conductor wrapped with polyethylene tapes impregnated with liquid helium. There are three insulated conductors in a common cryogenic envelope. The conductors are niobium on a backing of copper and are of 25.5 mm outer radius. The radius to the screen over the insulation is 37 mm. The following parameters govern the cable design:

working voltage (conductor to neutral) 76 kV
system impulse surge voltage to be withstood 650 kV
maximum allowable a.c. stress in insulation 10 kV/mm
maximum allowable impulse stress in insulation 82.5 kV/mm
rated current 17.5 kA
impregnated polyethylene tape tan $\delta \doteqdot 2.3 \times 10^{-6}$ at 4.2 K.

Examine the viability of the conductor radius and the insulation design. (Answers: E_r (impulse) 68.5 kV/mm, E_r (a.c.) 8 kV/mm, H 0.11 T, – Design is viable.)

Problem 7.5

A liquid nitrogen cooled cryoresistive cable comprises three insulated conductors in triangular formation in an aluminium pipe surrounded by thermal insulation. The nitrogen flows through ducts in the conductors to the far end of the cable and returns through the space effective area 0.038 m^2 between the insulated cores and the pipe. Calculate the current allowable through the conductors if the temperature rise at the far end of the cable is not to exceed 15 K over the inlet value. Using this current calculate the temperature rise of the return flow. Assume that the dielectric loss splits half into the conductor duct and half to the return flow. Relevant cable and coolant parameters are as follows:

Cable

Dielectric loss 6.8 W/m per conductor
Conductor loss $2.6 \times 10^{-6} I^2$ W/m per conductor
Heat inleak 5 W/m
Radius of conductor coolant duct 43 mm.
Length of cable 10 km

Coolant

Velocity in conductor duct 1 m/s
Velocity in space between cores and pipe 1.3 m/s
Mean density 700 kg/m^3; specific heat 2100 J/kg-K.

(Answer: 1820 A, $\Delta\theta$ 2.5 K.)

8

Costs

8.1 INTRODUCTION

Although technical matters have formed the major theme of this book, it is the most economic solution for a given transmission scheme that is required. This of course is always the case in electric power supply, but with underground circuits with their inherently high costs compared with the overhead equivalents it becomes crucial. No attempt is made to give in detail costs for specific cable types and installations, but the general approach will be outlined and a few figures given to show orders of magnitude. With increasing energy costs the capitalized cost of losses becomes more and more important, and this tends to favour the conventional underground cable compared with the higher-loss, overhead line. Expressions giving conditions for the costs of lines and cables to be a minimum were obtained by Kelvin[1] but unfortunately are not applicable to underground cables where the current is severely limited by thermal constraints. One method is to assess the cost of a scheme in terms of the cost per MW-year (or hour) of energy transmitted per km. This will be referred to as the Unit Cost. However often the cost per MVA-km is used in preference.

Two methods[2,3,4] are available to obtain costs:

(i) Capital cost plus the capitalization of the losses to their 'present worth' over the specified life of the installation.
(ii) Conversion of the capital cost to an annual payment plus the annual cost of losses.

The first method will be used here.

The present worth W_p of a recurrent annual cost over n years is a measure of the devaluation of this future cost to its present value, i.e. the sum which at the present time will provide at a given interest rate the yearly costs of losses over the allotted life span. It varies with the rate of yearly interest, as illustrated in Figure 8.1. It is seen that after the initial years the present worth increases only slowly and converges to the inverse of the interest rate in p.u. The interest charge includes depreciation and the value is usually increased by a further, say, 1 to 2% to cover maintenance and overheads. Also the rate depends on market conditions and can

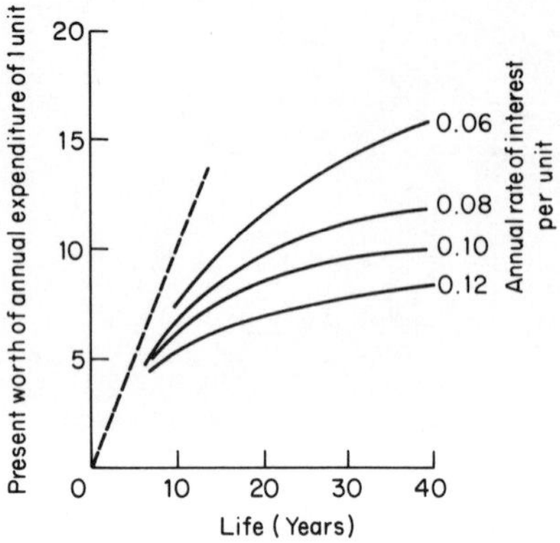

Figure 8.1 Present worth value as a function of time and interest rate.

vary considerably. However, when comparing different schemes a common rate is obviously assumed, e.g. 10%, and sensitivity studies are performed over a range of values. An economic life of 40 years is often assigned to cables; this is not necessarily the physical life (i.e. cable unfit for further service) which is usually longer and difficult to quantify. A life of 40 years and a rate of 8% (0.08 p.u.) results in a present worth (W_p) equal to 12 times the annual recurring cost.

8.2 LOSSES

In a cable system, constant losses (dielectric) and load-dependent losses exist. The latter depend upon the variation of load current with time and are quantified by means of the loss factor F′ which is a function of the load factor F and is defined as the I^2R loss over a given period divided by the I^2R loss if the load was maintained constant at its maximum value. The relationship, $F' = 0.2F + 0.8F^2$ typifies the load characteristic for many systems and is often used. Losses are supplied at the current day-to-day marginal generation cost, but usually an average value such as 0.02 \$ per kWh (in the USA) is taken. One kilowatt of peak loss represents an energy consumption of $8760F'$ kWh per annum, which over the life of the cable (present worth W_p) is capitalized to $8760 \times F' \times W_p \times$ cost per unit. The dependence of the loss factors F' on F is as follows: F 1.0, F' 1.0; F 0.76, F' 0.55; F 0.55, F' 0.34. Also, the provision of the generation capacity required for the loss is costed at, say, 60 \$ per kW and capitalized.

The unit cost of transmission per unit length is

$$10^{-6}\left(\frac{C + 3D \times 10^3 I^2 R}{T\sqrt{3}IV}\right) \quad \$/\text{MWh} \qquad (8.1)$$

where C = capitalized cost of the constant losses, D = present worth of supplying 1 kW of peak loss (load factor F), I = peak load current (kA), V = system voltage (kV), T = effective time load is supplied over the total period of operation = $F8760W_p$. The quantities D and T relate to the power system, and the cable system designer has no control over them. The value of current for minimum unit cost may be obtained from equation (8.1), and this value is such that the present worth of the losses equals the capital cost of the plant. However, the current so obtained is invariably well above the current determined by *thermal considerations*. Also the use of double or treble circuits for reasons of security must be accounted for, when each cable circuit must be thermally capable of carrying the full load.

The load factor and peak load may change over the considerable span of the cable life,and the unit cost is modified to the expresssion

$$\frac{C + \Sigma_1^n W_{pn}\ (\text{cost of losses in year } n)}{\Sigma_1^n W_{pn}\ (\text{energy supplied in year } n)}.$$

Whatever the technical merits of a particular cable type or scheme the final choice will be largely determined by relative costs. In the case of conventional cables a reasonable idea of costs is available. With novel schemes such as low-temperature cables the accurate predictions of costs is obviously more difficult. Certain commodity prices also fluctuate considerably, especially copper.

8.3 COMPARISON OF SCHEMES – USA

In the following sections the estimated costs of various schemes for bulk power transmission are given in the hope of presenting a reasonable comparison between alternative types of cable. Owing to the high inflation rates in many countries the temptation to include a large amount of data has been resisted. It is essential to define precisely the nature of the transmission link to be costed,[5] e.g. length, rural or urban situation, geographical features, and power levels. A detailed account[6] for high-capacity circuits was published in 1972 when various schemes were compared for three different settings.

The various parameters involved in the determination of the 'present worth' of a scheme in money units per MVA-km involve the following data, possibly in the ranges indicated:

rate of return and depreciation	6–12%
annual demand charge (for losses) – USA	\$20, \$60, \$100 per kW
energy charge (USA)	0.010, 0.020, 0.030 \$/kWh
load factor	75%, 100%
capital cost variation	75%, 100%, 125%
taxes and insurance	6%
operating and maintenance	1.8%

depreciation is based on a 40 year sinking fund with zero salvage value.

The process is complicated by the necessity to account for inflation. When comparing schemes similar route situations must be assumed. Typical examples are as follows:

(*a*) Urban route of length 8 km or less including rail and road crossings and with an assumed soil and rock content.

(*b*) Suburban/rural route including road, rail, and river crossings and combinations of publicly and privately owned land. The route is probably a straight line with lengths up to 80 km. Also in the rural setting will be special cases of short lengths of cable (say 300 m) undergrounding parts of overhead line circuits, e.g. across a major road.

The total cost will include the capital cost of the cable and associated equipment and its installation, civil works (trenching backfill, manholes, joint bays), and operating cost (losses, pumps, heat exchangers).

Often the ratio cost of overhead line to equivalent underground cable is quoted as shown in Figures 8.2 and 8.3. As cables are normally only used where overhead circuits are not possible owing to environmental and other reasons these ratios should be kept in perspective. Cost estimates for pipe-type (HPOF), self-contained (LPOF), and low temperature cable schemes in the USA have been based on the following; annual rate of return 10%, rate of inflation 0%, annual demand charge $60 per kW, charge for losses 20 mils per kWh, and load factor 100%.

Costs predicted for pipe-type HPOF schemes are for cables assumed forced cooled by circulating oil through the pipes. Heat exchangers are air cooled, avoiding the requirement for cooling water. For the 345 kV and 500 kV cables cellulose paper insulation is assumed, but for 765 kV a synthetic laminar tape (e.g. low loss polypropylene/paper is considered necessary (tan δ 0.0007 and ϵ_r 2.7). The spacing between refrigeration and pumping stations will vary because of pressure and temperature considerations. (With a 200 psi nominal minimum operating pressure the drop in the pipe is limited to 300 psi, 500 psi being the limit in the terminations.) Compensation is provided to achieve the maximum power transfer. The

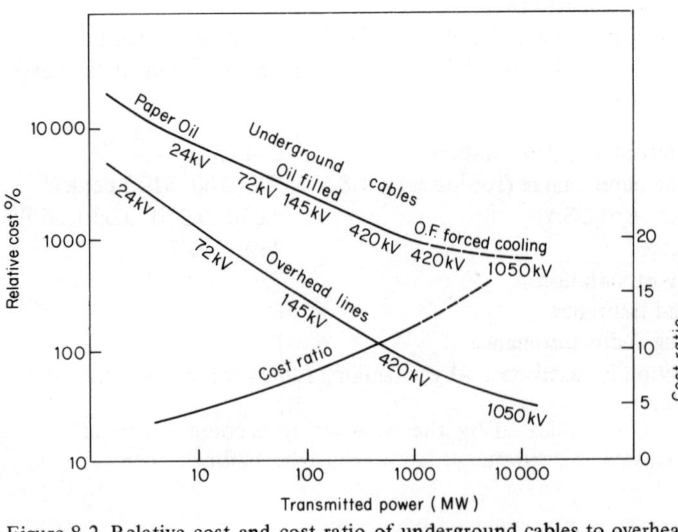

Figure 8.2 Relative cost and cost ratio of underground cables to overhead lines for various transmitted powers.[8] – European (Reproduced by permission of CIGRE, 1974.)

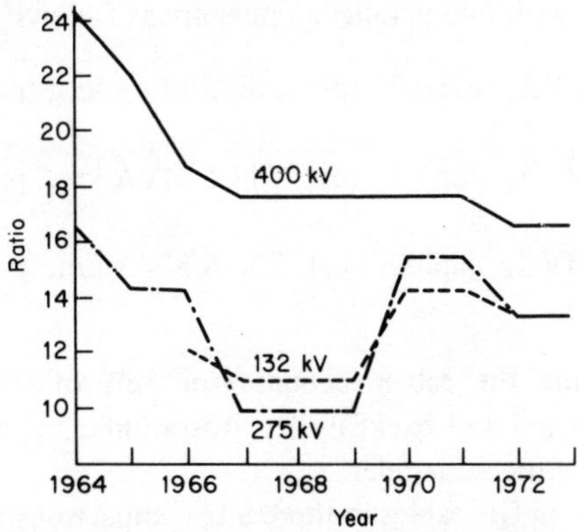

Figure 8.3 Ratio of costs of cables to overhead lines. Assumes 2 km length of circuit and includes cable terminations and compensation – British.[7] (Reproduced by permission of IEE.)

compensation units in urban areas are assumed totally enclosed SF_6 systems with only one disconnect switch, and are located on a common site with the refrigeration stations along the route. A maximum pulling length 535 m (1755 ft) is assumed, this being the distance between manholes. All the systems terminate in 345 kV substations. In a review of techniques[9] to reduce the costs of undergrounding, new trenching methods such as water-jet soil removal and mechanized spoil removal have been suggested. Also, the savings resulting in increased cable pulling lengths are given and shown in Table 8.1.

Estimated costs for hypothetical but typical pipe-type suburban/rural schemes

Table 8.1 Calculated savings resulting from 1/3 increase in cable pulling lengths

	138 kV	230 kV	345 kV
Manholes $/mile*	28 000	28 000	42 000
Splices $/mile*	10 000	12 000	21 000
Cable pulling**	12 000	12 000	18 000
Total/mile	50 000	52 000	81 000
Savings/mile @ 25%	12 500	13 000	20 250
Assumed annual installed mileage (conservative)	50	50	20
Estimated annual saving	625 000	650 000	405 000
Total saving/year		$1,680,000	

(Derived from table 6.3 of Reference 6.)
*Installed cost assuming 1/2 mile section lengths for 230 and 138; 1/3 mile 345 kV and assuming $6 000/cable pull for all voltages.
**Estimate including allowance for vacuum treatment.
(Reproduced by permission of IEEE.)

(length up to 80 km) with two parallel circuits are as follows[10] (1976 prices):

(*a*) 245 kV, 2250 MVA, capital cost 680 \$/MVA km, present worth 1620 \$/MVA km.
(*b*) 500 kV, 2500 MVA, capital cost 744 \$/MVA km, present worth 1625 \$/MVA km.
(*c*) 765 kV, 4500 MVA, capital cost 750 \$/MVA km, present worth 1360 \$/MVA km.

In the 500 kV scheme the cable accounts for 58% of the total cost, pipe and concrete 10%, excavation and backfill 7%, substation 12%, and refrigeration 7.7%. Smaller items make up the remainder.

For self-contained LPOF cables in the USA, comparisons of costs[10] assume that forced water cooling is used with one cable installed per pipe (surface cooling). As usual in the USA a duct system for self-contained cables is used. The cables have hollow core Milliken type segmented copper conductors, paper-tape insulation, semiconducting screening tapes, corrugated aluminium sheath, and extruded jacket. Unit and present worth costs based on a 40 year life and the base interest and cost figures already given, are as follows. Two parallel circuits are used in a suburban/rural situation (1976 prices):

500 kV, 2500 MVA, capital cost 1086 \$/MVA km, present worth 1895 \$/MVA km.
765 kV, 5000 MVA, capital cost 800 \$/MVA km, present worth 1465 \$/MVA km.

Suburban figures only are quoted as these give a fairer comparison with overhead lines. In urban areas, where overhead circuits cannot be used, cost of underground circuits are of course higher.

It is of interest to compare overhead line costs in the USA estimated on a similar basis to the above, i.e. some interest rate, energy charge, and load factor. The a.c. line costs are as follows[10] (1976 prices):

500 kV 2500 MVA (80 km), capital cost 165 \$/MVA km, present worth 540 \$/MVA km.
765 kV 5000 MVA (80 km), captical cost 190 \$/MVA km, present worth 424 \$/MVA km.
765 kV 5000 MVA (320 km), captial cost 140 \$/MVA km, present worth 290 \$/MVA km.

Costs for a d.c. line on a similar basis are as follows:

±400 kV 2500 MVA 80 km, capital cost 1040 \$/MVA km, present worth 1700 \$/MVA km.
±400 kV 2500 MVA 320 km, capital cost 300 \$/MVA kM, present worth 610 \$/MVA km.

Table 8.2 Comparison of self-contained LPOF cables – British practice

Conductor size, mm^2 (in^2)	1935(3.0)	1615(2.5)	1615(2.5)	1935(3.0)	2273(3.5)	2580(4.0)	2580(4.0)(d.c.)
Stress, MV/m	13.5	13.5	13.5	13.5	15.0	15.0	33.0 d.c.
Laying formation		flat			open trefoil	open trefoil	flat
Installation method	direct with special backfill	direct with special backfill	direct	direct	water-filled nonmetallic pipes	water-filled nonmetallic pipes	direct
Type of cooling	natural	natural	separate water pipes	separate water pipes	integral water pipe	integral water pipe	natural
Number of cables per phase	2	3	2	2	1	1	One cable per pole
Cable centres	0.45	0.25	0.25	0.25	0.40	0.40	0.40
Total width of cable reservation, m	17.4	23.4	16.2	16.2	8.4	8.4	7.6
MVA rating per circuit	2200	2640	2200	2640	2200	2640	2640 MW (d.c.)
Capital cost of scheme based on 1970 prices, £/MVA/km	156	181	159	140	125	113	53
Cost of reactive compensation, £/MVA/km	37	44	34	28	27	27	Terminal plant 3130
Total cost of scheme, £/MVA/km	193	225	193	168	152	140	3236

Notes: Milliken Conductors, cross-bonding for a.c., spacing between circuits 3 m, soil temperature – Summer 15 °C, Winter 10 °C, g_{soil} – Summer 1.2 deg C-m/W, Winter 1.05 deg C-m/W, cross country route length 5 km.
(Taken from Reference 7; reproduced by permission of IEE.)

It is interesting to note that the ratio of cable to line costs is much lower if present worth is considered. This is caused by the higher line loss coupled with high energy costs.

8.4 COMPARISON OF SCHEMES – BRITISH AND GERMAN

Cost figures (1975) for British LPOF self-contained installations have been summarized by Cherry.[7] Here the cables are direct buried into a trench with various cooling schemes. Table 8.2 gives parameters and cost based on cross-country routes for circuit ratings of 2.2 GVA and 2.64 GVA with a variety of installation and cooling methods. The ratios of underground cable to equivalent overhead lines costs based on British practice[7] are given in Figure 8.3 for each year since 1964. These costs for cables are significantly affected by the cost of materials, especially naturally-cooled schemes which are copper intensive. The price of aluminium has remained relatively stable compared with copper to the detriment of the underground cable compared with the overhead line. A detailed analysis of effects on cost of various proposed insulation materials has been given by Beale.[11]

A German study[12] in 1974 reviews the economics of high-power transmission using 110 kV paper–oil cables with internal water cooling in the conductor duct. The water enters the cable at 3 Bar and 30 °C and exits at 1 Bar and 80 °C. Costs are given for one cooling station with a separate water return pipe from the remote end of the cables. The following data apply:

Pumping and cooling station: interest 8%, repairs and maintenance 3% per annum, service life 20 years.

Cable and accessories: interest 8%, repairs and maintenance 1% per annum, life 40 years.

In Figure 8.4 the costs in DM (1974 values) are given for various circuit capacities, cooling duct diameters (d_h), conductor diameters (d_c). areas, and cooled section lengths. It is seen that for each conductor size there is an optimum cooled length.

Reliability

The majority of overhead line faults due to weather conditions are transient with no permanent damage. In view of this, quoted statistics on faults on cables and lines need be thoroughly studied. British experience[7] with LPOF cables indicates more faults per mile than with overhead lines and much longer repair times. This is illustrated in Table 8.4 which gives statistics for eight years up to 1975. Caution must be exercised with these figures because of the short lengths installed, and many problems have been specific to the new voltage levels.

8.5 HVDC CABLES

The cost of the conversion equipment is a major challenge to the use of d.c. cables. With the change from mercury arc valves to thyristors it is hoped that such

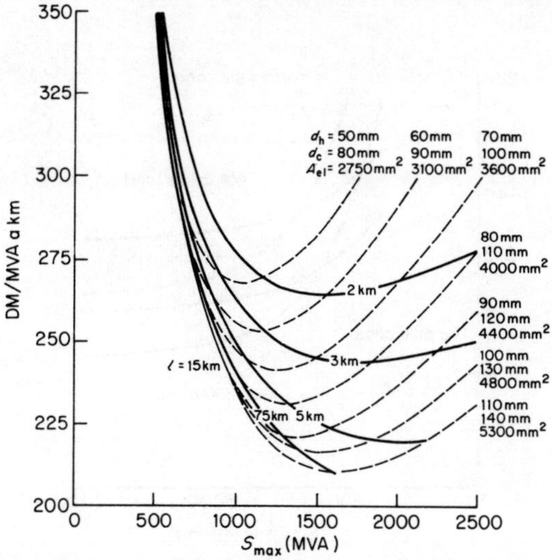

Figure 8.4 Costs of an internally water-cooled 110 kV cable related to MVA transmitted. Costs in Deutsche Marks (DM). Variation for given conductor area (A) and water channel diameter d_h – – – –. Variation for given length between cooling stations ———.[12] (Reproduced by permission of CIGRE from Report 21-09, 1974.)

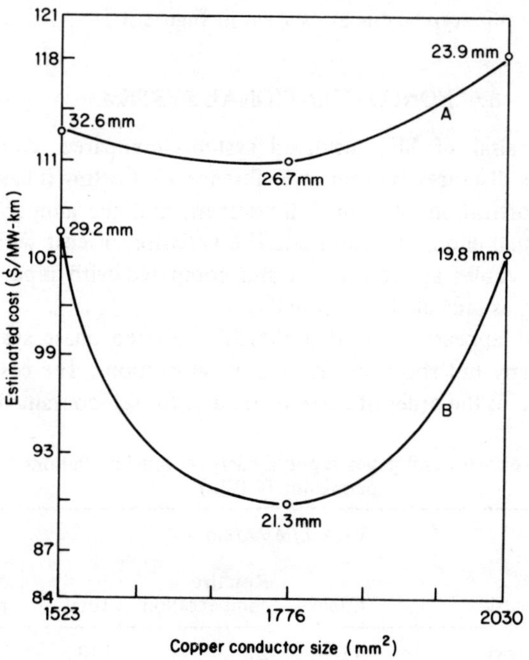

Figure 8.5 Estimated costs of a ± 600 kV 2.4 GW d.c. HPOF pipe-type cable. Maximum steady-state stress 34 kV/mm (A), 39 kV/mm (B). (From Bahder, G. *et al.*, *IEEE PAS-97*, pp. 2045–2055; reproduced by permission of IEEE.)

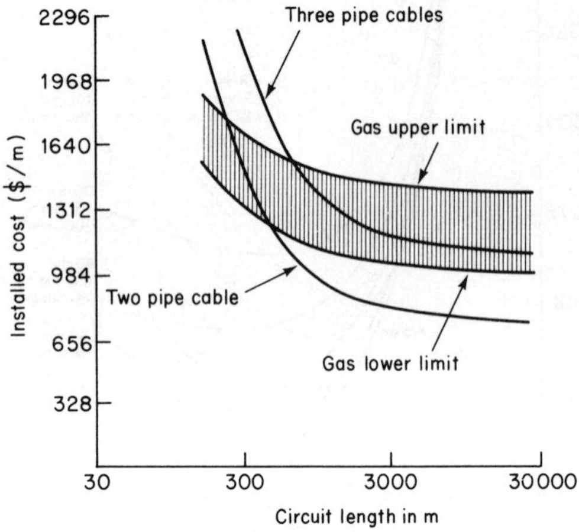

Figure 8.6 Installed cost of 345 kV, CGI, systems and equivalent pipe-type cable systems.[13] (Reproduced by permission of IEEE.)

equipment will become more competitive. Approximate figures suggest that d.c. underground circuits become competitive compared with a.c. for lengths greater than 174 km for two pairs of 500 kV d.c. cables of total capacity 2.64 GW. Costs are illustrated in Table 8.2 for a cross-country route of 5 km. Estimated costs of a ±600 kV d.c. HPOF pipe-type cable are shown in Figure 8.5.

8.6 NONCONVENTIONAL SYSTEMS

An economic appraisal of SF_6 insulated systems compared with conventional schemes for various distances is given in Reference 13. Costing is based on the sum of the annual amortization of capital investment and the annual cost of losses divided by the annual energy transmitted. The variation of cost with transmission length for 345 kV cables are considered and compared with pipe type-cables of equivalent capacity, as indicated in Figure 8.6.

British opinium[7] appears to be that the SF_6 isolated phase system would be uneconomical for any but short lengths, e.g. in substations. The cost of long runs they estimate to be in the order of three times that for self-contained LPOF cables.

Table 8.3 Comparative costs of oil/paper vs polythene/gas cable installations.[7] (Reproduced by permission of IEE.)

		Cost, £/MVA/km			
Cooling	Dielectric	Cable	Reactive compensation	Total	Number of cables per phase
Direct external	oil/paper	113	27	140	1
	polythene/gas	99	16	115	1
Natural	oil/paper	181	44	225	3
	polythene/gas	144	27	171	2

Table 8.4 Operational reliability of underground cables and overhead lines at 275 kV[7] (Reproduced by permission of IEE.)

Installation	Oil-filled cable circuits	Overhead lines
System voltage, kV	275	275
Total circuit-kilometre years	2350	27 000
Annual fault rate for 100 circuit km	2.5	1.17
Average outage time, h	268	23
Average nonavailability in hours per annum for 100 circuit km	65	2.7
Ratio: cable/overhead line 275 kV	25	

The reduction of diameters resulting from the use of higher gas pressure combined with a flexible corrugated tube design as discussed in Chapter 6 would decrease installation costs and may prove more attractive. A comparison of costs is given in Table 8.3. Costs for cryoresistive and superconducting systems have been mentioned in Chapter 7.

8.7 THE FUTURE

The amount of high capacity underground transmission required over the next few decades is difficult to quantify. It is doubtful whether past usage will be an accurate guide for two main reasons. Firstly, the rate of increase in the use of electricity in many developed countries has decreased because of greatly increased fuel costs. Secondly, it is difficult to foresee the strength of public opinion regarding visual amenity, and in particular whether this will increase to such an extent that increased electricity costs to cover a greater degree of undergrounding will be acceptable. In the long term, however, it is difficult to see any clear alternative to the more widespread use of electricity as a major energy vehicle as resources of natural gas and oil become deplete. Large improvements in fuel cell technology may influence a move towards energy transfer via hydrogen, but this is not at all clear at the present time. Even the widespread use of solar energy on a domestic level will not have a major effect on bulk electricity supply.

Given the premise therefore that large bulks of electric power are to be transmitted, the role which underground transmission will play is a important issue. The cost comparison figures are so clear as to require no argument. European figures[8] suggest that the use of all-underground circuits in the future (this does not include the conversion of existing lines) would entail additional investments corresponding to 50% of the total investment of a power utility. In view of this it must continue that overhead lines will form the major part of the transmission system in the absence of any radical change in underground technology.

Environmental considerations will dictate the use of underground circuits in urban and suburban areas and in certain rural areas of outstanding beauty. A further possible use would be transferring the power away from major generating facilities, which in the future could be of several GW capacity, to the overhead line system thus reducing severe environmental impact around the plant.

As has been mentioned with forced cooling the capacity per circuit of self-contained low-pressure oil-filled cable may be extended to 5 GVA at 750 kV. The costs, however, would be very high, especially those for reactive compensation. The cable to overhead line cost ratios increase with voltage when conventional cables are used. The ability to exploit pipe-type cables to these high capacities is more limited because of thermal considerations.

Of particular interest is the solid extruded cable with polyethelene dielectric. The advantages of this type of cable, both economic and technical, over impreg-

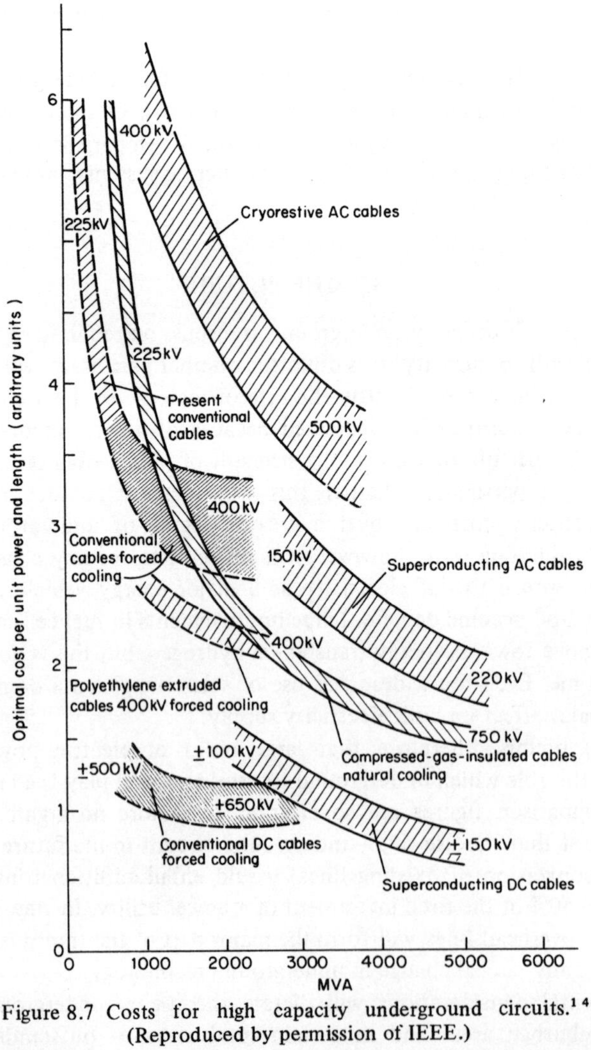

Figure 8.7 Costs for high capacity underground circuits.[14] (Reproduced by permission of IEEE.)

nated paper insulation cables have been mentioned. In particular, it is very applicable to intense cooling techniques. It is very likely that problems in the dielectric will be overcome by improved manufacture and materials and the voltage range extended. It may well supersede the paper–oil cable at medium and high capacities up to voltages of, say, 400 kV. Beyond this voltage taped synthetic materials are more promising because of the more controlled situation for dielectric in thin sheets. These will be impregnated with an insulating fluid such as a suitable oil or SF_6.

It is unlikely for high-voltage cables that copper and aluminium will be replaced, though sodium may show to economic advantage at low or medium voltages where the conductor weight is very large compared with the total cable weight. Flexible conductor areas of up to 3000 mm^2 are foreseeable using copper or aluminium.

Conventional paper–oil or extruded polyethylene cables would seem to have upper limits of about 2.5 GVA at 400 kV and 4 to 5 GVA at 750 kV assuming forced cooling and copper conductor areas of up to 2500 mm^2. The capacity of d.c. cables is higher than this with artificial cooling. In most countries it is unlikely that more than these capacity levels will be required by the end of the century. However, the economic incentive to develop alternative cable types is strong, along with a need to improve power system performance.

Compressed gas insulated cables may provide an economic solution, particularly in conjunction with SF_6 metal-clad substations. In Europe the power levels anticipated would not require consideration of superconducting cables for many years. However, in the United States a severe requirement for getting large powers into urban areas may make the use of superconducting cables attractive in the not so distant future. At high power transfers the cost of superconducting cables would be less than conventional cables. Also, superconducting cables possess good electrical system parameters. A summary of estimated costs[14] of the various options for high capacity underground transmission circuits is given in Figure 8.7.

REFERENCES

1. Lord Kelvin, 'On the economy of metal in conductors of electricity', *British Association Report*, p. 526 (1881).
2. JONES, B. *New approaches to the design and economics of ehv transmission plant*, Pergamon Press, Oxford, 1972.
3. NORRIS, T. E., 'Economic comparisons in planning electricity supply', *Proc. IEE,* **177**, 593–606 (1970).
4. BERRY, R. N., 'Economics of high voltage transmission by underground cables', *Proc. IEE*, pp. 573–590 (1948).
5. MORAN, J. A., 'Comparative cost analysis for cable power transmission systems', *Proc. IEEE Conference on Underground Transmission and Distribution, 1976*, pp. 442–445.
6. *Underground power transmission*, Electric Research Council, Project RP78–8, 1971.
7. CHERRY, D. M., 'Containing the cost of undergrounding', *Proc. IEE,* **122**, 293–300 (1975).

8. ANGELINI, A. M., 'The evolution of electric power transmission and the environment', *CIGRE*, 1974.
9. SMITH, W. G., 'Integrated system studies supporting underground electric transmission research and development', *Proc. IEEE Conference on Underground Transmission and Distribution, 1976*, pp. 1–4.
10. Shah K. R. *et al.*, 'Cost Components of high capacity transmission options' Report for EPRI (USA), EL-1065, Vol I and II, Project RP-568-1, 1979.
11. BEALE, H. K., 'Some economic aspects of extra-high-voltage underground cables', *Proc. IEE,* **112**, 109–118 (1965).
12. BIRNBREIER, H. *et al.*, 'High power cables with internal cooling', *CIGRE*, Report 21–09, 1974.
13. CRONIN, J. C. and MORAN, J. A., 'Economics of gas-insulated power transmission systems', *Proc. IEEE Conference on Underground Transmission and Distribution, 1974*, pp. 519–523.
14. DESCHAMPS, L., *et al.*, 'Prospects of development in France of electric power transmission', *Proc. IEEE Conference on Underground Transmission and Distribution, 1976*, pp. 459–464.

Appendix

Standards and Specifications

(IEEE Standard Techniques for High Voltage Testing, 1978. Distributed by J. Wiley and Sons Ltd.)

BRITISH STANDARDS

BS 801	Lead and lead alloy sheaths of electric cable.	1953
BS 1442	Galvanised mild steel wire for armouring cables.	1969
BS 4066	Flame-retardant characteristics of electric cables.	1969
BS 5099	Spark testing of electric cables.	1974
BS 5468	Crosslinked polyethylene insulation of electric cables.	1977
BS 5469	Hard ethylene propylene rubber insulation of electric cables.	1977
BS.6234	Polythene insulation and sheath of electric cables.	1969
BS 6360	Copper conductors in insulated cables and cords.	1969
BS 6746	PVC insulation and sheath of electric cables.	1976
BS 6791	Aluminium conductors in insulated cables.	1969
BS 6899	Specification for rubber insulation and sheath of electric cables.	1976

INTERNATIONAL ELECTROTECHNICAL COMMISSION PUBLICATIONS

IEC 141	Tests on oil-filled and gas pressure cables and their accessories.	
	141–1 Oil-filled, paper-insulated, metal-sheathed cables and accessories for alternating voltages up to and including 400 kV.	1976
	141–2 Internal gas pressure cables and accessories for alternating voltages up to 275 kV.	1963

	141–3 External gas pressure (gas compression) cables and accessories for alternating voltages up to 275 kV.	1963
IEC 183	Guide to the selection of high-voltage cables.	1965
IEC 228	Conductors of insulated cables.	1978
IEC 230	Impulse tests on cables and their accessories.	1966
IEC 287	Calculation of the continuous current rating of cables (100% load factor).	1969
IEC 331	Fire-resisting characteristics of electric cables.	1970
IEC 332	Flame-retardant characteristics of electric cables.	1970
IEC 502	Extruded solid dielectric insulated power cables for rated voltages from 1 kV to 30 kV.	1978
IEC 540	Test methods for insulation and sheaths of electric cables and cords (elastomeric and thermoplastic compounds).	1976
IEC 628	Gassing of cable and capacitor insulating oils under electrical stress and ionization.	1978

ASSOCIATION OF EDISON ILLUMINATING COMPANIES (USA)

AEIC 5–79	Specifications for thermoplastic and cross-linked polyethylene insulated shielded power cables rated 5 through 69 kV.
AEIC 6–79	Specifications for ethylene propylene rubber insulated shielded power cables rated 5 through 69 kV.
AEIC 2–67	Specifications for oil-impregnated insulated cable, high-pressure, pipe-type.

NATIONAL ELECTRICAL MANUFACTURERS ASSOCIATION (USA)

WC 7–1971	Crosslinked thermosetting polyethylene-insulated wire and cable for the transmission and distribution of electrical energy.
WC 8–1976	Ethylene propylene rubber-insulated wire and cable for the transmission and distribution of electrical energy.

Index